COUNT LIKE AN EGYPTIAN

COUNT LIKE AN EGYPTIAN

A HANDS-ON INTRODUCTION
TO ANCIENT MATHEMATICS

DAVID REIMER

Princeton University Press

Princeton and Oxford

Copyright © 2014 by Princeton University Press

Published by Princeton University Press, 41 William Street, Princeton, New Jersey 08540

In the United Kingdom: Princeton University Press, 6 Oxford Street, Woodstock, Oxfordshire OX20 1TW

press.princeton.edu

Library of Congress Cataloging-in-Publication Data

Reimer, David, 1962- author.
Count like an Egyptian / David Reimer.
pages cm
Includes index.
ISBN 978-0-691-16012-2 (hardcover : alk. paper) 1. Mathematics, Babylonian.
2. Mathematics--Egypt. I. Title.
QA22.R38 2014
513.2′11--dc23
2013036208

British Library Cataloging-in-Publication Data is available

This book has been composed in Times Ten

Printed on acid-free paper. ∞

Printed in the United States of America

1 3 5 7 9 10 8 6 4 2

CONTENTS

PREFACE

A number of years ago a colleague of mine looked over my class notes for a course in the history of mathematics. She pointed out that I had misinterpreted the ancient Egyptian method. At that time, Egyptian computation was only a small part of the class, but I wanted to make sure that everything was done properly. So I pulled all my history of mathematics texts off the shelf and reread the Egyptian sections. What I found hardly helped me at all. There were only a few trivial examples followed by abstract discussions filled with equations completely out of context.

I decided that I needed to learn Egyptian mathematics the old-fashioned way. I obtained a translation of the Rhind Mathematical Papyrus, the only known complete Egyptian math text, and worked my way through it as I would any modern textbook. I could always get correct answers to the exercises, but more often than not, the solutions in the papyrus were solved in a different way. Even worse, their solutions took far less work than mine. I would stare endlessly at the point where our solutions diverged and ask myself, what did they see that I could not?

Eventually I developed an instinct for the subject. By the second pass through the book, I could often match their solutions. By the third pass, I could occasionally better them, but I would always understand why they made the choices they did. I would also be aware of what knowledge I had and they didn't that enabled me to beat them. Surprisingly enough, at my peak, I found that I could perform most computations faster in Egyptian than I could with modern methods.

I attacked the subject both academically and pedagogically. Egyptian methods produce answers efficiently, but it's not clear on casual observation why it works as well as

it does. I spent a fair amount of time dissecting tables and algorithms in an attempt to understand this aspect of their system. I used pedagogy to attempt to get into the mind of the ancient mathematicians. Every teacher asks questions for a specific reason. If you look at adjacent questions in a math text, you will often see that although they are similar, the second one often requires one additional thought or method to solve. Noticing this, you can realize what the writer is trying to teach you with this question. By examining the difficulty of the different parts of the computation, you can begin to grasp what they want you to focus on and what is a trivial detail to be glossed over.

After doing this, I went back to my history of math texts thinking that my newfound knowledge would give me greater insights into the authors' discussions. However, upon rereading, I couldn't help but feel that they just didn't get it. They continually focused on insignificant details or challenging problems rather than on the themes central to the Egyptian system. They expressed their ideas in algebra, which often overly complicates simple ideas. Since algebra in its current form would not exist for more than three thousand years after the creation of the Rhind papyrus, using modern algebra was a bit off the point.

At about the same time, I had been working on my notes for the Greek section of my course. At first I alternated sections on math and historical background. However, as I delved deeper into the development of Greek philosophy, I noticed that the two were inextricably linked. The ideas of a philosopher would never explain how to solve a particular problem, but they would give great insight into why they considered the problem important. Further study revealed connections between Greek social structure, historical events, cosmology, and the changes in their mathematics. Slowly I learned to merge the cultural history and the math into single documents. The positive feedback from students and colleagues reading my notes convinced me that my efforts were well worth it.

This book is the result of these two influences. I've become convinced that you can't appreciate Egyptian mathematics without doing it. Just as you can't understand baseball by examining lists of dry statistics, you can't learn Egyptian math with a sampling of abstract discussions. As a former professor of mine once said, "Math is not a spectator sport; you need to play the game." In order to do this we need a complete working mathematical system. The truth is that we can't be sure how Egyptians performed some of their mathematics, because they often skipped steps or simply listed some things in table form with no discussion as to how they were derived. To remedy this, I've filled in the gaps giving possible methods. Often I will give more than one interpretation. I've tried to clearly state when something is known and when it is conjectured.

I've also done my best to give all of the sections some context consisting of light historical background. Giving someone a good reason to know something is as important as conveying the knowledge itself. It not only helps us appreciate the information but also makes us more likely to retain it. We need to set the material apart from the constant barrage of data overload we so readily discard.

While I'm fairly satisfied with my attempts in this regard, there are a few points late in the book where it was all but impossible to find meaningful context. Minor nuances in computational methods don't lend themselves to interesting anecdotes. In these cases, I've included some historical background in order to maintain the tone of the book. Often the connection to the material is metaphorical at best, but hopefully it will appeal to readers who appreciate the historical aspects of Egypt.

In order to make the book more accessible and appealing, I've kept the discussion light and the rhetoric conversational. I've included diagrams, humor, and anecdotes drawn from many historical periods. Like any good storyteller, I've accentuated the drama and made villains and heroes out of the characters within. I've also kept the mathematics light. You will need little beyond addition, subtraction, and the multiplication of whole numbers. Each section takes the reader in small, easily understood steps. This is not intended to be a book overstuffed with information but rather an easy read that slowly but surely leads you to a deep understanding of the material.

This book could be used as a supplementary text for a history of mathematics course. It could also serve as part of an interdisciplinary course. However, it is intended to be a light enjoyable read accessible to anyone at the junior-high level on up.

INTRODUCTION

ENGLISH IS STUPID

Far too many decades ago, I was sitting in a junior-high Spanish class. The topic of the hour was the conjugation of verbs, but one of them didn't follow the standard rules. This bothered my budding mathematical sensibilities, which required that all things follow well-regulated patterns. I asked the teacher why this word was different, and she simply responded "That's just the way it is," to which I replied, "Well, Spanish is stupid."

As her face turned red, I immediately realized my mistake. She was upset for a variety of good reasons, one of which was the comment that came from me. I was president of the Spanish club. This "club" was her euphemism for her detention hour where she sent students who didn't do their homework. I was president merely because I had logged in the most hours that year. She was also upset because no one likes some smart-aleck preteen to dismiss their life's work in a few words. But I suspect what bothered her most was my comment itself, which was, in fact, stupid.

She then began a long rant, not directed at me, but rather at the English language. She proceeded to list every English inconsistency that popped in her head. Consider the plural nouns "dogs," "mice," and "oxen." Can you see any consistent rule for pluralization? Now consider verb conjugations. The present-past tense pairs "am"/"was," "run"/"ran," and "eat"/"ate" have no pattern whatsoever. The spellings of the English language are even worse. English is consistently inconsistent. In retrospect, I can't even imagine how I learned the language. I suspect it would have been easier to invent a time machine, go back into the past, maim Daniel Webster, and leave a threatening note to any dictionary-writer wannabes to make their choices more carefully. As my teacher rightly pointed out, anyone who speaks English can't possibly criticize Spanish for its relatively few aberrations.

Fortunately, all horrible things have to come to an end, and the bell rang signaling the conclusion of class. I slipped out desperately trying to avoid eye contact with the teacher. Before the next meeting of Spanish club, I had nothing left to do but reflect on my experience. I learned two important lessons that day. The first is to think before you speak. The second is that it's very easy to perceive flaws in a system that's alien to your experience. At the same time, it's very difficult to give it a fair evaluation of your own, because familiarity often leaves us blind to our systems' limitations.

Egyptian mathematics has an alien feel to it. Most math historians refer to it as primitive or awkward. Even worse, many simply ignore it except for a passing reference. They look at this system and feel uncomfortable because it's so different. They perceive apparent "flaws" and move on. They don't understand Egyptian mathematics simply because they don't do it enough to truly appreciate it. To someone who's mastered it, Egyptian mathematics is beautiful. It scorns memorization and rote algorithms while it favors insight and creativity. Each problem is a puzzle that can be solved in many ways. Frequently, solutions will be surprising, something that never happens in the step-by-step drudgery that is modern computation.

There are a number of good reasons to learn Egyptian mathematics. Puzzle lovers will find it fun and challenging. History lovers will gain a new insight into the Egyptian mind-set. However, I believe the most important reason to study Egyptian mathematics is because it so different. We're taught throughout our entire education that math simply is. We learn laws and memorize steps, never questioning what is laid out before us, for if math is "fact," how could mathematics be wrong? When you're exposed to a different system, you're forced to reconsider the immutability of "the math."

Let's go back to the foreign language analogy. Ask an eight-year-old to conjugate a verb, describe a gerund, or identify the tense of a sentence, and they'll give you a blank stare. Yet they can speak as clearly as you or I. They've memorized their language system to the point where they don't need to understand what they're doing as they talk. It's only when they take a foreign language that they're forced to think consciously about language itself. Note that many years ago, I didn't even realize that language wasn't, could be, or should be consistent. It was only being exposed to a new one that forced me out of my comfort zone. In the same way, I believe, Egyptian mathematics shines a light on modern mathematics, providing insight into what we do and why we do it.

In fact, this book is a thinly disguised critique of modern mathematics. In the next-to-last chapter, I'll teach you Babylonian computation, the precursor to our system. Then I'll follow that chapter with a battle royal between the Egyptian and our modern system. We will consider which system is better and what exactly "better" means.

Despite some of my loftier goals, I've tried to make this book fun. I've included stories, analogies, and jokes in an attempt to bring the subject alive. I hope to show you mathematics within the context of the Egyptian world and not removed from it using the needless abstraction of most math books. The historical background is intentionally light, provided more for color than for substance. When possible, I've used historical context that motivates or helps explain the mathematics included. In some sections where there are no obvious connections, I have still included background material to maintain the tone of the book, going as far as using episodes in history as metaphors for mathematical ideas.

If you ask first-graders, "What's your favorite subject?" the vast majority will respond, "Math." The joy of mathematics has to be beaten out of us by endless drills and subjugation to seemingly arbitrary rules. Perhaps, by starting over, this time with Egyptian mathematics, you'll get a fresh chance to revive some of the delight you felt as a child. It's time to get back in touch with the math nerd in us all.

You can print the ruler and charts from the Website for the book (http://press.princeton.edu/titles/10197.html).

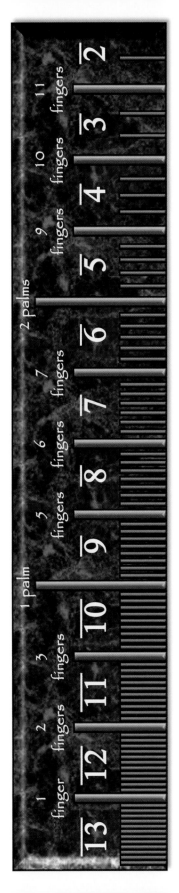

Egyptian Ruler

Tables

$$\times 2 \quad \div(2+1)$$
$$9 \;\rightleftarrows\; \overline{18} = \overline{6}$$
$$\times n \quad \div(n+1)$$

10	100	1000
∩	ℓ	𝍷
10,000	100,000	1,000,000
𝍷	𓆐	𓀀

Sum	Ans.
$\overline{6}\;\overline{6}\;\overline{6}$	$\overline{2}$
$\overline{25}\;\overline{15}\;\overline{75}\;\overline{200}$	$\overline{8}$
$\overline{50}\;\overline{30}\;\overline{150}\;\overline{400}$	$\overline{16}$
$\overline{25}\;\overline{50}\;\overline{150}$	$\overline{15}$
$\overline{7}\;\overline{14}\;\overline{28}$	$\overline{4}$

n	n/10
1	$\overline{10}$
2	$\overline{5}$
3	$\overline{5}\;\overline{10}$
4	$\overline{3}\;\overline{15}$
5	$\overline{2}$
6	$\overline{2}\;\overline{10}$
7	$\overline{3}\;\overline{30}$
8	$\overline{3}\;\overline{10}\;\overline{30}$
9	$\overline{3}\;\overline{5}\;\overline{30}$
$\overline{3}$	$\overline{15}$

Sum	Ans.
$\overline{14}\;\overline{21}\;\overline{42}$	$\overline{7}$
$\overline{18}\;\overline{27}\;\overline{54}$	$\overline{9}$
$\overline{22}\;\overline{33}\;\overline{66}$	$\overline{11}$
$\overline{28}\;\overline{49}\;\overline{196}$	$\overline{13}$
$\overline{30}\;\overline{45}\;\overline{90}$	$\overline{15}$

1 cubit = 7 palms
1 palm = 4 fingers
1 hekat = 320 ro

$\overline{3}$ Identities

$$\overline{2}\;\overline{6} = \overline{3}$$
$$\overline{3}\;\overline{3} = 1$$
$$2 \times \overline{3} = 1\;\overline{3}$$
$$\overline{3} \times \overline{3} = \overline{3}\;\overline{9}$$

2 × $\overline{\text{odd}}$ = table look up ↓

$\overline{n}$	$\overline{2n}$
$\overline{3}$	$\overline{3}$
$\overline{5}$	$\overline{3}\;\overline{15}$
$\overline{7}$	$\overline{4}\;\overline{28}$
$\overline{9}$	$\overline{6}\;\overline{18}$
$\overline{11}$	$\overline{6}\;\overline{66}$
$\overline{13}$	$\overline{8}\;\overline{52}\;\overline{104}$
$\overline{15}$	$\overline{10}\;\overline{30}$
$\overline{17}$	$\overline{12}\;\overline{51}\;\overline{68}$
$\overline{19}$	$\overline{12}\;\overline{76}\;\overline{114}$
$\overline{21}$	$\overline{14}\;\overline{42}$
$\overline{23}$	$\overline{12}\;\overline{276}$
$\overline{25}$	$\overline{15}\;\overline{75}$
$\overline{27}$	$\overline{18}\;\overline{54}$
$\overline{29}$	$\overline{24}\;\overline{58}\;\overline{174}\;\overline{232}$
$\overline{31}$	$\overline{20}\;\overline{124}\;\overline{155}$
$\overline{33}$	$\overline{22}\;\overline{66}$
$\overline{35}$	$\overline{30}\;\overline{42}$
$\overline{37}$	$\overline{24}\;\overline{111}\;\overline{296}$
$\overline{39}$	$\overline{26}\;\overline{78}$
$\overline{41}$	$\overline{24}\;\overline{246}\;\overline{328}$
$\overline{43}$	$\overline{42}\;\overline{86}\;\overline{129}\;\overline{301}$
$\overline{45}$	$\overline{30}\;\overline{90}$
$\overline{47}$	$\overline{30}\;\overline{141}\;\overline{470}$
$\overline{49}$	$\overline{28}\;\overline{196}$
$\overline{51}$	$\overline{34}\;\overline{102}$

2 × $\overline{\text{even}}$ = $\overline{\text{half}}$: 2 × $\overline{12}$ = $\overline{6}$

$\overline{n}$	$\overline{2n}$
$\overline{53}$	$\overline{30}\;\overline{318}\;\overline{795}$
$\overline{55}$	$\overline{30}\;\overline{330}$
$\overline{57}$	$\overline{38}\;\overline{114}$
$\overline{59}$	$\overline{36}\;\overline{236}\;\overline{531}$
$\overline{61}$	$\overline{40}\;\overline{244}\;\overline{488}\;\overline{610}$
$\overline{63}$	$\overline{42}\;\overline{126}$
$\overline{65}$	$\overline{39}\;\overline{195}$
$\overline{67}$	$\overline{40}\;\overline{335}\;\overline{536}$
$\overline{69}$	$\overline{46}\;\overline{138}$
$\overline{71}$	$\overline{40}\;\overline{568}\;\overline{710}$
$\overline{73}$	$\overline{60}\;\overline{219}\;\overline{292}\;\overline{365}$
$\overline{75}$	$\overline{50}\;\overline{150}$
$\overline{77}$	$\overline{44}\;\overline{308}$
$\overline{79}$	$\overline{60}\;\overline{237}\;\overline{316}\;\overline{790}$
$\overline{81}$	$\overline{54}\;\overline{162}$
$\overline{83}$	$\overline{60}\;\overline{332}\;\overline{415}\;\overline{498}$
$\overline{85}$	$\overline{51}\;\overline{255}$
$\overline{87}$	$\overline{58}\;\overline{174}$
$\overline{89}$	$\overline{60}\;\overline{356}\;\overline{534}\;\overline{890}$
$\overline{91}$	$\overline{70}\;\overline{130}$
$\overline{93}$	$\overline{62}\;\overline{186}$
$\overline{95}$	$\overline{60}\;\overline{380}\;\overline{570}$
$\overline{97}$	$\overline{56}\;\overline{679}\;\overline{776}$
$\overline{99}$	$\overline{66}\;\overline{198}$
$\overline{101}$	$\overline{101}\;\overline{202}\;\overline{303}\;\overline{606}$

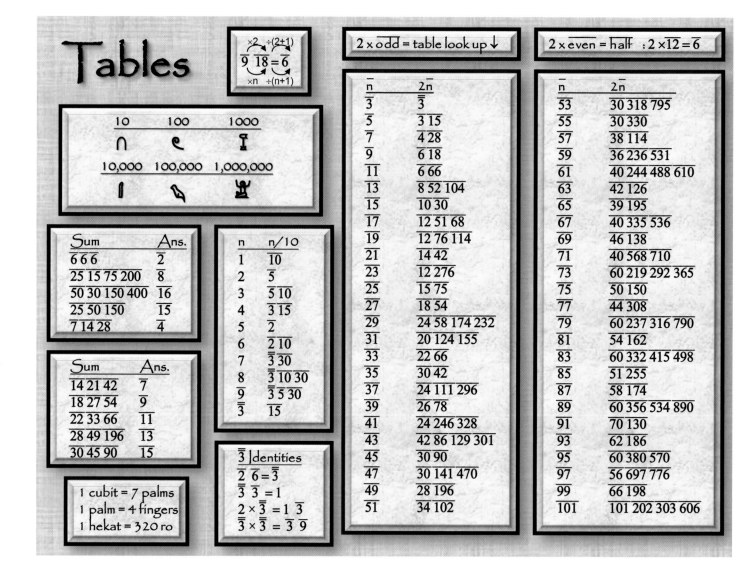

BABYLONIAN
MULTIPLICATION TABLETS

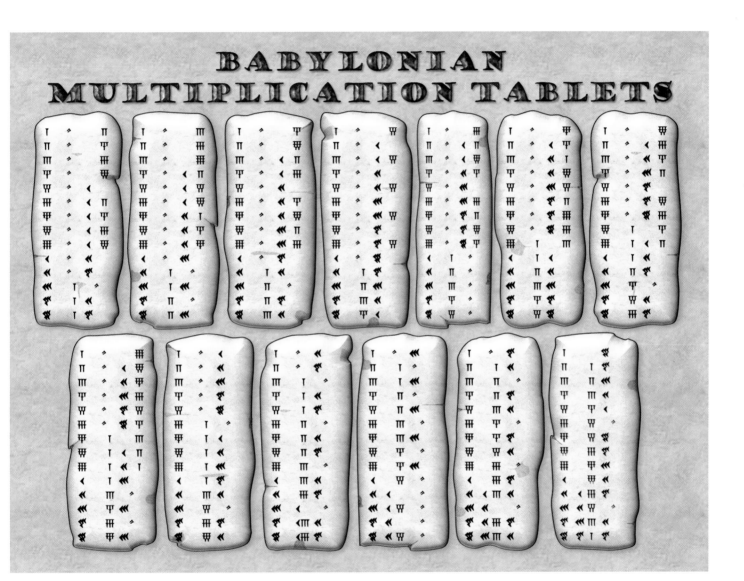

COUNT LIKE AN EGYPTIAN

1
NUMBERS

THE WORDS OF THE GODS

Hieroglyphic Numbers

In the primal waters at the dawn of time, the Egyptian god Ptah brought himself into being. This bearded god had skin blue as the night sky, and he carried a scepter whose form combined the Egyptian symbols of stability, dominion, and life. In his heart, Ptah conceived of the world, and his tongue turned his thoughts into words. At the sound of his voice, the universe changed. The amorphous eight gods of the Ogdoad, including the primeval waters, darkness, chaos, and the invisible power, came together. There they formed the primeval mound, the first piece of the earth. The act drained the power of the Ogdoad and the mound became their tomb, but their sacrifice created the birthplace of the sun, the father of the Egyptian pantheon.

This mound was the center of the earth, which the Egyptians believed resided right in the middle of their nation. The Egyptians called the central part of the world *the Mansion of the Life Force of Ptah*, which the ancient Greeks translated as *Aigyptos*, the origin of our word "Egypt."

The magic of Ptah's words created the world, and words in ancient Egypt had real power. This was especially true for hieroglyphics, which the Egyptians called *the words of the gods*. These artistic writings, along with other magical diagrams, cover the walls of their tombs and temples. But hieroglyphs are more than mere writing. When Egyptians wanted just to write, they used the hieratic script, a simplified form of hieroglyphs. They used the hieroglyphs only when their words needed a small portion of the same

Ptah, the god of creation.

power that Ptah had used to create the world. They used the magic of words to protect themselves from the evil that was in the spirit world.

Such spells usually took the form of either monologues or stories. In monologues, Egyptians spoke directly to the gods, and they would plead with a god for his or her assistance. However, the words contained so much power that these spells contained threats directed at the gods. The magic in the monologues' words was apparently strong enough to prevent divine retribution for their harsh words. Similarly, words infused stories, the second form of a magical spell, with divine power. Hence telling a tale of a god healing another god had the power to heal.

The diagrams that accompanied the hieroglyphs were also magic. Spells granted them the ability to come to life to serve the dead or protect the living. One such spell, *the opening of the mouth*, allowed the spirits both of the dead and the divine to enter or leave a mummy, statue, or drawing. Ptah's name literally translates into the words "the opener," interpreted precisely in this sense. Ptah, in fact, was the patron god of the craftsmen who built and decorated the tombs and temples of ancient Egypt.

These craftsmen had to create the images according to precise specifications because of their mystical nature. Important objects needed to have more magic and hence needed to be drawn bigger. They also had to be drawn with attention to mathematical proportion so they wouldn't come to life misshapen and malformed. These "magical blueprints" required that all the parts were carefully detailed. Hence the figures took on odd poses to clearly depict each essential body part. Many of the poses also possessed symbolic value and in turn conveyed different occult powers. Egyptians were quite capable of accurately drawing figures in natural postures, but these images were not art but, rather, detailed specifications for their afterlife.

Words had so much power that they were often dangerous, even to their users. The bad parts of a magical story could harm someone as easily as the good parts could help. So when a tale included an evil event such as a murder, it often skipped these parts or made a vague reference to such events. Even the symbols used to make up words presented some danger. Imagine the frustration you'd feel if your soul woke up shortly after your funeral only to be chased around your tomb by the spirit of a venomous snake. This would have happened because some craftsman didn't take the proper precautions when writing a word containing the *j* sound, whose symbol takes the form of a cobra. A better-trained craftsman would have drawn the snake sliced up or impaled for the safety of the deceased.

There is no mathematics written in hieroglyphs, but numbers are used for the occasional date or quantity. They use a straight vertical line, I, to represent the number one. This is no surprise since virtually every culture uses a similar symbol to represent 1 just as we do. This practice is tens of thousands of years old and far predates writing, which is a mere five thousand years old. It seems to have been started by hunter-gatherers who used notched bones or sticks to record quantities. While it's easy to

The hieroglyphic symbol for the *j* sound can become a venomous snake in the afterlife.

cut a straight line with a knife across a piece of wood, a curved shape, like our 2, would be needlessly difficult. So, when a denizen of the ice age needed to remember the number 5, he or she would make five straight cuts into a stick. The Egyptians carried on this practice in their writing. Hence, the Egyptian 3 appears as III, just like three notches on a bone.

Unlike their contemporaries, such as the Mesopotamians, the Egyptians didn't group their 1s in specific patterns. For example, 4 could be written in one line as IIII, or in two rows of two as II. This is consistent with their other hieroglyphs, since the layout was concerned more with the aesthetic look of the word than with a systematic layout. For example, the word "day" could be written ⬭👄🐦. These three symbols represented a hut, a mouth, and a quail chick and made the sounds of *h*, *r*, and *w* respectively. Because the first two symbols were short compared to the picture of the quail, it was often written as below, filling up the space on the temple wall more uniformly.

The numbers 1 and 3 had special use in Egyptian hieroglyphs. As we've seen, the symbol 🐦 can represent the

The Egyptian word "day."

sound made by the letter *w*, but it could also represent an actual quail. In order to help the reader distinguish between the two, the Egyptians wrote a symbol identical to the number 1 below the drawing when they wished to identify the object and not the sound. Similarly they could pluralize the object by writing the number 3 below it. For example, the following depicts both the singular and plural of fish.

The system of writing numbers as a bunch of 1s has a serious flaw. Look at the number ||||||||||||||||||||. It's far from

A fish and many fish.

obvious that this is the number 21. Too many lines blur together making them difficult to count. The Egyptians, like most ancient cultures, used symbols to represent groups larger than 1. For example, they used ∩, a picture of a cattle hobble, to represent the number 10. Using the ∩ and the | symbols, they represented numbers up to 99. For example, the number 21 could be written ∩∩|.

For larger numbers they used the symbols ℮, 𝙸, 𝟘, and ⟍ to represent 100; 1,000; 10,000; and 100,000 respectively. The pictures represent a coil of rope, a lotus flower, a finger, and a tadpole. So we can write the number 251,342 as follows:

$$⟍ ⟍ 𝟘 𝙸 𝙸 𝙸 𝙸 𝙸 𝓘 ℮ ℮ ℮ ∩ ∩ ∩ | |$$

The pictures used for the numbers may give clues to how the words were pronounced. Words in ancient Egypt were usually spelled without vowels. If we used a similar system, we could use the symbol, ⌂, to represent the word "bell." However, it could also be used to write "ball" and "bull." The symbol for cattle hobble, ∩, is composed of three consonants: *m, j* and *w*. So the Egyptian word for 10 could sound something like "mojaw," "mijow," and so on. We need to remember that Egyptian is a dead language and no one is really sure how any of the symbols sounded. Egyptologists have made intelligent guesses based on their knowledge of the ancient language Coptic,

which evolved from Egyptian. But this problem is compounded when we realize that even if we knew all the written sounds, we still don't know what vowels go in between. So when we see the coiled rope symbol, ℮, we believe the consonants are *s, h,* and *t* and can only guess the vowel. While most of us can think of a few interesting words using these letters, each adjacent pair has an unknown vowel sound between them suggesting 100 is pronounced something like "sehet" or "sahot."

The number for a million is depicted by a man holding his hands in the air, 𝍠. Egyptians used this word to represent extremely large numbers in exactly the same way we do when we say we've done something "a million times." It's difficult to say what the pose means. Some have suggested it's the arms of man outstretched, overwhelmed when confronted with the concept of infinity. It's also reminiscent of the pose the air god Shu takes while holding up the sky, restraining her from embracing her lover, the earth, and crushing all things between heaven and earth. The symbol, 𝍠, also stands for each of the Heh gods. These are the spirits of the Ogdoad, who died to form the earth and coincidentally help Shu hold up the sky.

The number 1 million was used repeatedly in the Egyptian mythos. Perhaps the most important example is the barque of millions. A barque was a boat a god used to sail across the sky, which according to the Egyptians, was made of water. The barque of millions was the sun god Ra, which was navigated by the god Thoth and his wife Ma'at across the sky each day. The "millions" refers to all the souls who had achieved salvation and manned the barque as it descended into the nether world each night. Together with Seth, the god of strength and violence, they defended the boat on its journey to the dawn, when the sun god would be reborn anew in order to shine another day.

THOTH, SCRIBES, AND BUREAUCRACY

Hieratic Numbers and Addition

On the day an ancient Egyptian was born, Thoth, the god of scribes and wisdom, would change into his ibis form. He needed this form so he could fly from his barque, the moon, down to the earth, where he would carry out the will of the gods. Unseen by human eyes, Thoth would find one of the bricks of the house in which the baby was born and write down the day the child was fated to die. When that day would finally arrive, the soul of the mortal would once again encounter Thoth in the Hall of Osiris. Here, in the land of the setting sun, which formed the barrier between heaven and hell, the soul would be given final judgment. Regardless of the outcome, Thoth would dutifully record the result.

In order to fully appreciate the importance of Thoth and the scribal class in Egypt, we need to understand the central role of bureaucracy in Egypt. Contemporary movies about the ancient world seem to invariably include scenes of a vast marketplace where the characters are offered a wide array of treats and forbidden goods. This imagery is based on a modern misconception. The economies of ancient civilizations were, by and large, controlled by a central government. The state provided everything its ruling class thought the citizens needed, and the former took what they considered to be their share. When not working for the government, individuals would occasionally exchange goods and services with each other for a few items the government wouldn't provide. Relatively speaking, it was a small part of the economy, which otherwise was dominated by Egypt's ruling class.

The pharaoh, governors, and high priests who controlled the government needed an army of bureaucrats to manage the economy. The scribes of Egypt performed this function, documenting every aspect of ancient life. Just like Thoth, they were there from a person's birth to their death, recording all. Scribes were on the farms, in the storehouses, and on the factory floors. They were even on the battlefields, recording the details of the fight and tabulating the casualties by counting thousands of hands severed from the dead.

The constant need to keep records on every aspect of Egyptian life was a huge drain on the time of the scribes. Hieroglyphics consist of detailed pictures that take a long time to properly write. Apparently, the ancient scribes didn't have the time or patience to make their records with hieroglyphics, so they invented hieratic. This is essentially a cursive form of hieroglyphics, but it is different enough to be considered its own script.

The god of wisdom and scribes, Thoth.

Some of the hieratic symbols are recognizable from their hieroglyphic roots. For example, the hieratic number 2 evolved from the two straight lines of the hieroglyphics. When an ancient scribe painted the first line, he wouldn't lift his brush all the way before painting the second line, making the motion a little faster. This would have the effect of connecting the two vertical lines near the bottom.

Our number 2 was created in much the same way long ago in India. The only significant difference is that they

The transition of the hieroglyph 2 to the hieratic 2.

started with two horizontal lines. The curve of the 2 is nothing more than the backstroke to begin the second line. The number 3 evolved in much the same way.

As the figures grew more complicated, the Egyptian scribes reduced parts of the hieroglyphic symbols to

The evolution of the Hindu 2 into the modern 2.

simple strokes. Consider the hieroglyphic 7. Normally it could be written in four vertical strokes on top of three more. The impatient scribe would paint all four as one horizontal brush stroke and zigzag back for the next row. He could not make a stroke for the second row because it would be unclear how many ones it contained. So he would jiggle his brush representing two and follow it with a slash down representing the third one. While the figure hardly looks like the original seven lines, all that really matters is that the scribes recognized this symbol and were able to easily distinguish it from similar figures.

The ancient scribes of Egypt, like any other accountants, eventually needed to find the total of the values of

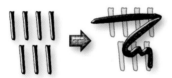

The transition of the hieroglyph 7 to the hieratic 7.

the objects they inventoried. This is perhaps the most difficult subject for me to write about. I can easily explain how they calculated the volume of an unfinished pyramid, multiplied mixed numbers, and created fractional identities, but I can't be sure how they added 15 and 12. Both of the ancient math scrolls that have been found regard addition as being too simple to detail. Hence the work of the solutions is not shown. Only the answers are given. With no written record, we can do little but guess.

On the surface, the problem doesn't seem that bad. Consider the following addition of 82 and 54. The Egyptians

would have written the numbers in hieratic, but I'm using hieroglyphics just because they're more recognizable.

The above sum is easily added in one's head. We can first combine the ones, |, adding 2 and 4 to get 6. Then we

can compute the tens, ∩, adding 8 and 5 to get 13. We can interpret this as ten 10s and three 10s. Since ten 10s is 100, or ℰ, the answer is ℰ ∩∩∩ ||||||.

We can't automatically assume that the ancient Egyptians used this method. Although there are hints in ancient texts, there is no direct evidence as to how they added. We also don't see their work when they added lists of fractions often having different denominators. I can't imagine doing such problems in my head. If we know they didn't do all additions in their head, how can we be sure they did any in their head?

We've fallen into the trap of mathematical familiarity. Subconsciously we think to ourselves, "They must do their math the way we do, because our way is the right way." Of course it's foolish to think that anything we do is the so-called right way. It's often simply one choice of many.

In Mesopotamia at that time, there is some evidence that the people used tokens to perform calculations. In order to explain this in modern terms imagine that you have 82 cents in one hand and 54 in the other. In the first hand you have 8 dimes and 2 pennies. Note the relationship between the 8 dimes and the eight ∩ in the scribe's number. Similarly in the second hand you have 5 dimes and 4 pennies. All you need to do to add these numbers is to pour the coins from one hand into the other. So you now have 13 dimes and 6 pennies in one hand. You now decide you have too many coins, so you replace 10 dimes with a one-dollar bill. The addition is now complete. You now write ₵ representing the dollar, ∩∩∩ representing the three dimes, and ||||||| the six pennies. Note that we never actually added any digits. We simply pushed the piles together and made a currency exchange.

I'm not advocating that the Egyptians used tokens. You could rightly argue that there is no physical evidence of tokens being used in ancient Egypt. I could counter by saying that there is no need for physical tokens. They might have done their mathematics on a dust board. This is roughly the equivalent of doing math on a dirty car window. They could have "placed tokens" by making marks with their fingers and "picked up tokens" by smearing dirt over the marks they wished to erase. I'm simply pointing out that there are other ways, beside the modern methods, to solve problems. In the absence of evidence, speculation is fine, but we have to understand it for what it is. Ignorance has never stopped me in the past, so let's add some Egyptian numbers.

EXAMPLE: Add 𓆼∩∩||||| and 𓆼𓆼∩∩∩∩∩∩∩|||||.

Let's proceed as we would when adding modern numbers. First we add the 1s. This is more difficult because of the way I wrote the numbers. Note that the lines that denote the 1s blur together and are difficult to count. Ancient people recognized this and generally never wrote more than four identical number symbols in a single row. My modern word processor just doesn't seem to appreciate this difficulty, refusing to do anything but go left to right. If we count them carefully, we will get six 1s in the first number and five in the other. When added these make 10 and 1, or equivalently, ∩|. We treat the ∩ as we would a carry.

We now add the two 10s to the seven 10s along with the carry to get ten 10s. This presents us with two minor problems. First of all, the entire 10, ∩, gets "carried." So we don't write down any ∩. The second problem is a little more subtle. We're used to seeing the hundreds digit next to the tens digit. When we see 324, we assume the 3 represents the hundreds because it's before the 2, which is the tens digit. However in the Egyptian number 𓆼∩∩|||||, the 𓆼 is not a hundreds digit but actually represents 1000.

Today we would avoid this confusion by writing 1026, putting a 0 between the tens and thousand. However, when we write zero ₵, we simply don't write anything. So if we weren't paying attention, we might be tempted to add an extra 𓆼 as our carry because it's the next digit in the number. However ten 10s is ₵, not 𓆼. Finishing our addition, we add the one and two lotus flowers to get 𓆼𓆼𓆼, giving us a final answer of 𓆼𓆼𓆼₵|. Here's the work of the above problem without all the discussion. Note that I've left room for the missing hundreds and wrote the carries above in the modern way.

SOLUTION:

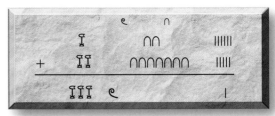

Now try this one on your own.

PRACTICE: Add |||| ₵₵₵₵₵₵| + |||||||𓆼𓆼₵₵₵₵₵₵₵₵||.

ANSWER: 𓂋𓆼𓆼𓆼₵₵₵|||

Although we can't be sure that the Egyptians viewed addition in this way, it seems fairly simple. It's not hard to understand why they didn't devote any precious book space to this elementary subject. Multiplication, on the other hand, is different.

BREAD, BEER, AND PESU

Whole-Number Multiplication

Ask an ancient Egyptian, "What's for dinner?" and they'd probably reply, "Bread and beer." If they were feeling like going wild, they would add some honey or butter. They did eat meat, particularly beef, but this was generally reserved for special occasions, at least for the labor class. The average Egyptian could scrounge up some other food items. They might have a duck to lay the occasional egg or a goat for some milk. They might do some moonlighting on the weekend to earn a few specialty items like fish or figs, but since they were paid in grain, bread and beer were served daily.

You might think that the Egyptians were raging alcoholics, drinking beer with every meal, but their beer was watered down. The purpose of the alcohol was to kill the bacteria in the river water, not to get drunk. The grain to brew your own beer came from the same grain used to bake your own bread. So if you drank too much, you might go hungry. This was no small concern in the ancient world, where having a little belly fat was considered a sign of wealth.

The Egyptians carefully measured the amount of grain that went into a loaf of bread or jug of beer. They measured grain in *hekat*, a hekat being about a gallon. They measured bread in *pesu*, the number of loaves that could be made out of 1 hekat of grain. This is different from most modern systems in that the larger the pesu, the smaller the loaf of bread, since the more loaves you made out of 1 hekat, the smaller they would have to be. Perhaps it's easiest to think of pesu as a fraction. A 10-pesu loaf is a *tenth* of a hekat in grain and so on. There are some modern equivalents to this system. For example a quart is a *quarter* of a gallon. This makes for some unusual math since we usually associate larger numbers with greater quantities. With pesu, the reverse is true. For example, two 10-pesu loaves is not 20 pesu but is equivalent to a 5-pesu loaf.

We might ask, why would the Egyptians use this system? Once again we have to set aside modern methods and ask ourselves whether is there a good reason for the Egyptians to have chosen methods different from our own. While it is more difficult to add loaves of bread in the pesu system, perhaps there are problems that are easier in this system. If I want 2 hekat of 25-pesu bread, I need 50 loaves. This is obvious in the pesu system since each hekat has 25 loaves, so 2 hekat have $2 \times 25 = 50$ loaves. In our system, this problem would be trickier. We would rate a 25-pesu loaf as having 0.04 hekat of grain. To figure out how many loaves make 2 hekat we would have to calculate $2 \div 0.04$, to obtain 50. Which is easier, multiplying by a whole number or dividing by a decimal? The answer is obvious.

This problem naturally leads us to the Egyptian method of multiplication. Consider the following problem:

EXAMPLE: How many loaves of 10-pesu bread can be made out of 7 hekat?

We know the answer is 10×7, or 70. The Egyptians had a unique method of solving such a multiplication problem. They'd begin by writing the numbers 1 and 7 in the first row of their solution. I'm going to also write 10 above the 1 for reasons you'll soon understand. They then proceeded to double the row repeatedly forming new rows as follows:

10	
1	7
2	14
4	28
8	56

We stop doubling when the first column reaches 8 because if we double it again, it will reach 16 which is bigger than 10.

Our goal now is to find numbers in the first column that add up to 10. Although it's easy to find them for such a simple problem, let's do it in a systematic way that will work for more complex problems. Start with the number 8 on the bottom and place a checkmark at the end of the row. Now go up to the next row, 4. If we add 8 and 4 we get 12, which is too big since we want 10. So we ignore the 4 row and proceed to the next row up. Now we add 8 and 2, which is 10. This is of course not bigger than 10 so we check the 2 row. If we didn't notice that we're already at 10, we could check the last row but of course adding 1 to our total which is now 10 will give us too much. Our work now looks like this:

10	
1	7
2	14 ✓
4	28
8	56 ✓

Finally we add up the second column, only using the checked rows. This gives us 14 + 56, or 70, the solution to 10×7. The final work appears below.

SOLUTION:

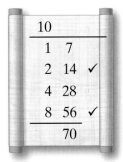

Let's try another one.

EXAMPLE: How many loaves of 25-pesu bread can you make out of 3 hekat?

Since there are 25 loaves in each of 3 hekat, we have to determine 25×3. As in the first problem, we write 1 and 3 in the first row and double it until the first column is as large as it can be without being greater than 25.

	25	
1	3	
2	6	
4	12	
8	24	
16	48	

Check the 16 row and add as we go up, remembering to skip rows that make the number larger than 25. Note that 16 + 8 is 24, which is fine, but 24 + 4 is too big. So is 24 + 2. Finally we note that 24 + 1 gives us the 25 we need.

	25	
1	3	✓
2	6	
4	12	
8	24	✓
16	48	✓

We end by adding up the checked entries of the second column getting 75, which is the solution of 25×3.

	25	
1	3	✓
2	6	
4	12	
8	24	✓
16	48	✓
	75	

Try this one on your own. You should get 156.

PRACTICE: Multiply 13 by 12 as an Egyptian would.

The method is remarkably fast and easy when you get used to it. Many people are surprised that it works, but it's not hard to see why, provided you look at it in the right context. Consider the following multiplication of 6 and 5:

	6	
1	5	
2	10	✓
4	20	✓
	30	

Instead of writing the numbers, I'm now going to represent them with tokens, specifically pennies and nickels. The first row is 1 and 5, which I will represent with one penny and one nickel. Note that the cash value represents the number. Instead of doubling the numbers, I'm going to double the number of coins. So the second row is two

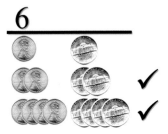

Beginning the multiplication of 6 and 5 using "coins."

pennies and two nickels and the third row is four pennies and four nickels.

Note that each row has exactly the same number of pennies as nickels. This means that the cash value in the right column is five times the value in the left. Finally, note that the cash values are precisely the values in the preceding numerical example.

We want a 6 in the left column, so we check the rows that begin with two and four pennies. When we add these two rows, we get six pennies in the first column and six nickels in the other. Notice that this sum also has the same number of coins in each column. So getting six pennies in the first column means that there will be six nickels in the second. Clearly this means that there will be a cash value of six times the value of a nickel, or 6×5, which is precisely the multiplication we wished to solve.

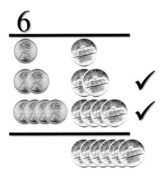

Each value in the second column is five times that of the first.

The above thought experiment would work on any multiplication of positive whole numbers. For example, if we wanted to multiply 11 by 25, we would simply mimic the above process with quarters, getting the left column to add up to 11 pennies. The bottom of the right column would then have 11 quarters, which has a value of 11×25. Similarly we can multiply by any number, such as 7, although we would have to imagine a 7-cent coin.

SIMPLICITY

Whole-Number Division

My eldest daughter just finished the third grade. Her strongest subject is mathematics, which couldn't make her math-professor father more proud. She did have one

trouble spot: memorizing her multiplication tables. This doesn't surprise me, because in a very real sense, memorizing is not mathematics.

We live in a world that obeys the laws of mathematics. We know that $2 + 2 = 4$, not because we memorized this but because every time we had two things and got two more, we ended up with four. Our common sense has been fine tuned by living in a mathematical world. For example, how do we know multiplication by 0 gives 0? I know that two cartons of eggs contain 24 eggs. I can transfer this knowledge into the mathematical fact that 2×12 is 24. Now that I need to know the value of 0×12, I can simply ask myself, how many eggs are in zero cartons? The answer is clearly 0 so I know 0×12 is 0.

Our brains have evolved to deal with mathematics. They even have special locations devoted to mathematical ideas and knowledge. If I were to surgically remove the part of your brain where your knowledge of letters and written words are stored, you would of course no longer be able to write. Oddly enough you would still be able to write down numbers. This is because your subconscious brain recognizes that number knowledge is different and stores their symbol shapes in a separate part of the brain than the location for the symbols of language. Mathematical understanding is different from our knowledge of language. I can't explain why the word "dog" means dog; it just does. We memorize it, and people who speak other languages learn different, arbitrary, words for dog. Mathematical knowledge is not subjective. We don't arbitrarily decide the value of $2 + 2$.

For some reason, our brains don't recognize our multiplication tables as mathematics. So when my daughter is confronted by the fact that $7 \times 8 = 56$, her mind regards this as arbitrary memorization and essentially stores this knowledge in the same place it keeps things like advertising jingles. Of course $7 \times 8 = 56$ is not quite as catchy as a tune sung by a cute five-year-old on the subject of hot-dogs. So we tend to remember one and forget the other.

Memorization in place of understanding seems to have a detrimental effect on early education. Consider the English number system. With the forties we have the numbers "forty-one," "forty-two," and so on. There's an obvious pattern. The word "four" is followed by the syllable *ty*, which is short for "ten" and then it's followed by a number. So "forty-two" literally means "four tens and two." It's very simple. We have the same pattern with the fifties, sixties, and so on. However, in the tens, we run into a problem. We say "eleven" and "twelve" when we should be writing "onety one" and "onety two." Likewise, the teens still don't follow the standard pattern. Notice that fourteen doesn't follow the same pattern as twenty-four and thirty-four. In contrast, the Chinese word for 11 essentially translates to "ten one." The Chinese system follows a fixed pattern for these numbers while ours does not. As a result the mathematical development of English-speaking students slows down compared to the Chinese just at the time when they hit these numbers. Excessive memorization inhibits mathematical growth.

After my daughter finally commits her multiplication table to memory, she'll begin to tackle the methods of long multiplication and long division. These are fairly complicated processes. They may seem easy to some of us now, but that's because some dedicated teacher forced us to attempt problems over and over until we memorized the algorithm.

You might be asking what this has to do with Egyptian math. What I want to do now is begin to compare our system to theirs. Consider the following long multiplication and the corresponding long division. As you look at them, think about how many memorized facts are required.

$$
\begin{array}{r}
56 \\
\times\ 37 \\
\hline
392 \\
168 \\
\hline
2072
\end{array}
\qquad
\begin{array}{r}
37 \\
56\ \overline{)\ 2072} \\
168 \\
\hline
392 \\
392 \\
\hline
0
\end{array}
$$

In this multiplication, students need to have memorized 7×6, 7×5, 3×6, and 3×5. They also have to adjust the answer to 7×5 by the carry of 7×6, and the same is true for 3×6 and 3×5. As they do each digit multiplication, they need to carefully build the number going from right to left. When they start the 3 multiplication, they

must remember to go down a line and shift left by one. Finally, they need to compute a sum.

The long division is trickier. They first have to see that 56 does not go into 2 or 20 and then guess how many times 56 goes into 207. This guess is far from obvious, especially to a child in grade school. They take their guess of 3 and multiply it by 56, using the knowledge of 3×5 and 3×6 and modifying the first by the carry of the second. They also must know to place the answer 168 so its rightmost digit, 8, lies directly underneath the rightmost digit of 207. They then subtract the two answers and repeat the above steps, guessing how many times 56 goes into 392. Finally, they must stop when they get 0, knowing that the process ends when they get any number less than 56.

When you looked at the above multiplication and division, it may not have seemed complicated, but note that when I describe the system in words just how involved it seems to be. It's like riding a bicycle. It seems easy because we've blocked out the memory of how many times we fell as we waited for our muscles to learn how to make it "easy."

Now look at the same multiplication and division done the Egyptian way.

37		
1	56	✓
2	112	
4	224	✓
8	448	
16	896	
32	1792	✓
	2072	

	2072	
1	56	✓
2	112	
4	224	✓
8	448	
16	896	
32	1792	✓
37		

The first thing that should catch your eye is that the multiplication and the division look almost exactly the same. That's because they are basically the same. Each element in the second column is 56 times larger than the corresponding number in the first column. Hence when we make 37 in the first column, we get 37×56 in the second. Reversing the same logic, every number in the first column is equal to the number in the second divided by 56. So if we make 1792 in the second column, we get 1792 ÷ 56 in the first.

Let's look at an example.

EXAMPLE: Compute 133 ÷ 7 as an Egyptian would.

We start by putting 1 and 7 in the first row. Above the second column, put 133. Continually double each row until the second column approaches but does not exceed 133. In this example we stop when we get to 112 since 112×2 is 224, which is greater than 133.

	133
1	7
2	14
4	28
8	56
16	112

Start at the bottom in the right column and check off the 112. Now we add the numbers going upward, making sure the sum doesn't exceed 133. So we try 112 + 56, which is 168. Since this number is bigger than 133, we go to the next row up. Then we try 112 + 28, which is 140 and still too big. When we try 112 + 14 we get 126, which is fine, so we check the 14 row. Finally we try 126 + 7, which is 133, our number, and we check off the 7 row. Now we add up the elements in the first column in the checked off rows giving 16 + 2 + 1, or 19, the solution to 133 ÷ 7.

SOLUTION:

	133	
1	7	✓
2	14	✓
4	28	
8	56	
16	112	✓
19		

Sometimes we get remainders. Consider the following problem:

EXAMPLE: How many hekat are there in 81 loaves of 15-pesu bread?

Recall that to determine the number of loaves you multiply the number of hekat of grain by the pesu. To find the number of hekat, you divide the number of loaves by the pesu. So we want to compute 81 ÷ 15. Once again we make a table, with the first row being 1 and 15. We label the second column with an 81. Now we double the first row until we're about to pass 81 in the second column and add rows going up.

	81	
1	15	✓
2	30	
4	60	✓
5	75	

Notice that as we add going up we get to 75 and are forced to stop. This is because 15 goes into 75 but not 81. We're 6 short. This seems like a problem, but it really isn't. The 5 we get as our answer is 75 ÷ 15. When we divide 81 ÷ 15 we still get 5 but with 6 remaining, which is sometimes expressed as "5 remainder 6." Try these two. The second will have a remainder.

PRACTICE:

a) Divide 187 by 17
b) Divide 100 by 21.

ANSWERS:

a) 11
b) 4 remainder 16

Let's return to the lesson of this section. In order to learn modern long multiplication, we need to memorize a

10-by-10 multiplication table. Then we need to learn how to multiply a single digit by a longer number, adding carries to later multiplications as we go. For a long multiplication we have to do this repeatedly, lining up numbers in a shifting pattern. Finally these numbers need to be added. Long division is even worse. We must continually guess how many times some ugly number like 37 goes into an even uglier number like 207. Once this guess is made, we need to multiply our answer by 37, which takes a fair amount of time, as we see above. Then we must line this up just right and subtract. This procedure has to be repeated multiple times.

Now let's compare this process to the Egyptian system. First of all, the only number you need to multiply by is 2. In fact, you don't really need to know this because it's the same as adding a number to itself. You do this a few times, add parts of the columns, and you're done. Division is almost exactly the same as multiplication. It's all simple addition.

I teach Egyptian mathematics at my college. I can teach my students how to do Egyptian whole-number multiplication and division in less than five minutes. They do it faster than I can and almost error free. Now compare that process to the two or so years my daughter will spend in class using drills and rote memorization in order to learn her arithmetic. I suspect that somewhere in the heavens, Thoth, the scribal god, is looking down at us with a smirk on his face thinking, "Who's primitive now?"

2
FRACTIONS

THE SHELTERING ARMS OF THE DESERT

Egyptian "Fractions" and Decimals

Egyptian society was concurrent with a number of great Mesopotamian empires; however, those empires rarely lasted more than a couple of generations, being brought down by barbarians or a warlord with a desire for an empire of his own. Although it was a great civilization, Mesopotamia suffered from constant turmoil and abrupt change. Although equally as large, Egypt was more of a nation than an empire. For almost three thousand years, Egypt remained relatively stable. There were a few "intermediate" periods consisting of either internal strife or foreign influence, but these were small compared to the thousands of years of internal peace.

Egypt owed much of its stability to its unique geography. The land simply kept the violent outsiders away. In order to understand ancient Egypt, however, we must realize that modern borders meant little in the ancient world. Egypt was split into two parts: the Nile valley and the northern delta. The valley is hundreds of miles long but never more than about ten miles wide. A map of this part of Egypt would look like a long string winding its way through the desert. The western desert comprised the harsh sands of the Sahara Desert. The eastern desert was a stony, mountainous landscape. Neither side could support enough people to seriously threaten the Egyptians who lived on the banks of the Nile, and both sides presented a sweltering, waterless barrier to outside invaders.

The Egyptians were also relatively protected in the south. Most nations have trouble defending borders that span up to a thousand miles long. In order to protect the entire country, they would need to disperse their armies

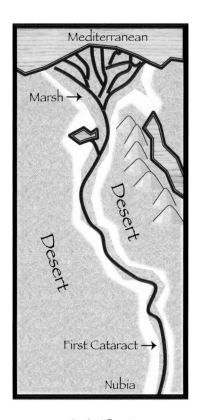

Ancient Egypt.

into small, vulnerable groups. The southern border of Egypt was only a few miles wide because it consisted only of the width of the Nile and its banks. It was easy to concentrate troops at this one point. Egyptians had little fear of an armada sailing downriver from the south. The Nile has a series of rapids, called cataracts, along its southern portion. Large boats could not navigate these waters and would have to be portaged over land to get around them.

The Egyptians wisely built forts at these points to harass any fleet that attempted this tactic.

The northern delta was also fairly safe. This triangular region was also protected on the east and west by desert. The delta did have some vulnerability due to its long border with the Mediterranean, but this was not as bad as it might seem. During most of Egypt's history, sea travel was a risky business. Ship-building technology had not progressed far enough at this time. Any king would be taking a great risk by placing the army charged with protecting his kingdom in a series of not-so-seaworthy ships. One storm could wipe out the bulk of his armed forces.

Even if an invading force managed to sail a fleet to the delta, it would immediately encounter problems. Ships that can survive in the sea are not very good on rivers and would be at a serious disadvantage against Egypt's river fleet. If the force attempted to disembark and proceed on foot, they would run into other problems. The delta consists of many small islands surrounded by the branches of the Nile. This area is swampy and not conducive to moving heavy military equipment. If an invading army conquered one of the islands, they would have to load all of their equipment back in their boats and sail to another. Holding an island would force them to leave men behind, splitting their troops. In any case, the time such operations took would enable the Egyptians to mount a large-scale response. Clearly, Egypt was safe until seafaring and military technology advanced to the point of overcoming these obstacles.

According to one of the Egyptian theologies, Ptah formed the world out of chaos, which still surrounded it. Egypt was the center of this world, and as you moved away from Egypt, the closer you came to this disorder. The violent and barbaric ways of the outsiders probably seemed natural to the Egyptians' world view because the foreigners bordered the primal chaos. Egyptians had little respect for and wanted little to do with non-Egyptians. In many ways, this view was justified despite its seemingly racist overtones. Imagine how a pharaoh whose position was based on a thousand-year-old tradition would view a Mesopotamian king who recently secured his position through violence and plunder. Egypt had little contact with the outside world and had little desire to change that situation. As a result, Egypt developed in its own way and did things radically differently from the rest of the world. This is not to say that there was no foreign influence on the Egyptians, but they had the ability to reject or accept ideas as they saw fit, and they were predisposed to reject them. Nowhere is this more apparent than in the way they dealt with fractional values.

The Mesopotamians had a system remarkably close to our decimal system except that they used base sixty. When we want to express a half, we write 0.5 in decimal form because 5 is half of 10. When the Mesopotamians wanted to write one-half, they wrote something like 0.30, since 30 is half of 60. Actually, they didn't have zeros or a "point" symbol, so they would just write a symbol that represented 30. In any case, this is why today half an hour is 30 minutes. We're still writing the "decimal" value of time in ancient Mesopotamian. We essentially use the Mesopotamian system today in part because they were not isolated from the rest of the world of which we are a part.

The Egyptian method seems decidedly strange to the modern mind. They only used parts, like a fourth or a tenth. The hieroglyphic representation of a fourth was simply the number four under the symbol of a mouth. The mouth symbol makes an *r* sound in the Egyptian language. It's possible that they added *er* to the end of a number, just as we add *th* to the end of a number like ten to form a part, like a tenth.

One-fourth written in hieroglyphs.

So a fifth could be written ⅢⅢ and a thirteenth as ⋒Ⅲ. We will use the modern shorthand of placing a bar over the number, so $\overline{4}$ will be used to represent an Egyptian fourth. Today these numbers are called *unit fractions*. A fourth is the fraction ¼. The word "unit" refers to the 1 in the numerator of the fraction. This is a misnomer because Egyptian fractions are not fractions in the modern sense. There is no 1 in the numerator because there is no numerator.

When modern mathematicians are confronted with Egyptian fractions, they often shake their heads in disbelief. Remember that Egyptian notation could express fractions only where we would put a 1 in the numerator. There simply was no way to write ⅖. When Egyptians wanted to express this value, they had to write $\overline{3}\,\overline{15}$. This is ⅖, since ⅓ + ¹⁄₁₅ is ⅖. Who would possibly want to express non-integer values as the sum of fractions? The answer is simple. We would, and we do it every day.

Every math student eventually learns the approximation 3.14 for π. What does the representation 3.14 mean? If you're sufficiently familiar with the base system, you should recognize that this is the sum of three parts: a unit, a tenths, and a hundredths part. So when we write 3.14 we really mean this:

$$3.14 = 3 + \tfrac{1}{10} + \tfrac{4}{100}$$

But this can be simplified to the following:

$$3.14 = 3 + \tfrac{1}{10} + \tfrac{4}{100} = 3 + \tfrac{1}{10} + \tfrac{1}{25} = 3 + \overline{10} + \overline{25}$$

Just as we ignore the plus signs between the parts of our decimal representations, so did the ancient Egyptians. They might represent the quantity 3.14 as $3\ \overline{10}\ \overline{25}$ or in hieroglyphs as ⲓⲓⲓ ⲛ̇ ⲛ̇ⲛⲓⲓⲓⲓⲓ. I think part of the problem modern mathematicians have in appreciating Egyptian mathematics is the phrase we use to describe their numbers, "unit fractions." We compare their system to our system of fractions instead of to our decimal system, with which it has far more in common.

To truly comprehend the Egyptian system of fractions, we must deeply understand the properties of our decimal system that make it so effective. Imagine that you won the lottery. Someone calls you up on the phone and tells you that you won 28 million and …. At this point your phone drops the signal. The caller was going to say $28,732,593, but they only got off two of the eight digits.

Yet only hearing 28 million, you have a very good approximation of what you've won. Technically, the number is the sum

$$
\begin{array}{r}
20,000,000 \\
8,000,000 \\
700,000 \\
30,000 \\
2,000 \\
500 \\
90 \\
+\ \underline{\hspace{6em}3} \\
\end{array}
$$

The sum of the last six terms is dwarfed by the first two numbers, so we don't really need to add all the terms to get a good idea of the total value.

The same holds true for decimals. Consider the following approximation for the square root of two, 1.414214. To an untrained eye, this expression is overly complicated. It's the sum of an integer and six different fractions, each with a different denominator. It consists of parts measured in millionths, parts too small to intuitively grasp. Yet when we look at it, we immediately see a number close to one and a half, or a little more than 1.4. The beauty of the decimal system is that it gives as little or as much information as you need. I can view the representation of the approximation of the root as *a little more than 1.4 or a little more than 1.41 or a little more than 1.414*. With each phrase I'm forced to tolerate more complexity in exchange for more accuracy. The best part is that the choice is ours. The rapidly declining significance of the place values gives us power to adjust our number interpretations to our needs. It gives us a quick approximation together with an accurate estimation.

The Egyptian system does exactly the same thing. Their representation of 3.141, $3\ \overline{10}\ \overline{25}\ \overline{1000}$, can be quickly assessed as a little more than $3\tfrac{1}{10}$. If they need more accuracy, they can include more "digits." For contrast, let's compare this system to our fractional system. Try to guess the rough value of $\tfrac{4586}{1310}$. Can you come up with an estimate? Even if you can, do you have any idea how close your approximation is? However, in Egyptian this number is $3\,\overline{2}\,\overline{1310}$, which is obviously a number less than a thousandth away from 3 ½. Clearly the Egyptian system has more in common with our place-value system. Just like our decimal system, there's an easy balance between accuracy and estimation.

One common observation of the Egyptian system of fractions is that they never would write the same fraction twice within one number. So you never see a number like $7\,\overline{5}\,\overline{5}$. This so-called rule is more likely the misinterpretation of a more general rule of thumb. Think of our decimal representation of this number:

$$7\,\overline{5}\,\overline{5} = 7 + \tfrac{1}{5} + \tfrac{1}{5} = 7 + 0.2 + 0.2 = 7.4$$

Our placement system refuses to accept repeated digits, which is perhaps the reason we take this for granted. If we tried to force $7\,\overline{5}\,\overline{5}$ into our decimal system we might get something like this: 7.2.2. We would then interpret this as a number close to 7.2, but it isn't. It's actually 7.4, and 7.2 is a bad approximation. Similarly the Egyptians would not tolerate $\overline{5}\,\overline{5}$. It's not close to a fifth. It's twice as big as a fifth. To be off by 100% in an approximation is terrible. The Egyptians would write this fraction as $\overline{3}\,\overline{15}$. The fraction is close to a third, not a fifth. It's a third and a little bit more. It's important that $\overline{15}$ is significantly smaller than $\overline{3}$ because it is a refinement of an approximation. So they know it's basically a third, and if they need more accuracy, they can add the fifteenth.

The rule applies to more than just equal fractions. The Egyptians would never write $\overline{5}\,\overline{6}$, since a sixth is too close to a fifth to be a refinement. This number is not close to a fifth—it is almost twice as much. So they would instead write $\overline{3}\,\overline{30}$. We can easily check to see that these represent the same value as follows:

$$\overline{5} + \overline{6} = \tfrac{1}{5} + \tfrac{1}{6} = \tfrac{6}{30} + \tfrac{5}{30} = \tfrac{11}{30}$$
$$\overline{3} + \overline{30} = \tfrac{1}{3} + \tfrac{1}{30} = \tfrac{10}{30} + \tfrac{1}{30} = \tfrac{11}{30}$$

As a result, Egyptian fractions are fairly easy to read. But isn't it difficult to work with fractions that have radically different denominators? We will see that it is true only if you stick to the modern method, but not if you calculate like an ancient Egyptian.

THE BEST THING SINCE SLICED BREAD

An Intuitive Model for Egyptian Fractions

One of the most common duties of an ancient Egyptian scribe was to pay the workers. This was not always easy because they often received shares for pay and not a set salary. A scribe would receive some amount of food and have to divvy it out fairly according to each worker's share value. Not every job had equal shares, just as today not every job has an equal salary. For the purpose of this section, however, we'll assume everyone has a share value of 1.

Assume that seven workers get a loaf of bread to share. This isn't very realistic because they were usually paid in grain, but let's accept it for argument's sake. The scribe realizes that one loaf divided between seven men means that each gets a seventh, which the scribe would record as $\overline{7}$ in his records.

One loaf of bread cut up to feed seven workers.

So far everything is working out well. However, the next day the workers get four loaves of bread. The scribe could cut each loaf of bread into seven pieces and give each worker four, but it seems like too much cutting and the workers won't appreciate all the small pieces. The scribe gets an idea and decides to cut each loaf in half. Now he has eight pieces, enough to give each worker one and he has one left over.

The loaves cut in half make one piece for each of the seven workers with one left over.

Being an honest scribe, he doesn't keep the one remaining piece for himself but cuts it into seven pieces, one for each worker. The size of the smaller pieces is a seventh of a half, which is a fourteenth, since 7×2 is 14.

When the eighth piece is cut into seven, the smaller pieces can be handed out to the workers.

So each worker gets a half loaf and a fourteenth of a loaf, which the scribe records as $\overline{2}\ \overline{14}$.

Note that we've just performed our first Egyptian division involving fractions. We now know that $4 \div 7 = \overline{2}\ \overline{14}$. Although this was not their typical method of division, it does show how natural the Egyptian system is. It's also possible that their method of fractions was formed from similar considerations.

Let's try another division.

EXAMPLE: Divide 3 by 16 using "sliced bread."

Before we can begin, we need at least 16 slices so each worker can get one. If we were to cut each loaf into four pieces, we would get only 12 slices. Mathematically, this tells us that $3 \div 16$ is smaller than $\overline{4}$, so we can't write the division as $\overline{4}$ plus some other fractions. If we cut each loaf into 6 slices, we get 18 slices, which is more than enough

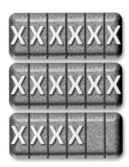

Three loaves to be divided between 16 workers. After cutting them in sixths, each worker gets one, leaving two slices.

for the 16 workers. We'll perform the first slice and mark off the 16 slices that will be distributed.

We now know that each worker gets $\overline{6}$ of a loaf with 2 slices left over. If the scribe now cuts the remaining 2 slices into eight pieces, he gets 16 smaller slices, which is one for each worker. The size of these smaller slices is $\overline{48}$ since we cut an eighth of a sixth and 8×6 is 48. This makes ³⁄₁₆ equal to $\overline{6}\ \overline{48}$. The final answer is as follows:

SOLUTION: $\overline{6}\ \overline{48}$

The remaining two slices get cut into eighths and distributed to the 16 workers.

PRACTICE: Divide 2 by 5 using "sliced bread." Make your first cuts as thirds.

ANSWER: $\overline{3}\ \overline{15}$

PRACTICE: Divide 4 by 18 using "sliced bread." Make your first cuts as fifths.

ANSWER: $\overline{5}\ \overline{45}$

Note that some divisions can't be done in two slices. Consider the following problem:

EXAMPLE: Divide 4 by 5 using "sliced bread."

Four loaves cut in half with each of five workers getting one slice. Three halves are left over.

If we cut the loaves in two, we'll get eight slices. We hand out five of them to each of the five workers, leaving three halves.

We can treat these three remaining halves just as we would three whole loaves. In order to divide three "loaves" into five, we can cut each in half. Note that these

The three remaining halves can be cut in half. Five of the six pieces are distributed between the workers.

are already halves so half of a half is a fourth. Now the three halves become six pieces of size $\overline{4}$, of which we can give one to each worker leaving one of the fourths.

We can now cut the remaining fourth into five pieces, giving each worker a piece of size $\overline{20}$. This makes ⅕ equal to $\overline{2\ 4\ 20}$.

The remaining quarter is cut in five pieces and given to the workers.

ANSWER 1: $\overline{2\ 4\ 20}$

The above example shows us that divisions might require three or more fractions in the solution. This should not come as a surprise since Egyptian fractions are closer to our decimals than to our fractions. You should realize that just as decimals may require a lot of digits, like 8.77928347723, so Egyptian math may require many fractions.

Note that in the above problem we did not need to cut all of the original four pieces in half. By cutting the first

This time only three of the loaves are cut in half.

three pieces in half we would get six halves, enough to give one to each of the five workers.

We can now cut both the half and the remaining whole into five pieces each. So now each worker gets a half, a fifth of a half, and a fifth, or equivalently $\overline{2\ 10\ 5}$.

The half and the whole loaf are each cut into fifths. Each worker gets two slices, one small and one large.

ANSWER 2: $\overline{2\ 5\ 10}$

Note that we got two different answers for the same problem. This means that Egyptian fractions can be written in more than one way. Specifically we see for the above problem that $\overline{4\ 20}$ is the same as $\overline{5\ 10}$.

PRACTICE: Divide 5 by 7 in two ways using "sliced bread."

ANSWER: $\overline{2\ 6\ 24\ 168}$ or $\overline{2\ 7\ 14}$

You might argue that having more than one way to write a number is extremely awkward, but as I mentioned in the introduction, we see difficulties in alien systems far more easily than we notice them in our own. In our number system all of the following are exactly the same.

$$1.75 = 175\% = \frac{7}{4} = 1\frac{3}{4} = \frac{63}{36}$$

At least the Egyptian system is consistent with the way it portrays numbers even though they are not exactly the same.

FILLING THE VOID: THE FRACTION $\overline{\overline{3}}$

Once each year the star Sirius would disappear behind the sun. Toward the end of this period, the Egyptians would scan the skies at dawn, searching for its return. The first day of its reappearance marked the beginning of a new year. This sign from the gods foretold the coming of the inundation—soon the Nile would flood, ending the harvest season. All of Egypt would then be covered by water except for the settlements on the banks and the higher islands of the delta. There wasn't much for the Egyptians to do except wait for the flood to recede, and yet this was the most crucial time in the Egyptian year.

The flood carried with it the two most valuable resources in Egypt, water and topsoil. Obviously water is as good as gold in a land surrounded by desert. The Egyptians built barriers to trap the flood waters and use them for irrigation in the drier seasons. Water was so precious that restricting its flow onto your neighbor's farm was an offense punishable by damnation during your soul's final judgment in the Hall of Osiris.

The flood waters also brought the most fertile topsoil found in the Mediterranean. During its four-thousand-mile journey, the waters of the Nile picked up nutrients and then scattered them throughout Egypt, enabling a huge civilization to flourish in the desert. The color of this rich soil lent the very name used by the Egyptians to describe their home, the Black Land.

The size of the flood varied from year to year. Some farmland was replenished annually. Other areas received the life-bringing waters only in high-flood years. The scribes of Egypt assessed the value of farmland based on how likely it was to be inundated. Too many years of low waters could result in famine for a culture that was overly dependent on the bounty of the Nile. However, if the flood was too high, their homes on the banks would have been threatened. The Egyptians needed to know the extent of a year's flood, so they invented the nilometer. This measuring device is essentially a stone stairway that descended to the Nile. As the waters rose, individual stairs would be covered in water, and marks on the nilometer would give the depth of the water. Hence the nilometer was essentially a giant ruler built into the Nile.

Reading a ruler tells us a lot about the way decimals are used to make approximations. Consider measuring a pencil with a ruler. In the diagram below, the ruler shows us that the pencil is somewhere between 8 and 9 inches. If we approximate the length as 8 inches, we'll be off by at most 1 inch.

If we want more accuracy, we could look at the smaller ruler marks indicating tenths of an inch between 8 and 9. Below we can see that the tip of the pencil falls between the sixth and seventh tick mark between 8 and 9. So the pencil length is between 8.6 and 8.7 inches. If we

The length is estimated by the marks the point falls between.

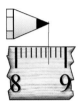

The pencil is between 8.6 and 8.7 inches.

approximate the pencil length as 8.6, we will be off by at most 0.1 inches.

There are a couple of things we should note. The first is that the distance between the ruler's marks determines the accuracy of the measurement. In fact, this distance is exactly the maximum error. The latter can be calculated by subtraction. The above pencil point falls between the 8.6 and 8.7 mark, and hence the maximum error is 8.7 − 8.6 = 0.1. We should also note that in a decimal measurement system, the distance between marks is uniform. This makes the error the same no matter where on the ruler our measurement occurs. This is not true for Egyptian fractions.

If we made a ruler marked with Egyptian fractions, it would look like the following diagram. In the middle would be the mark for $\overline{2}$ since it's half way. Similarly, one-third of the way over, we would find $\overline{3}$, and so on. The marks would become more tightly packed for the smaller measurements. They get so close that eventually we would have to stop marking the ruler so they didn't overlap.

The pencil being measured below is $\frac{3}{16}$ of a foot long. The ruler shows it being longer than a sixth of a foot but

The marks are close below $\overline{2}$, and hence the measurement is more accurate.

less than a fifth. If we convert $3/16$ into Egyptian fractions, we get $\overline{6}\,\overline{48}$. The Egyptian number verifies that it is a little, specifically a forty-eighth, more than a sixth.

Someone measuring the pencil with the Egyptian ruler can tell that it's more than a sixth but can't exactly tell by how much. Since the marks are not uniformly distributed, it's not as easy to know the largest possible error as it would be with a more conventional ruler. However, we can calculate the error by finding the distance between the $\overline{6}$ and $\overline{5}$ marks using subtraction. Since I haven't taught you yet how Egyptians subtract fractions, we'll use modern methods.

$$\overline{5}-\overline{6}=1/5-1/6$$

The common denominator is 30 and we can get both to be 30 by multiplying the $1/5$ by 6 and the $1/6$ by 5 giving

$$\frac{1\times 6}{5\times 6}-\frac{1\times 5}{6\times 5}=6/30-5/30=1/30=\overline{30}$$

Note that the error of a measurement between $\overline{5}$ and $\overline{6}$ is $\overline{30}$ and 30 is 5×6. It's not difficult to show using algebra that this is always true provided the Egyptian fractions are adjacent numbers. Hence the error measurement between $\overline{10}$ and $\overline{11}$ would be at most $\overline{110}$. A mathematician would phrase this as "the error of an Egyptian fraction is roughly the square of the smallest term."

We actually didn't even need to subtract the two values; we could just have easily noticed that the two lines on the ruler are "really close." They're all really close on the left side of the ruler, and hence all measurement on this half would be fairly accurate. However, there seems to be a problem with the right side. The following pencil is $5/6$ of a foot long. The best our ruler can do now is to estimate it as more than $\overline{2}$. In fact we can write $5/6$ as $\overline{2}\;\overline{3}$ in Egyptian fractions. Unfortunately, a half is a bad approximation for $5/6$. If we treat the $\overline{2}$ as the approximation and the $\overline{3}$ as the error term, the error is $66\tfrac{2}{3}\%$ of the estimate, which is too big to be considered accurate.

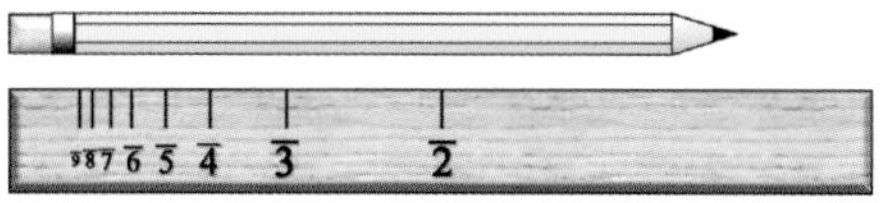

Objects that extend into the right half of the ruler are subject to bad approximations. The error is the distance from the $\overline{2}$ to the pencil tip.

This seems to fly in the face of our interpretation of Egyptian fractions as being a system of arbitrarily good approximations. However, the Egyptians solved this problem, and they did it in the most obvious way. Thinking in terms of rulers, the difficulty arises because there are no marks on the right-hand side of the ruler. So the Egyptians included an extra a mark or two.

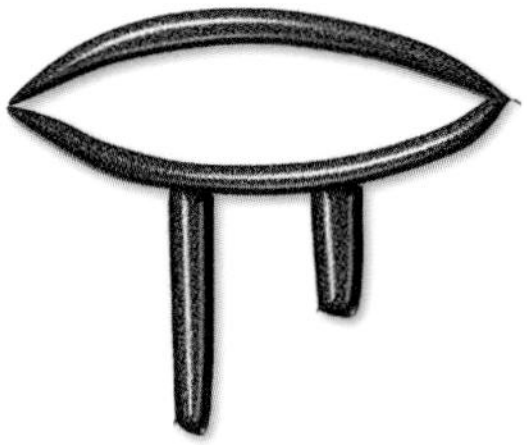

The hieroglyphic 2/3.

By far, the more common of the two fractions added to their "ruler" is $2/3$, symbolized by the following hieroglyph. We will use the modern transcription, $\overline{\overline{3}}$, for this symbol.

The choice of symbol is truly inspired. It looks like the mouth and is used to denote a fraction over the number 1 and "a half." This is in fact what $2/3$ is. In modern terms we get

$$\overline{1\tfrac{1}{2}}=\overline{(3/2)}=1/_{3/2}=2/3=\overline{\overline{3}}$$

We'll look at this relation in more detail later. Right now let's see the impact it has on approximations in measurement. Here's our previous pencil measurement using a ruler with the $\overline{\overline{3}}$ mark added.

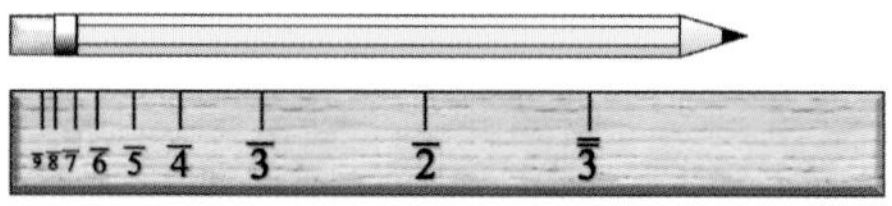

The distance between the $\overline{\overline{3}}$ mark and the pencil tip is smaller than that of the $\overline{2}$ mark. Hence the error in measurement is reduced.

Now the ruler reads $2/3$ and a little more, and the $2/3$ is a much better approximation than the $1/2$ we had previously.

The Egyptians would express the length of the pencil as $\overline{\overline{3}}$ $\overline{6}$ rather than $\overline{2}$ $\overline{3}$. If we treat the length as roughly $\overline{\overline{3}}$ with an error of $\overline{6}$, we get a 25% error, which is much better than the 66⅔% error we got with $\overline{2}$ $\overline{3}$.

You might argue that the gap on the ruler to the right of the $\overline{\overline{3}}$ is still a bit large. If you do, you'll find that a fairly small group of ancient Egyptians agree with you. It's extremely rare, but there are instances of a special symbol for ¾ being used. Most Egyptians seem to have felt that the ⅔ symbol was sufficient, and we will restrict ourselves to $\overline{\overline{3}}$, since the mathematical rules involving it are clearly spelled out by ancient texts.

When I first learned of the $\overline{\overline{3}}$ symbol, I was uncomfortable. Mathematicians like consistency and order. This symbol, being unique, bucked the rules and had to be treated with special operations. However, as I became more proficient with Egyptian mathematics, I began to understand that this gave their mathematical system added flexibility. In order to begin to appreciate the versatility of Egyptian math, we'll need to learn its basic operations.

3
OPERATIONS

MEMORIZATION AND TRIANGLES

Multiplication by $\overline{2}$

"Beginning of the teaching, explaining to the heart, instructing the ignorant, to know all that exists, created by Ptah, brought to being by Thoth, the sky with its features, the earth and what is in it, the bend of the mountain, and what is washed by the primeval waters ...." So begins the Onomasticon of Amenemipet.

Egyptian children destined for the scribal class entered school at the age of five. They had to start early because they had so much to learn. In the modern world, we take reading for granted. We have to learn only twenty-six letters associated with a few dozen sounds. The Egyptians had at least six hundred hieroglyphs to memorize, many of which had multiple meanings. Students had to spend many years in the temple schools, called the houses of life, learning to read and write.

Despite its grandiose title, the Onomasticon of Amenemipet is nothing but a list of Egyptian words grouped by meaning. Although we don't know exactly how the teachers used the Onomasticon, we do know that students copied texts, presumably to learn the symbols. The list consists of 610 words. The first 62 describe natural features like the sky, water, and earth. The next 167 words give the titles of gods, spirits, royalty, and officials. It continues by describing people, foreign lands, towns, architecture, agriculture, and food. Another version found on a damaged scroll also includes plants and trees.

The Onomasticon of Amenemipet is similar to most texts of the ancient world in that it doesn't discuss its subject in depth but rather is a mere list of information. Even a subject as involved as medicine appears as a list of symptoms, diagnoses, and treatments. Similarly, the papyrus of Ahmose appears, at least on the surface, to be little more than a list of math problems. These lists suggest that education in the ancient world consisted of a lot of rote memorization.

Although much of what they learned in those days is different from what we learn today, some things never seem to change. "One-half base times height" is perhaps the earliest geometric formula burned into our brains. I've taught senior citizens who haven't had a math class in many decades who remember that equation even though they've forgotten what it's for. Just in case you have had a memory lapse, it's the area of a triangle.

The "base times height" part of the equation calculates the area of a rectangle. This equation is apparent, at least for whole numbers. If we take a 2-by-3 rectangle, we can easily see that it consists of two rows of three 1-by-1 squares. Since a 1-by-1 square has the area 1 by definition, the total area is the number of squares. But since two rows of three objects has $2 \cdot 3$ squares, the area is just 6, the product of the height and the base.

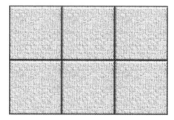

A 2-by-3 rectangle is made of two rows of three unit squares.

The "one-half" is simply an acknowledgement that a right triangle is half a rectangle, since the rectangle can be split in two down the diagonal.

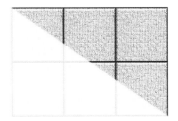

A triangle is half a rectangle.

The above argument can be made to work for any triangle, with some modification. Because the relations are so visually clear, it's not surprising that the Egyptians knew of this equation.

The calculation of the area of a triangle would have been trivial to an Egyptian. Their methods of multiplication and division are based upon repeated doubling of a number. Without a thought, they would know that double 7 is 14, but they would just as easily know that half of 14 is 7. This is called an *inverse relationship*. We learn to perform one operation; in this example, double 7 is 14. To find the inverse relationship we then pose the same expression as a question, what do we double to get 14? Of course the answer is 7. Since we reverse our knowledge of doubling to calculate half, the two operations are inverses.

Egyptians used this knowledge in their multiplications. For example, if I wanted half of 14, I could multiply it by ½ as follows.

Half of 14 is particularly easy since 14 is an even number. If instead we asked for half of an odd number, like 17, the process would not be that much harder. Since 17

is $16+1$, we can start by taking half of 16, which is 8. We can now treat the remaining 1 separately and realize that twice $\overline{2}$ is 1, so half of 1 is $\overline{2}$. When we combine our two answers we get 8 $\overline{2}$, which is half of 17.

The notion of breaking up 17 into 16 and 1 would be completely natural to an ancient Egyptian. Consider the following multiplication of 17 and 3. See how the division splits 17 into 16 and 1, precisely as we did above.

17	?	
1	3	✓
2	6	
4	12	
8	24	
16	48	✓
	51	

This trick of breaking up a number into pieces was probably helpful to the Egyptian students learning this method, especially when they moved on to larger numbers. It's particularly easy for numbers with all even digits.

EXAMPLE: Find half of 684 by breaking up the number.

SOLUTION:

$$684 = 600 + 80 + 4$$
$$\overline{2} \times 684 = 300 + 40 + 2$$
$$= 342$$

We can write 9 as $8+1$. So $\overline{2}$ of 9 is $4+\overline{2}$. We can still apply this trick if 9 is in the tens digit. For example, consider 94. The 9 in this number is really 90. Just as we thought of 9 as 8 and 1, we can think of 90 as 80 and 10. The 10 can be combined with the 4 making 14. Hence 94 can be rewritten as 80 and 14. So half of it is 40 and 7 making 47.

EXAMPLE: Find half of 76.

SOLUTION:

$$76 = 60 + 16$$
$$\overline{2} \times 76 = 30 + 8 = 38$$

PRACTICE: Find half of 58.

ANSWER: 29

This trick can be used for an odd digit anywhere in a number. The leftover 1 simply makes a "10" in front of the next digit. The next example uses the same 58 from the above practice problem, but here it's 58 hundred. Notice the similarity of computation.

EXAMPLE: Find half of 5832 by breaking up the number.

SOLUTION:

$$5832 = 4000 + 1800 + 20 + 12$$
$$\overline{2} \times 5832 = 2000 + 900 + 10 + 6$$
$$= 2916$$

PRACTICE: Find half of 258 by breaking up the number.

ANSWER: 129

The last thing you need to know is that if there is a 1 left over in the final digit, it turns into a $\overline{2}$.

EXAMPLE: Find half of 371 by breaking up the number.

SOLUTION:

$$371 = 200 + 160 + 10 + 1$$
$$\overline{2} \times 371 = 100 + 80 + 5 + \overline{2}$$
$$= 185\,\overline{2}$$

PRACTICE: Find half of 8743 by breaking up the number.

ANSWER: 4371 $\overline{2}$

PRACTICE: Repeat for 951.

ANSWER: 475 $\overline{2}$

With a little practice, you should be able to take half of a number in your head and just write down the answer. From now on we will take half just as we double, that is, without showing any work. We will now use this skill to calculate the area of a triangle in two multiplications. The following problem finds the area of a triangle. It first determines the value of half the base using the above method. The second multiplication multiplies this number by the height.

EXAMPLE: Find the area of a triangle with a base of 37 and a height of 12.

SOLUTION: Find half the base.

	$\overline{2}$	
1	37	
$\overline{2}$	18 $\overline{2}$	✓
	18 $\overline{2}$	

Now multiply this by the height.

	18 $\overline{2}$	
1	12	
2	24	✓
4	48	
8	96	
16	192	✓
$\overline{2}$	6	✓
	222	

And 222 is the area.

In the above multiplication we doubled the 1 until we were about to pass 18. To get the $\bar{2}$, we went back to the first row, which contains 1 and 12, and took half to get the last line, $\bar{2}$ and 6.

EXAMPLE: Find the area of a triangle that has a base of 13 and a height of 7.

SOLUTION: Find half the base.

$$\begin{array}{cc}
\bar{2} & \\
\hline
1 & 13 \\
\bar{2} & 6\,\bar{2} \quad \checkmark \\
\hline
& 6\,\bar{2}
\end{array}$$

Now multiply this by the height.

$$\begin{array}{cc}
6\,\bar{2} & \\
\hline
1 & 7 \\
2 & 14 \quad \checkmark \\
4 & 28 \quad \checkmark \\
\bar{2} & 3\,\bar{2} \quad \checkmark \\
\hline
& 45\,\bar{2}
\end{array}$$

And $45\,\bar{2}$ is the area.

PRACTICE: Find the area of a triangle that has a base of 9 and a height of 10.

PRACTICE: Repeat for a base of 23 and a height of 11.

ANSWERS: 45 and $126\,\bar{2}$

CIRCLES IN THE SAND

Fractions Built from Powers of 2

Every culture has wrestled mathematically with the dimensions of the circle. It is a surprisingly difficult problem to figure out how far it is around a circle when given how far across it is. However, anyone with a tape measure and a tree can quickly determine that the circumference is a little more than three times the diameter.

The length around a can is about three times longer than its diameter.

Most cultures use a rough estimate of a number we call π in their calculations. This approximation varies from culture to culture. Some used 3, others $^{22}/_{7}$, and we use the familiar 3.14. None of them are correct; in fact it's impossible to be correct. Modern mathematicians have proven that π is irrational, which means that it can only be described by an infinite expansion. It wasn't until the time of Archimedes, around 200 BCE, that there was an explicit method to calculate π to any desired accuracy. He also seems to be the first mathematician who understood that the same value was related to the area of a circle and the volume and surface area of a sphere.

We don't know what value the Egyptians used. The Babylonian culture was roughly concurrent with that of the Egyptians, and the two did have some contact with each other. So, for the sake of argument, we will use the Babylonian value of $3\,\bar{8}$, which they multiplied by the diameter to obtain an estimate for the circumference of a circle. This value of an eighth is particularly easy in Egyptian mathematics because it can be obtained by taking half of a half of a half. Consider the following. We start with a whole and cut it in two. Since two pieces make up a whole, each piece is size $\bar{2}$.

Next we slice each of the two halves in half. This makes four pieces, giving fourths.

Finally the four fourths are each cut in half, making eight eighths.

This allows us to easily multiply by an eighth in much the same way we multiplied by a half in the triangle problems. Consider taking an eighth of 10.

EXAMPLE: Calculate $\overline{8} \times 10$.

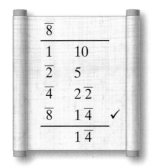

In order to estimate the circumference of a circle, we need to multiply by $3\,\overline{8}$. Very little changes from the above example except that we also need to multiply by 3.

EXAMPLE: Estimate the circumference of a circle of diameter 12.

SOLUTION: Multiply 12 by $3\,\overline{8}$.

$3\,\overline{8}$		
1	12	✓
2	24	✓
$\overline{2}$	6	
$\overline{4}$	3	
$\overline{8}$	$1\,\overline{2}$	✓
	$37\,\overline{2}$	

So the circumference is $37\,\overline{2}$.

Since we're dividing by 2 more than once, we will occasionally get fractions that must be halved, as in the following problem.

EXAMPLE: Find the circumference of a circle of diameter 13.

SOLUTION: Multiply 13 by $3\,\overline{8}$.

$3\,\overline{8}$		
1	13	✓
2	26	✓
$\overline{2}$	$6\,\overline{2}$	
$\overline{4}$	$3\,\overline{4}$	
$\overline{8}$	$1\,\overline{2}\,\overline{8}$	✓
	$40\,\overline{2}\,\overline{8}$	

So the circumference is $40\,\overline{2}\,\overline{8}$.

Note that when we divided $3\,\overline{4}$ by 2 we got $1\,\overline{2}$ from the 3 and $\overline{8}$ from the $\overline{4}$. The two fractions were added simply by placing the two next to each other. From an Egyptian standpoint $1\,\overline{2} + \overline{8} = 1\,\overline{2}\,\overline{8}$ is perfectly valid as an answer. It's interpreted as *a little more than one and a half*.

PRACTICE: Find the circumference of a circle of diameter 14.

PRACTICE: Repeat for diameter 31.

ANSWERS: $43\,\overline{2}\,\overline{4}$ and $96\,\overline{2}\,\overline{4}\,\overline{8}$

Although we can easily see how the ancients might have come up with the equation for the area of a triangle, it's not clear how they arrived at their approximation for the circumference multiplier. There are a number of ways it could have been discovered, but each of these is just speculation. One often-suggested method can be accomplished with ropes. Go into the desert and drive a stake into the ground. Take a length of rope around the stake and pull both ends until they are taught. Hold another stake at the ends and walk around the first stake using the second stake as the pencil of a compass. Doing this you will trace out a circle in the sand whose diameter is roughly the length of the rope.

Tracing out a circle in the sand with a rope whose length is the diameter of the circle.

The gap is filled, making 3 and an $\overline{8}$.

Take three lengths of rope equal in length to the original rope and lay them out along the circle. You'll find that they almost complete the circle, leaving a small gap. The three ropes are the 3 found in our 3.14 representation of π and the Babylonians' 3 $\overline{8}$.

Three ropes of diameter length almost complete the circle.

In order to fill the gap, we need a smaller piece of rope. One way of obtaining a shorter length is to repeatedly fold the rope in half, cutting it at the middle. If three cuts are made this way, the rope segment roughly fills the gap. As we've seen, dividing something in half three times gives you an eighth of the original.

A rope cut in half three times is an eighth of the original length.

Hence we have derived the Babylonian approximation 3 $\overline{8}$ for the circle.

While the above method is interesting, it shouldn't be taken too seriously. Any measurement around a circle, say by counting bricks around a tower, when divided by the measured diameter, will produce a number around 3 $\overline{8}$. We're simply showing that it *can* be done using primitive tools without actually knowing *how* it was done.

Consider an architect building a grain silo whose foundation is pictured below. The diameter consists of 8 bricks and the circumference of 25.

A diameter of 8 bricks and circumference of 25.

The architect could ask himself, what would I need to multiply the diameter of 8 by to get the circumference of 25? This is just 25 divided by 8, and if the computation were completed, he would get 3 $\overline{8}$, the value used by Babylonians.

	25	
1	8	✓
2	16	✓
$\overline{2}$	4	
$\overline{4}$	2	
$\overline{8}$	1	✓
3 $\overline{8}$		

The value 3 $\overline{8}$ found by dividing the circumference, 25, by the diameter, 8, as measured in bricks.

Of course, in general it's unlikely he would get exactly $3\,\overline{8}$, but if he consistently got values like $3\,\overline{8}\;\overline{50}$ or $3\,\overline{8}\;\overline{90}$, he might quickly settle on an approximation of $3\,\overline{8}$. So, as you can see, there are many ways to obtain this approximation.

In this section we've dealt with $\overline{8}$ primarily because it arises naturally in the Egyptian calculation of circumferences. However, our method works just as easily with many other fractions that can be obtained by repeated halving, such as $\overline{4}$ or $\overline{32}$.

EXAMPLE: Multiply 20 by $1\,\overline{4}\;\overline{16}$.

$1\,\overline{4}\;\overline{16}$	
1	20 ✓
$\overline{2}$	10
$\overline{4}$	5 ✓
$\overline{8}$	$2\,\overline{2}$
$\overline{16}$	$1\,\overline{4}$ ✓
	$26\,\overline{4}$

PRACTICE: Multiply 48 by $1\,\overline{8}\;\overline{32}$.

ANSWER: $55\,\overline{2}$

The fractions that work well are found by taking half of 1 repeatedly to get $\overline{2},\,\overline{4},\,\overline{8},\,\overline{16}$, and so on. These are the "powers of 2" that naturally arise in Egyptian mathematics. Of course most numbers are not powers of 2, and we'll need to learn to work with them.

AHMOSE'S TABLE

Doubling Fractions

Ahmose the scribe lived around three and a half thousand years ago. His name translates into something like "Moon Born," presumably in deference to the scribal god, Thoth. During the fourth month of the inundation, when Egypt was covered in the waters of the Nile, Ahmose diligently copied an ancient scroll. There was little to do until the waters receded, so it was the perfect time to catch up on chores, like the preservation of an old papyrus.

The original scroll dated from the glorious Twelfth Dynasty, which was around the nineteenth century BCE. The scroll was produced during a peaceful and prosperous time that lasted about three hundred and fifty years. The educated in the national bureaucracy dominated the local nobility of this age, thereby reducing internal power struggles. The sons of the aristocrats were enlisted into the ranks of the educated elite, providing the nobles with a path to success that did not involve conspiracy or rebellion. It's perhaps no surprise that our two principal mathematical texts both originate from around the time of the supremacy of the intellectuals.

This period stands in direct contrast to Ahmose's day, when Egypt was split into fragments, many of which were dominated by foreign rulers. Ahmose dates his scroll as the thirty-third year of the reign of Apophis, a Canaanite king. Apparently even foreign kings needed the scribes and mathematical wisdom of ancient Egypt to manage their kingdoms. A few years after Ahmose copied his famous scroll, the Egyptian kings rebelled against their foreign masters, and Apophis and many of the leaders of the insurrection were killed. Having finally mastered the military technology of the outsiders, the rebels won, ushering in the last great period of Egyptian hegemony.

Ahmose's scroll is now known as the Rhind Mathematical Papyrus, named after a modern collector who purchased it. The scroll bears the title *Accurate Reckoning, the Entrance into the Knowledge of All Existing Things and All Obscure Secrets*. Modern school children would probably be disappointed to learn that this is a title to what is only an introductory book on fractions, but the topic is extremely important. The word "Accurate" in the title probably refers to the use of fractions to make exact calculations. It's not hard to see that $^{156}\!/_{15}$ is roughly 10, but an "accurate" value is $10\,\overline{3}\;\overline{15}$. Hence it is not possible to accurately calculate without fractions. Such a skill was necessary to run a kingdom. In a world where only a few percent of the people could read, let alone calculate, these mathematical tools would make a scribe indispensable.

What's the first thing an ancient Egyptian would need to know about fractions? In order to multiply or divide, an ancient scribe would need to know how to double them. This is exactly the first subject covered in Ahmose's papyrus. Part of this process is easy. Consider the fraction $\overline{3}$. We know that three of these are 1 and hence three blocks of area $\overline{3}$ is the same as one of size 1.

If we divide each $\overline{3}$ in two, we get six pieces that still make up the 1 block in size. Hence, these pieces are of size $\overline{6}$ and we can conclude half of $\overline{3}$ is $\overline{6}$.

We can repeat this argument for any number. For instance we could show that half of $\overline{4}$ is $\overline{8}$ and half of $\overline{5}$ is $\overline{10}$. The pattern is simple. You just double the number under the bar to take half.

EXAMPLE: What is half of $12\ \overline{5}\ \overline{30}$?

SOLUTION: $6\ \overline{10}\ \overline{60}$

PRACTICE: What is half of $17\ \overline{6}\ \overline{27}$?

ANSWER: $8\ \overline{2}\ \overline{12}\ \overline{54}$

We double a fraction by simply reversing the process. Hence twice $\overline{6}$ is $\overline{3}$, twice $\overline{8}$ is $\overline{4}$, and twice $\overline{10}$ is $\overline{5}$. So if you want to double a fraction, you just take half of the barred number. Note that this only works on even fractions. We could say that twice $\overline{5}$ is $\overline{2.5}$ but this would not be an acceptable fraction to an ancient Egyptian.

We can now use this in a standard multiplication. Consider the following example.

EXAMPLE: Multiply $7\ \overline{12}$ by 5.

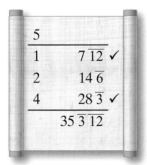

Try this one on your own.

PRACTICE: Multiply $3\ \overline{20}$ by 6.

ANSWER: $18\ \overline{5}\ \overline{10}$

Now try to apply the ideas of the previous section to solve the following practice problem. Be sure to put $3\ \overline{4}$ on the top of the left column.

PRACTICE: Multiply $6\ \overline{14}$ by $3\ \overline{4}$.

ANSWER: $19\ \overline{2}\ \overline{7}\ \overline{14}\ \overline{56}$

The original author of the Rhind Mathematical Papyrus obviously thought that doubling even fractions was too simple; hence, it wasn't included in the text. Odd fractions, however, were another matter entirely, and this is where the scroll begins.

Let's consider $2 \times \overline{5}$. The trick to finding the value of this multiplication is to understand that it's a division problem. Imagine you're an Egyptian craftsman and you know that one loaf of bread is being divided up between five workers. You don't need a scribe to determine that you get a fifth; in other words, $\overline{5}$ of a loaf. The next day you don't see how much bread is being distributed but the scribe hands you two pieces of bread, each $\overline{5}$ in size. So you received a fifth twice, or $2 \times \overline{5}$. You realize that since you have twice as much bread as you did on the first day, the scribe must have started out with twice as much bread today. Since yesterday he had one loaf of bread, today he started with two loaves of bread. Hence he divided two loaves between five workers, so your share is $2 \div 5$. Clearly this means that $2 \times \overline{5} = 2 \div 5$. In general, twice any fraction is the solution to 2 divided by the number under the fraction bar.

Ahmose's papyrus begins by dividing 2 by all the odd numbers from 3 to 101, and hence it is in essence a table of 2 times an odd number. But it's much more than a table since it justifies each division with a multiplication. For example, when it tells us that $2 \div 5$ is $\overline{3}\ \overline{15}$, it immediately multiplies $\overline{3}\ \overline{15}$ by 5, showing that the result is 2. For simplicity, we will use a condensed table. If we need to know what $2 \div 11$ is, we could simply look down the first column until we hit $\overline{11}$, look to the right, and find our solution of $\overline{6}\ \overline{66}$. This table also appears in the Computation Tables at the beginning of the book. You will need it when you do Egyptian math, so feel free to copy it for your own personal use and keep it handy.

3	$\overline{\overline{3}}$	21	$\overline{14}\ \overline{42}$
5	$\overline{3}\ \overline{15}$	23	$\overline{12}\ \overline{276}$
7	$\overline{4}\ \overline{28}$	25	$\overline{15}\ \overline{75}$
9	$\overline{6}\ \overline{18}$	27	$\overline{18}\ \overline{54}$
11	$\overline{6}\ \overline{66}$	29	$\overline{24}\ \overline{58}\ \overline{174}\ \overline{232}$
13	$\overline{8}\ \overline{52}\ \overline{104}$	31	$\overline{20}\ \overline{124}\ \overline{155}$
15	$\overline{10}\ \overline{30}$	33	$\overline{22}\ \overline{66}$
17	$\overline{12}\ \overline{51}\ \overline{68}$	35	$\overline{30}\ \overline{42}$
19	$\overline{12}\ \overline{76}\ \overline{114}$	37	$\overline{24}\ \overline{111}\ \overline{296}$

39	$\overline{26}\ \overline{78}$	55	$\overline{30}\ \overline{330}$
41	$\overline{24}\ \overline{246}\ \overline{328}$	57	$\overline{38}\ \overline{114}$
43	$\overline{42}\ \overline{86}\ \overline{129}\ \overline{301}$	59	$\overline{36}\ \overline{236}\ \overline{531}$
45	$\overline{30}\ \overline{90}$	61	$\overline{40}\ \overline{244}\ \overline{488}\ \overline{610}$
47	$\overline{30}\ \overline{141}\ \overline{470}$	63	$\overline{42}\ \overline{126}$
49	$\overline{28}\ \overline{196}$	65	$\overline{39}\ \overline{195}$
51	$\overline{34}\ \overline{102}$	67	$\overline{40}\ \overline{335}\ \overline{536}$
53	$\overline{30}\ \overline{318}\ \overline{795}$	69	$\overline{46}\ \overline{138}$

71	$\overline{40}\ \overline{568}\ \overline{710}$	87	$\overline{58}\ \overline{174}$
73	$\overline{60}\ \overline{219}\ \overline{292}\ \overline{365}$	89	$\overline{60}\ \overline{356}\ \overline{534}\ \overline{890}$
75	$\overline{50}\ \overline{150}$	91	$\overline{70}\ \overline{130}$
77	$\overline{44}\ \overline{308}$	93	$\overline{62}\ \overline{186}$
79	$\overline{60}\ \overline{237}\ \overline{316}\ \overline{790}$	95	$\overline{60}\ \overline{380}\ \overline{570}$
81	$\overline{54}\ \overline{162}$	97	$\overline{56}\ \overline{697}\ \overline{776}$
83	$\overline{60}\ \overline{332}\ \overline{415}\ \overline{498}$	99	$\overline{66}\ \overline{198}$
85	$\overline{51}\ \overline{255}$	101	$\overline{101}\ \overline{202}\ \overline{303}\ \overline{606}$

Doubling Odd-Fraction Table

We will reference this table every time we have to double an odd fraction. Consider the following example. When going from the second to the third line we will need to double $\overline{9}$. To do this we simply look at the above tables to find that double $\overline{9}$ is $\overline{6}\ \overline{18}$.

EXAMPLE: Multiply $5\ \overline{18}$ by 9.

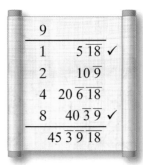

ANSWER: $45\ \overline{3}\ \overline{9}\ \overline{18}$

This answer may seem unwieldy but that's because it's not simplified. We will soon learn how to quickly reduce our solution to $45\ \overline{2}$, or in modern terms $45\frac{1}{2}$. Try this one.

PRACTICE: Multiply $3\ \overline{\overline{60}}$ by 16.

ANSWER: $48\ \overline{5}\ \overline{15}$

We can now multiply any fraction by a whole number. However, our method is slow and cumbersome. There's still a bit more to learn in order to be as proficient as an ancient Egyptian.

PYRAMIDS AND SEKED

Row Switching

The pharaoh Sneferu had a big problem. It seems that the pyramid for the founder of the Fourth Dynasty was falling down while it was under construction. This was a relatively new problem because at the time, the pyramids were only half a century old. Before that time, a pharaoh was buried in a *mastaba*, a rectangular building over an underground tomb.

In the Third Dynasty, Imhotep, vizier to the pharaoh Djoser, had had an inspired idea. By essentially stacking mastabas of decreasing size, he oversaw the construction of the first Egyptian pyramid, which served as a physical recreation of the primordial mound, where the universe was created at the dawn of time. It also served as a ramp by which the deceased pharaoh was to ascend to the heavens. By choosing to build it out of limestone instead of the mud brick of the mastabas, Imhotep set into motion the building of monuments that have survived the four-and-a-half-thousand-year journey to the present. For his efforts, Imhotep was deified, becoming a god of wisdom and healing.

About fifty years later, Sneferu made an improvement to Imhotep's innovation. Why not make the sides of the pyramids flat? So he began construction of his Shining Pyramid of the South. The workers were instructed to slant the walls at an angle that the Egyptians referred to as a *seked of five*. However, when they were about half done, the pyramid began to fall apart. But why?

Consider an hourglass. As the sand falls, it piles up in a cone shape. As the grains continue to accumulate, the cone gets larger but the slope of the side essentially remains the same. Because of the material they're made of, the grains have a natural angle of accumulation. You may think that a many-ton block of limestone is a little more stable than a grain of sand, but you would be wrong. We perceive the stone block as strong because it's immune to the forces we typically encounter. However, stack a few million similar blocks on top of it, and the forces they generate can push it aside as easily as you can blow sand from your hand.

Sneferu's pyramid was just a little too steep, and the laws of physics began to kick in. In order to rectify the situation, the architects needed to increase the seked, which, as we will see, is equivalent to lowering the slope. They raised it to just under $7\,\overline{2}$. This alteration has led modern archaeologists to call the Shining Pyramid of the South, the Bent Pyramid. Apparently not satisfied, Sneferu had another pyramid built entirely with a seked of approximately $7\,\overline{2}$, and eventually he built a fourth with a seked of exactly $5\,\overline{2}$, almost the steepest such a pyramid can be and not crumble.

Today we calculate slope using the mnemonic "rise over run." Consider the half pyramid pictured below.

Notice that it has a height of 4 units and a width of 5. We would declare the rise to be 4 and the run to be 5. Then rise over run would be 4 over 5, or equivalently, 4/5, which is the slope.

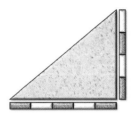

The rise is 4 units, the run is 5, and hence the slant has a slope of 4/5.

The Egyptian seked is similar to our slope but instead of using rise over run, the Egyptians started with *run over rise*. So our slope of ⅘ would be interpreted as ⅘. The Egyptians took an extra step and multiplied the run over rise by 7. I will tell you shortly where this number comes from. On the surface, this seems like an unusual step, unnecessarily complicating the value; however, this view stems from our shallow understanding of the concept of slope.

Our first encounter with slope usually occurs in grade school in reference to *x-y* graphs. In these classes we are taught equations to find the slope, use it to graph lines, and so on. To most students slope is an abstract concept, never extending beyond their graph paper.

However, slope has a very important real-world interpretation. Consider the fair exchange of dimes and nickels. We can exchange 1 dime for 2 nickels. We can express this as a pair of numbers starting with the number of dimes followed by the equivalent number of nickels (1,2). We can depict this pair by drawing a point one square over and two up from the bottom left of a graph. Similarly we can change 2 dimes for 4 nickels and plot (2,4) on our graph. We can then draw a line though these two points. This line will cross through every point representing a fair exchange. For example, if the diagram below were extended, the line would run through (5,10), telling us that 5 dimes is 10 nickels. If we calculate the rise and run of the line between the two points, we get 2 and 1. This gives us a slope of ⅔, which is 2.

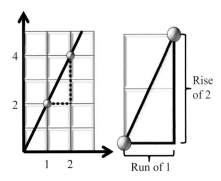

The points (1,2) and (2,4) have a line of slope 2 between them.

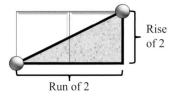

A pyramid with a slope of ½ has a seked of 14.

Although it's not too difficult to calculate the slope by examining a graph, the work we did above was completely unnecessary. We're exchanging dimes for nickels and we know there are 2 nickels per dime. Therefore the slope is 2. Think about how we found the slope above. We looked at the relation between two points. The horizontal move of 1 represents 1 dime. The corresponding vertical move of 2 represents the 2 nickels we get for the 1 dime. When we divide 2 nickels by 1 dime, we are calculating how many nickels we get per dime.

This works for any slope. If we were to graph inches on the vertical axis and feet horizontally, the slope would be 12 since there are 12 inches per foot.

"Per" is the operative word in slopes. Every time you see this word in a quantitative context, you're considering slope. For example, miles per hour, seconds per minute, and dollars per yen can all be viewed as the slopes of different graphs. When we try to interpret the Egyptian seked, we must realize that they're measuring *something* per *something else*.

Consider a pyramid with slope of one-half. Every unit up the pyramid one goes, one moves over two units. In order to find the seked, we divide the run by the rise and then multiply our answer by seven. This gives $(^2/_1) \times 7$, which gives a seked of 14.

To understand this value, we have to know something about Egyptian units of length. A cubit is roughly the typical distance from one's elbow to the tips of one's fingers. A palm is the width of the palm of a hand. The Egyptian measurement system had 6 palms in a cubit. This is roughly accurate. My palm is about 3.4 inches across. My

"cubit" is about 20 inches long. When we multiply 3.4 by 6 we get 20.4 inches, which is essentially what the Egyptians said it should be.

Horus's cubit and palm.

A royal cubit is 7 palms in length. Why did the pharaohs need their cubits longer than those of the common people? Perhaps they were built like orangutans, or perhaps it's just an elitist thing. In any case, when building the pyramids, royal cubits were used to construct the royal tomb.

Now imagine you're a worker positioning a new 1-cubit block on the next level of a half-built pyramid. You need to know how far back you should push the block to keep the pyramid at 14 seked. Consider the diagram below. The top left corner of the block must be right on the slope of the pyramid's slanted edge. We can see from our diagram that the block needs to be pushed back 2 cubits in order to make this happen. The block needs to be positioned precisely, so we probably want to measure this distance in a smaller unit. The Egyptians used palms. So the 2 cubits becomes $2 \times 7 = 14$ palms. This is precisely the seked of the pyramid.

If the block were 2 cubits high, it would have to go back twice as far, hence 28 palms back. The ancient scribe

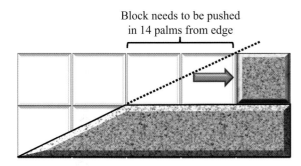

Block needs to be pushed
in 14 palms from edge

Fourteen-seked pyramids position blocks 14 palms per cubit high.

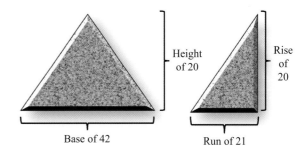

Height
of 20

Rise
of
20

Base of 42

Run of 21

The rise is the height and the run is half the base.

could easily calculate this by multiplying the height of the block by the seked of the pyramid, obtaining 2×14, or 28, palms. Notice that it's 14 palms across per cubit up. The "per" is precisely why we can think of seked as a slope.

EXAMPLE: How many palms would you need to push a $1\,\overline{2}$ block back if the seked was 12?

SOLUTION: We have 12 palms per cubit and $1\,\overline{2}$ cubits, so we must multiply them, giving

$$
\begin{array}{c|cc}
12 & & \\
\hline
1 & 1\,\overline{2} & \\
2 & 3 & \\
4 & 6 & \checkmark \\
8 & 12 & \checkmark \\
\hline
& 18 &
\end{array}
$$

Hence the block goes back 18 palms.

PRACTICE: How many palms would you need to push a $3\,\overline{4}$-cubit block back if the seked were 7?

ANSWER: $22\,\overline{2}\,\overline{4}$ palms

In Ahmose's scroll, we're introduced to seked in a problem to find the slope of a given pyramid. Let's do a similar problem. Consider a pyramid of height 20 and base of 42. The first step is to find the rise and run of the slope. Ahmose accomplished this by arithmetically splitting the pyramid in half, dividing the base by 2. This gives a rise of 20 and a run of 21.

The next step in calculating the seked is to divide the run by the rise of the pyramid. The division of 21 by 20 starts as follows:

$$
\begin{array}{c|c}
? & 21 \\
\hline
1 & 20
\end{array}
$$

Note that we only need 1 more to reach 21. If we divide 20 by 2 repeatedly, we get 10, 5, $2\,\overline{2}$, and $1\,\overline{4}$. We will not get 1. We could keep dividing by 2, getting $\overline{2}\,\overline{8}$, $\overline{4}\,\overline{16}$, and $\overline{8}\,\overline{32}$, but we would not be able to get these pieces to add up to 1, and the division would get uglier and uglier.

The Egyptians had a trick up their sleeve that I'll refer to as "switching." They returned to the top line and switched the 1 and the 20. In doing so they put a bar over each. Note that $\overline{1}$ is just 1.

$$
\begin{array}{c|cc}
? & 21 & \\
\hline
1 & 20 & \checkmark \\
\overline{20} & 1 & \checkmark \\
\hline
1\,\overline{20} &
\end{array}
$$

Finally we have to multiply by 7 in order to convert to palms.

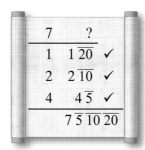

7	?
1	1 $\overline{20}$ ✓
2	2 $\overline{10}$ ✓
4	4 $\overline{5}$ ✓
7 $\overline{5}$ $\overline{10}$ $\overline{20}$	

We can use the above switching method to perform any integer division. Consider the following division of 32 by 5. You proceed doubling until you are about to exceed the number over the second column. Then you determine how much you're short, switch the first row to get a 1 in the second column, and then double it to obtain the remainder. Remember you will have to use a copy of the doubling of odd fractions that can be found in the previous section and in the Computation Tables at the beginning of the book.

EXAMPLE: Divide 32 by 5.

		?	32	
1)		1	5	
2)		2	10 ✓	(1) × 2
3)		4	20 ✓	(2) × 2
4)		$\overline{5}$	1	(1) ↔
5)		$\overline{3}$ $\overline{15}$	2 ✓	(4) × 2
		6 $\overline{3}$ $\overline{15}$		

Note that after we reached 20, we could add the 10 and 20 to get 30. Knowing we now needed 2 more, we went to the first row and switched it (↔). Once the 1 was doubled, we got the 2 we needed.

PRACTICE: Divide 23 by 10.

ANSWER: 2 $\overline{5}$ $\overline{10}$

PRACTICE: Divide 68 by 12.

ANSWER: 5 $\overline{3}$ or 5 $\overline{2}$ $\overline{6}$

We can now apply our new method of division to determine if a pyramid is safe. We will apply the following steps:

- Let the rise be the height and the run be half the base.
- Calculate run/rise.
- Multiply by 7 to determine the seked.
- If the answer is 5 $\overline{2}$ or more, it's safe.

EXAMPLE: Is a pyramid with a base of 44 and a height of 40 safe?

- Rise = 40, Run = half 44 = 22.
- Run/Rise = $^{22}/_{40}$

?	22
1	40
$\overline{2}$	20 ✓
$\overline{40}$	1
$\overline{20}$	2 ✓
$\overline{2}$ 20	

- 7 times Run/Rise

7	?
1	$\overline{2}$ 20 ✓
2	1 $\overline{10}$ ✓
4	2 $\overline{5}$ ✓
3 $\overline{2}$ $\overline{5}$ $\overline{10}$ $\overline{20}$	

- The seked is less than 5 $\overline{2}$ and so the pyramid is unsafe.

PRACTICE: Is a pyramid with height of 24 and a base of 52 safe?

ANSWER: Yes, since its seked, 7 $\overline{3}$ $\overline{6}$ $\overline{12}$, is greater than 5 $\overline{2}$.

As we will see, switching is a powerful tool. In general, the switch of a number just adds or removes the fraction bar. Be careful not to try to switch sums; the switch of $\overline{4}\ \overline{7}$ is not $4+7$. There is no good way to switch any sum.

THE NIGHT WATCHMEN'S SHARES

Multiplying by $\overline{\overline{3}}$

The afterlife offered pharaohs an awe-inspiring existence among the gods. From their tombs within their pyramids, the souls of the deceased kings could read the spells written on the walls, which simultaneously described and enabled their ascent into the heavens. One such spell is found in the tomb of the pharaoh Unas. His cataclysmic ascension caused the stars to tremble and rain to fall from the sky. Some gods shook in fear, others fled. Unas transformed into the Bull of Heaven, consuming all that was produced by the gods and the gods from the Lake of Flame. Unas then took his throne, sitting back to back with the earth god, Seb. In front of him, men and gods alike were sacrificed in order that Unas could consume them and thus take their power and establish himself as a god.

About five hundred years later, the pharaoh Senusret II, like most other Egyptian kings, wanted the same heavenly reception for himself. So he set aside a patch of arid land in Lahun between the Nile and the oasis-like Fayum Basin, and a town was constructed nearby to service Senusret's mortuary complex. The area had temples for two gods: Senusret himself and falcon-headed Sopdu. For Senusret, the area's pyramid, town, and temple provided for his ascent into heaven and for his sustenance thereafter. For the temple's night watchmen, it provided steady employment.

Unlike their use in modern times, the purpose of these ancient temples was not to provide a place of worship but to cater to the needs of the gods, who, often in the form of small statuettes, had to be clothed, fed, washed, and entertained. The temples had large staffs and land holdings to provide for the workers who served the god. The food, tools, and temple treasures had to be protected, so one of the temples of Lahun hired two guards to keep watch at night when everyone else had left.

None of the temple staff received a fixed salary; instead they had fixed "shares." The size of your share would determine what portion of the temple's wages you received. The head lay priest received a 3 share, while the head reader received a 6 share. While their wages could vary from pay period to pay period, a 6 share would always receive twice as much as a 3 share. The temple director got a whopping 10 share, almost a quarter of the 42 shares for the whole temple.

On payday, a scribe would mathematically calculate the value of a single share and then multiply this value by the number of shares an individual worker earned. Say, for example, the temple received 84 loaves of bread. Since there are 42 total shares, each share would be worth 2 loaves. The head lay priest, with his 3 share, would receive 3×2, or 6, loaves of bread. The head reader and the temple director would receive 12 and 20 loaves respectively because they had shares of 6 and 10.

It's difficult to tell how either of the night watchmen felt about his measly $\overline{\overline{3}}$ share, which was not even a whole one. The typical priest earned six times his salary. Even the scribe, who only showed up to work occasionally to record a few numbers, make a couple of calculations, and write a letter or two, made twice a watchman's salary. Perhaps the watchman knew that most of the staff were educated, a rare and valuable trait in the ancient world, and deserved their pay. Perhaps he took solace in the fact that his local god Sopdu, one of the oldest gods, was also a watchman, guarding the eastern desert from invasion and protecting the Sinai turquoise mines. Perhaps he thought that his job of standing around in the cool desert night was far better than slaving away in the fields under the hot Egyptian sun, the task most uneducated Egyptians were assigned.

The scribe of Lahun had a different problem with a watchman's salary. He had to be able to multiply a 1-share value by two-thirds and do it quickly because he had many other workers who needed their pay computed. Oddly enough, ancient scribes multiplied by $\overline{\overline{3}}$ all the time, never showing any work, as if it were as easy as multiplying or dividing by 2. Some historians are shocked by this apparently difficult computation. Others assume that the Egyptian scribes must have had tables to perform such an onerous task. While nobody knows how they multiplied by $\bar{\bar{3}}$, it is in fact quite easy to do. With a little practice, you can find two-thirds of large numbers quickly and in your head.

In order to understand how to do this, we must think like an Egyptian. Consider the following division of 9 by 6.

?	9
1	6 ✓
$\bar{2}$	3 ✓
$\overline{12}$	

A scribe working the above problem starts with a 6 in the right column and is required to construct a 9. He knows he needs 3 more, which can easily be obtained by dividing the 6 by 2. Essentially, the scribe found 9 by adding 6 and its half. What you have to notice is that $\bar{\bar{3}} \times 9 = 6$. It turns out that adding a number to its half is the inverse of taking two-thirds in the same way that doubling is the inverse of taking half. When I want half of 30, I ask myself, what number do I need to double to get 30, and I realize the answer is 15. Similarly when I want to know what $\bar{\bar{3}}$ of 30 is, I ask myself, what number do I need to add to its half to get 30? Since 20 and its half, 10, is 30, I know $\bar{\bar{3}}$ of 30 is 20.

EXAMPLE: What is $\bar{\bar{3}}$ of 18?

SOLUTION: Since $12 + 6 = 18$, $\bar{\bar{3}}$ of 18 is 12.

PRACTICE: What is $\bar{\bar{3}}$ of 15?

ANSWER: 10

You might ask, why does this work? It works for much the same reason you can multiply 22 by $\bar{2}$ if you ask yourself, what do I multiply by 2 to get 22? A modern mathematician would say that 2 and $\bar{2}$ are reciprocals and that multiplying by one is the same as dividing by the reciprocal. The reciprocal of $\bar{\bar{3}}$ is the fraction ⅔ upside down, or ³⁄₂, which is *1 and its half*.

The Egyptians didn't have to resort to arithmetic rules to know that this trick worked. It's obvious if you stop

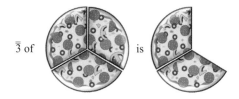

Finding two-thirds of a pizza.

looking at numbers and start looking at things. Consider a small pizza cut into three slices. If you take $\bar{\bar{3}}$ of the pizza, you get two slices.

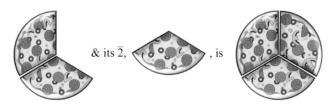

Finding two-thirds of three slices by showing two slices and its half are three slices.

Half of our two slices is one slice. So if we take our slices "and its half" we get three slices, the original pie. Hence two slices is two-thirds of a pie.

When using this trick, the first thing you notice is that it is easiest to find a *number and its half* if the *number* is even. So an ancient Egyptian scribe could make a table of

a # + its $\bar{2}$		is
2	+ 1	3
4	+ 2	6
6	+ 3	9
8	+ 4	12
10	+ 5	15
12	+ 6	18
14	+ 7	21
16	+ 8	24
18	+ 9	27

A table of numbers followed by itself plus its half. Note that the left column is $\bar{\bar{3}}$ of the right column.

such numbers. The table above consists of almost all the entries a mathematician would need.

It's doubtful a scribe would need such a table. There are only nine entries. If the scribe forgot an entry like the "14 21" row, he simply needed to add 14 to its half, 7, in his head to get 21. On the other hand, a scribe would use these nine entries so much that he would almost certainly have them memorized in the same way we know three quarters is 75 cents without ever performing the calculation 3×25.

Right now you should be skeptical. Although I have shown you a table to find two-thirds of nine numbers less than 30, can we easily find the $\overline{\overline{3}}$ of numbers greater than 30?

EXAMPLE: Find $\overline{\overline{3}}$ of 39.

Note that while this number is not on the table, it's composed of a 3 and a 9, both of which are on the table. The $\overline{\overline{3}}$ of 3 is 2 since 2 and its half is 3. The $\overline{\overline{3}}$ of 9 is 6 since 6 and its half is 9. Hence the $\overline{\overline{3}}$ of 39 is 26. In a modern sense, we've split 39 into $30 + 9$.

So we get:

SOLUTION:

$$\overline{\overline{3}} \times 39 = \overline{\overline{3}}(30 + 9) = (\overline{\overline{3}} \times 30) + (\overline{\overline{3}} \times 9)$$
$$= 20 + 6 = 26$$

EXAMPLE: Find $\overline{\overline{3}}$ of 15,618.

SOLUTION:

$$\overline{\overline{3}} \times (15,000 + 600 + 18) = 10,000 + 400 + 12$$
$$= 10,412$$

Now try these.

PRACTICE: Find $\overline{\overline{3}}$ of 24 and of 39,156.

ANSWERS: 16 and 26,104

The above problems were selected so that the numbers on the table were immediately apparent. For most problems it's slightly more difficult. Consider $\overline{\overline{3}}$ of 168. The starting digit, 1, is too small to take two-thirds of. If we look at the first two digits, we notice that 16 doesn't appear on the table. In such a case we simply take the largest table entry under 16, which in this case is 15. When we remove the 15 from 168, a 1 is left where the 6 was, leaving us 18. What we've essentially done is to rewrite 168 as $150 + 18$, both of which are easy to take two-thirds of. Since 10 and its half is 15, and 12 and its half is 18, the answer is $100 + 12$, or 112.

EXAMPLE: Find $\overline{\overline{3}}$ of 261.

SOLUTION: $\overline{\overline{3}} \times 261 = \overline{\overline{3}}(240 + 21) = 160 + 14 = 174$

PRACTICE: Find $\overline{\overline{3}}$ of 192 and 294.

ANSWERS: 128 and 196

In order to find two-thirds of larger numbers, you do the same thing, you just do it more times. So for example, if I need to find two-thirds of 82,134, I break it up piece by piece.

$$\begin{aligned} 82134 &= 60000 + 22134 \\ &= 60000 + 21000 + 1134 \\ &= 60000 + 21000 + 900 + 234 \\ &= 60000 + 21000 + 900 + 210 + 24 \end{aligned}$$

Now when we take two-thirds of this we get

$$\begin{aligned} \overline{\overline{3}} \times 82134 &= \overline{\overline{3}}(60000 + 21000 + 900 + 210 + 24) \\ &= 40000 + 14000 + 600 + 140 + 16 \\ &= 54756 \end{aligned}$$

You might think the above process is tedious, but it only appears to be so because I wanted the solution to be easy to follow. Notice that I wrote 60,000 over and over. If I were solving the problem for myself. I simply would have written 40,000, which is two-thirds of 60,000, and then recorded the remainder 22,134. Then in my head I would have pulled out the 21,000, written 14,000, and recorded the remainder 1,134, and so on. I'm going to continue doing it the "slow" way because it's easier to follow. When you get used to this method, try doing it in your head.

EXAMPLE: Find $\overline{\overline{3}}$ of 741,834.

SOLUTION:

$$\begin{aligned} 741834 &= 600000 + 141834 \\ &= \cdots + 120000 + 21834 \\ &= \cdots + 21000 + 834 \end{aligned}$$

$$= \cdots + 600 + 234$$
$$= \cdots + 210 + 24$$
$$= \cdots + 24$$

So $\overline{\overline{3}} \times 748{,}134$ is

$$\overline{\overline{3}}(600000 + 120000 + 21000 + 600 + 210 + 24)$$
$$= 400000 + 80000 + 14000 + 400 + 140 + 16$$
$$= 494556$$

With numbers this large, it's easy to make a mistake. However, it's also easy to check. In the following check, we multiply our answer by $1\,\overline{2}$. Since we got 741,834, the original number, our solution is correct.

CHECK:

$1\,\overline{2}$	?
1	494,556 ✓
$\overline{2}$	247,278 ✓
	741,834

Checking that 494,556 and its half is 741,834.

PRACTICE: Find $\overline{\overline{3}}$ of 36,168 and 574,329.

ANSWERS: 24,112 and 382,886

You might also wonder if this method would seem palatable to an Egyptian. We don't know how they performed these calculations, and there's always the danger of imposing modern ideas and methods on ancient mathematicians. The question we have to ask is, would such a method be in line with the way Egyptians viewed math? What we're essentially doing is performing a multiplication by breaking it into pieces that are easy to take two-thirds of. But this is basically what Egyptian multiplication is about. Consider the following multiplication of 6 by $\overline{2}$.

In the next scroll, we broke 6 into 4 and 2. When we multiplied 741,834 by $\overline{\overline{3}}$ in the previous example, we did essentially the same thing. The only real difference is we didn't generate a row by manipulating previous rows. We

6	?
1	$\overline{2}$
2	1 ✓
4	2 ✓
	3

simply use the nine numbers, 3, 6, 9, … , 27, whose two-thirds are easy to calculate. While we can't prove that the Egyptians used this method, it's certainly far from inconceivable that they did something similar.

741,834	?
1	$\overline{\overline{3}}$
600,000	400,00 ✓
120,000	80,000 ✓
21,000	14,000 ✓
600	400 ✓
210	140 ✓
24	16 ✓
	494,556

The $\overline{\overline{3}}$ method written as an Egyptian multiplication.

We need to know one more thing in order to easily multiply any whole number by two-thirds. Consider the following problem:

EXAMPLE: Find $\overline{\overline{3}}$ of 226.

SOLUTION: $226 = 210 + 16 = 210 + 15 + 1$

At this point we run into a small problem. The number 1 isn't one of the numbers on our table. All we know is how to take two-thirds of multiples of 3 from 3 to 27. The number 1 is too small. We may be briefly stumped because we don't know how to determine the value of $\overline{\overline{3}} \times 1$, but suddenly we recognize that, obviously, $\overline{\overline{3}} \times 1$ is $\overline{\overline{3}}$. So, our solution appears as follows:

SOLUTION:

$$226 = 210 + 16$$
$$= 210 + 15 + 1$$
$$\overline{\overline{3}} \times 226 = \overline{\overline{3}}(210 + 15 + 1)$$
$$= 140 + 10 + \overline{3}$$
$$= 150\,\overline{3}$$

The only other number we might end up with beside 1 is 2. In this case we're just asking what $\overline{\overline{3}} \times 2$ is. Its value is $1\,\overline{3}$. This, once again is clear if we think about $\overline{\overline{3}}$ doubled as slices of pizza.

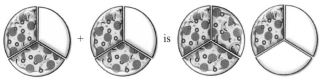

$\overline{\overline{3}} + \overline{\overline{3}}$ is $1\,\overline{3}$ since it's just four slices of pizza.

Here's an example using this identity.

EXAMPLE: Find $\overline{\overline{3}}$ of 4178.

SOLUTION:

$$4178 = 3000 + 1178$$
$$= \cdots + 900 + 278$$
$$= \cdots + 270 + 8$$
$$= \cdots + 6 + 2$$

$$\overline{\overline{3}} \times 4178 = \overline{\overline{3}}(3000 + 900 + 270 + 6 + 2)$$
$$= 2000 + 600 + 180 + 4 + 1\,\overline{3}$$
$$= 2785\,\overline{\overline{3}}$$

PRACTICE: Find $\overline{\overline{3}}$ of 9,454 and 41,627.

ANSWERS: 6302 $\overline{\overline{3}}$ and 27,751 $\overline{3}$

We now know how to easily find two-thirds of any whole number. Next we will learn how to find two-thirds of any number.

BEER—GOOD TO THE LAST DROP

Taking $\overline{\overline{3}}$ of Fractions

A night watchman at the temple of Lahun was paid in bread and beer. Beer in the ancient world was typically drunk watered down, ensuring that there was enough alcohol to kill the bacteria in the drinking water. Knowing this, it still surprises me that he received more than twice as many jugs of beer than loaves of bread. To know how strong the beer was, we would need to know its pesu, that is, the number of jugs one hekat of grain would make. The higher the pesu of the beer, the lower the grain, and hence the alcohol, content. I suspect that either the beer had a high pesu or that the night watchmen were not very effective.

We've already discussed pesu in terms of loaves of bread, but that's not what we're concerned with here. In this section, we're going to find out how to calculate a watchman's share. If the shares were whole numbers, we could use what we learned in the previous section to calculate his two-thirds share. The problem for the scribe who had to allocate the temple staff's pay was that the shares were rarely easy numbers to work with.

At the temple there were a total of 42 shares, a $\overline{\overline{3}}$ share of this was a night watchman's share. On payday the temple had to distribute 115 $\overline{2}$ jugs of beer. So to calculate the value of a 1 share, he divided the 115 $\overline{2}$ jugs by 42. This could be done as follows.

?	115 $\overline{2}$
1	42
2	84 ✓
$\overline{2}$	21 ✓
$\overline{4}$	10 $\overline{2}$ ✓
2 $\overline{2}\,\overline{4}$	

The "usual reader" at the temple received a 4 share. So his salary was calculated by multiplying a 1 share, 2 $\overline{2}\,\overline{4}$, by 4. This is easily done using Egyptian methods as follows:

4	?
1	2 $\overline{2}\,\overline{4}$
2	5 $\overline{2}$
4	11 ✓
	11

When we calculate a night watchman's share, we must multiply $2\,\overline{2}\,\overline{4}$ by $\overline{\overline{3}}$. This means we need to know how to multiply fractions by two-thirds. Fortunately, it's not very hard, especially for even fractions. The trick is in realizing that Egyptian fractions work exactly the opposite way that whole numbers do. For example, when we want to double the fraction $\overline{6}$, we take half of it to get $\overline{3}$. Using modern notation, we can express this reverse relationship as follows:

$$2 \times 3 = 6 \text{ hence } 2 \times \overline{6} = \overline{3}$$

Let's apply this idea to multiplication by two-thirds. Since 4 plus its half is 6, we know $\overline{\overline{3}} \times 6$ is 4. This now gives us

$$\overline{\overline{3}} \times 6 = 4 \text{ hence } \overline{\overline{3}} \times \overline{4} = \overline{6}$$

But since 6 was found by adding 4 and its half, two-thirds of a fraction can be found by taking the number under the bar and adding its half. So, since 10 and its half is 15, we know that $\overline{\overline{3}} \times \overline{10} = \overline{15}$.

EXAMPLE: Find $\overline{\overline{3}}$ of $\overline{8}$.

SOLUTION: $\overline{\overline{3}} \times \overline{8} = \overline{8+4} = \overline{12}$

PRACTICE: Find $\overline{\overline{3}}$ of $\overline{30}$.

ANSWER: $\overline{45}$

We can now calculate the number of jugs of beer a night watchman received. We just multiply $\overline{\overline{3}}$ by $2\,\overline{2}\,\overline{4}$. We've already seen that $\overline{\overline{3}} \times 2$ is $1\,\overline{3}$. We now know that $\overline{\overline{3}} \times \overline{2}$ is $\overline{3}$ and $\overline{\overline{3}} \times \overline{4}$ is $\overline{6}$. So the answer would be $1\,\overline{3}+\overline{3}+\overline{6}$, which simplifies to $1\,\overline{\overline{3}}\,\overline{6}$. A scribe would calculate it as follows:

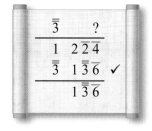

When calculating two-thirds of an odd fraction, we run into the same problem we faced when trying to double an odd fraction. In theory we can say that two-thirds of $\overline{5}$ is $\overline{5+2.5}$, or $\overline{7.5}$, but no Egyptian would express a number that way. They didn't use a table as they did for multiplication by 2. Instead they used a rule that said *to take $\overline{\overline{3}}$ of a fraction, add its double and its six* (i.e., its sextuple). For example, to take two-thirds of $\overline{5}$, we "double" $\overline{5}$ to get $\overline{10}$ and take 6 times $\overline{5}$ to get $\overline{30}$. Putting these together we get $\overline{\overline{3}} \times \overline{5} = \overline{10}\,\overline{30}$.

EXAMPLE: Take $\overline{\overline{3}}$ of $\overline{7}$.

SOLUTION: $\overline{\overline{3}} \times \overline{7} = \overline{2 \times 7}\,\overline{6 \times 7} = \overline{14}\,\overline{42}$

PRACTICE: Take $\overline{\overline{3}}$ of $\overline{11}$.

ANSWER: $\overline{22}\,\overline{66}$

This method is taken straight out of Ahmose's papyrus, but how did the ancient Egyptians find this rule? As always in Egyptian mathematics, if you look at it the right way, it's obvious. First, we need to realize that when we "double" an odd fraction like $\overline{5}$ to get $\overline{10}$, we're really taking half of it. Similarly, when we take "its six," we're cutting it into sixths. Consider a pizza cut in six pieces. Its half is three slices and its sixth is one slice. Added together, this is four of the six slices, which in turn is two-thirds of the pie. Hence, to take two-thirds of something you can just take half and a sixth of it.

Half a pizza and its sixth is two-thirds of a pie.

We can now take two-thirds of any normal fraction since the "its double and its six" rule will work for any fraction. So if we want to take $\overline{\overline{3}}$ of $\overline{6}$, we could use it to get $\overline{12}$ $\overline{36}$. However, we could just as easily use the even-fraction rule to get $\overline{6+3} = \overline{9}$. While the two answers look different, they are the same number. In general, it's probably best to use the even rule when possible because it gives one fraction instead of two. So we will generally only use the "its double and its six" rule on odd fractions. The following example uses both rules.

EXAMPLE: If a 1 share is $\overline{3}$ $\overline{4}$, find a night watchman's share.

SOLUTION: $\overline{\overline{3}} \times (\overline{3}\ \overline{4}) = (\overline{\overline{3}} \times \overline{3}) + (\overline{\overline{3}} \times \overline{4}) = \overline{6}\ \overline{18} + \overline{6} = \overline{3}\ \overline{18}$

For the following practice problem, remember that $\overline{30}$ $\overline{30}$ is just double $\overline{30}$, which of course is $\overline{15}$.

PRACTICE: If a 1 share is $\overline{5}$ $\overline{20}$, find a night watchman's share.

ANSWER: $\overline{10}$ $\overline{15}$

The only fraction we haven't figured out how to take two-thirds of is ⅔ itself. We can use the above rule to find this value. First we take half of $\overline{\overline{3}}$, which is $\overline{3}$. Then we take a sixth of $\overline{\overline{3}}$, which is a third of its half, or $\overline{9}$. Hence $\overline{\overline{3}} \times \overline{\overline{3}}$ is $\overline{3}$ $\overline{9}$.

On payday the 1 share for bread was $1\ \overline{\overline{3}}$ loaves. In order to calculate a night watchman's share, the scribe multiplied $\overline{\overline{3}}$ by $1\ \overline{\overline{3}}$ to get $\overline{\overline{3}}\ \overline{3}\ \overline{9}$. Since $\overline{\overline{3}}\ \overline{3}$ is 1, a watchman received $1\ \overline{9}$ loaves of bread. The following example uses the above value for $\overline{\overline{3}}$ of $\overline{\overline{3}}$ and the rule for $\overline{\overline{3}}$ of an even fraction.

EXAMPLE: If a 1 share is $\overline{\overline{3}}$ $\overline{20}$, find a night watchman's share.

SOLUTION: $\overline{\overline{3}} \times (\overline{\overline{3}}\ \overline{20}) = (\overline{\overline{3}} \times \overline{\overline{3}}) + (\overline{\overline{3}} \times \overline{20}) = \overline{3}\ \overline{9}\ \overline{30}$

Try this one. When you simplify your answer, remember that to add an odd fraction to itself you must consult Ahmose's table.

PRACTICE: If a 1 share is $\overline{\overline{3}}$ $\overline{6}$, find a night watchman's share.

ANSWER: $\overline{3}\ \overline{6}\ \overline{18}$, or equivalently, $\overline{2}\ \overline{18}$

We can now take $\overline{\overline{3}}$ of every Egyptian number. The following example will use the method on whole numbers, $\overline{\overline{3}}$, odd fractions, and even fractions. In the example we will simplify using $\overline{\overline{3}} + \overline{3} = 1$.

EXAMPLE: If a 1 share is $25\ \overline{\overline{3}}\ \overline{9}\ \overline{40}$, find a night watchman's share.

SOLUTION:

$$\overline{\overline{3}} \times (25\ \overline{\overline{3}}\ \overline{9}\ \overline{40}) = (\overline{\overline{3}} \times 24) + (\overline{\overline{3}} \times 1) + (\overline{\overline{3}} \times \overline{\overline{3}}) + (\overline{\overline{3}} \times \overline{9}) + (\overline{\overline{3}} \times \overline{40})$$
$$= 16 + \overline{\overline{3}} + \overline{3}\ \overline{9} + \overline{18}\ \overline{54} + \overline{60}$$
$$= 17\ \overline{9}\ \overline{18}\ \overline{54}\ \overline{60}$$

While simplifying the following example, remember that $\overline{\overline{3}} + \overline{3}$ is $\overline{\overline{3}}$ and that the two $\overline{30}$s simplify to $\overline{15}$.

PRACTICE: If a 1 share is $32\ \overline{\overline{3}}\ \overline{15}$ $\overline{20}$, find a night watchman's share.

ANSWER: $21\ \overline{\overline{3}} \times \overline{9}\ \overline{15}\ \overline{90}$

The last problem we will consider here is the number of jugs of beer a watchman received. A 1 share was worth $2\ \overline{\overline{3}}\ \overline{10}$. This will involve three separate rules: one for whole numbers, one for $\overline{\overline{3}}$, and one for even fractions. We can calculate a night watchman's share as follows:

$$\overline{\overline{3}} \times 2\ \overline{\overline{3}}\ \overline{10} = (\overline{\overline{3}} \times 2) + (\overline{\overline{3}} \times \overline{\overline{3}}) + (\overline{\overline{3}} \times \overline{10})$$
$$= 1\ \overline{3} + \overline{3}\ \overline{9} + \overline{15}$$
$$= 1\ \overline{3}\ \overline{9}\ \overline{15}$$

Oddly enough, the scribe records an equivalent answer of $1\ \overline{2}\ \overline{3}\ \overline{90}$. Does this mean our method of taking two-thirds is wrong? The answer is no. This is the first of many examples we will see of the creative solutions of the Egyptians. Up to this point I've been treating Egyptian mathematics as rote. This is as far from the truth as you can get. The ancient scribes were always thinking about the numbers they worked with, searching for tricks to simplify their calculations. To understand where the solution $1\ \overline{2}\ \overline{3}\ \overline{90}$ came from, we must examine the previous calculations. Someone called the Wtw priest had a 2 share, which the scribe had calculated as a value of $5\ \overline{2}\ \overline{30}$. When he needed to calculate the salary of the Thur guardians, he had to determine how much a $1\ \overline{3}$ share was worth. He presumably did this by realizing that two-thirds of 2 is $1\ \overline{3}$. So he could calculate the Thur guardian's share by multiplying the Wtw priest's share by $\overline{\overline{3}}$. This multiplication used the method for taking $\overline{\overline{3}}$ of a whole number and two applications of $\overline{\overline{3}}$ of even fractions. Doing this he got

$$\overline{\overline{3}} \times (5\ \overline{2}\ \overline{30}) = \overline{\overline{3}} \times 5 + \overline{\overline{3}} \times \overline{2} + \overline{\overline{3}} \times \overline{30}$$
$$= 3\ \overline{3} + \overline{3} + \overline{45}$$
$$= 3\ \overline{\overline{3}}\ \overline{45}$$

Now that the scribe had calculated the $1\ \overline{3}$ share, he apparently knew that half of this was a share. So he obtained a night watchman's share by taking half of a $1\ \overline{3}$ share, getting

$$\overline{2} \times (3\ \overline{\overline{3}}\ \overline{45}) = \overline{2} \times 3 + \overline{2} \times \overline{\overline{3}} + \overline{2} \times \overline{45}$$
$$= 1\ \overline{2} + \overline{3} + \overline{90}$$
$$= 1\ \overline{2}\ \overline{3}\ \overline{90}$$

The farther we delve into Egyptian mathematics, the more we will see how they rarely solved problems in straightforward, obvious ways. They formed an unusual combination of precision and creativity. On one hand, they felt compelled to calculate a night watchman's salary to the few drops of a ninetieth of a jug, yet at the same time they attained their precision in inventive ways.

THE PERFECT WOMAN

Division by 3

No matter what they tell you, size matters, at least in the pictures on the wall of an Egyptian tomb. In fact, everything matters. If you look at a figure, its size, its pose, and the objects it carries all have symbolic meaning. Size determines importance. The owner of the tomb usually dwarfs most of the other figures painted on the walls, but there's often an exception to this rule: wives are often depicted next to their husbands drawn to the same scale, indicating they are equal in status to their spouses.

While females in Egypt were not completely treated as equals to their male counterparts, Egyptians, compared with much of the ancient world, particularly Greece and Rome, were extremely liberal. Women are portrayed side by side, even with the most powerful of men. They held positions in many temples. There is even evidence that the wives of scribes could read, and this was in an era where only a couple of people in a hundred could do so. Perhaps women's greatest power was their ability to take men who had wronged them to court, a right that helps guarantee other rights.

Although wives are drawn with equal status near their husbands in the tombs of Egypt, there's something unusual about their appearance. Almost all of the women look exactly the same. They are all young and beautiful, and even their facial features are by and large indistinguishable.

It's not hard to understand why this is so. Imagine you're an Egyptian man overseeing the decoration of your family's tomb at the time when the artisans are painting your wife's portrait. As you glance at her accurately portrayed belly, you begin to feel uncomfortable. Your belief system tells you that one day this painting will come to life, acting as a host for your love's immortal soul. You suddenly come to the realization that you may spend your eternal afterlife dodging the question, does my tomb painting make me look fat? After a brief conversation with the artist, a few brush strokes are able to accomplish what years of dieting could not. Confronted with the world's cheapest form of cosmetic surgery, you correct every perceived flaw in your wife, yourself, and every other member of your family. In a similar way, you surround your portrait with pictures of a perfect home, fields, livestock, and other possessions. What you end up with is not a family portrait but rather an idealized version of Egyptian family life, including the form of the "perfect" woman.

It was important to the ancient Egyptians that figures of equal status have equal heights. Even seated characters sat on raised chairs to keep their heads at equal height to the standing figures. In order to ensure this, the Egyptian artists drew their figures on a grid. This grid would usually be painted over or carved away in the finished product. However, when a pharaoh died young, the artwork on the walls of their tombs was often left unfinished. On some of these, the grid lines are plainly visible.

The typical figure stood 18 grid lines tall, from the bottom of its feet to the hairline of its forehead. The reason the artists didn't measure from the top of the head was that this was often obscured by hair styles and hats. Equal-status figures had to look each other in the eye. It didn't matter if one wore a hat that was taller than the other.

Eighteen grid blocks for a human figure may at first seem like an odd choice for an Egyptian. Suppose craftsmen needed to paint a character 12 palms high. They would have to divide this number by 18 in order to calculate the height of one grid block. In order to understand the consequence of 18 grids, we need to be aware of a simple trick of division. Consider the division $50 \div 10$. We know that the answer is 5 because 10 is easy to divide by. However, we could have arrived at the answer another way. We could have first divided by 2, getting $50 \div 2$, or 25, and then we could have divided by 5, getting $25 \div 5$, which is 5. This trick works because 10 is 2×5.

Similarly, to divide by 18, we could first divide by 3, then by 3 again, and finally by 2 since $18 = 3 \times 3 \times 2$. This means that in order to use this trick, an Egyptian needs to know how to divide by 3. We know it's easy to divide by 2 as an Egyptian. If they had used 16 grids, they could partition the height by dividing by 2 four times since $16 = 2 \times 2 \times 2 \times 2$. Yet they still chose 18. Although we can't be sure why they selected 18 over 16, this would not have caused an Egyptian much difficulty because, as we will see, it's not hard to divide by 3 as an Egyptian.

The trick is to understand that dividing by 3 is the same as taking a third. Remember that Egyptians knew well how to take two-thirds of a number. In order to obtain a third, they merely needed to take half of this number, an operation they performed incessantly. For example, if we needed to divide 15 by 3, we could take $\overline{\overline{3}}$ of 15. This is 10, because 10 and its half is 15. Then we could divide the 10 by 2 to get 5, the answer to $15 \div 3$. We could write this as follows:

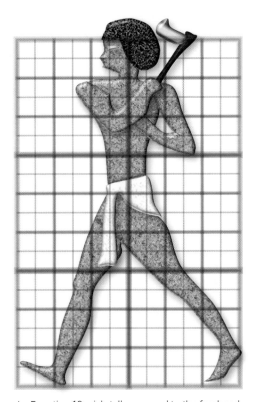

An Egyptian 18 grids tall measured to the forehead.

$\overline{3}$	?
1	15
$\overline{\overline{3}}$	10
$\overline{3}$	5 ✓
	5

$15 \div 3$ in the Egyptian style.

Using this method, they could divide by 3 in two easy operations. Consider the following division of 29 by 3. Remember 18 and its half is 27, leaving 2. To take of 2 we just remember that it's $1\,\overline{3}$, giving a total of $18 + 1\,\overline{3}$ or $19\,\overline{3}$.

$\overline{3}$	?
1	29
$\overline{\overline{3}}$	$19\,\overline{3}$
$\overline{3}$	$9\,\overline{2}\,\overline{6}$ ✓
	$9\,\overline{\overline{3}}$

$29 \div 3$ in the Egyptian style.

The final answer was calculated using the well-known Egyptian identity $\overline{2}\,\overline{6} = \overline{\overline{3}}$. Remember the pizza example in which one-half a pizza and a sixth is two-thirds of a pizza? A true Egyptian mathematician would have known this and would have replaced $\overline{2}\ \overline{6}$ with in the first division.

The Egyptians already knew how to divide by three. So why didn't they just do the problem in the "standard" way? Because it simply takes longer, and Egyptians were always looking for insightful shortcuts to simplify computation. Consider the following standard division of 29 by 3.

?	29	
1	3	✓
2	6	
4	12	
8	24	✓
$\overline{3}$	1	
$\overline{\overline{3}}$	2	✓
$9\,\overline{\overline{3}}$		

29 ÷ 3 in the nonstandard Egyptian style.

Notice that the second division is longer than the first. This difference gets worse the larger the number that must be divided by 3. In addition to being longer, the second problem requires more careful thought, whereas in the first they simply take two-thirds and then they take half. Here's another example using fractions.

EXAMPLE: Divide $49\,\overline{6}$ by 3 using the Egyptian trick.

SOLUTION:

$\overline{3}$	?	
1	$49\,\overline{6}$	
$\overline{\overline{3}}$	$32\,\overline{\overline{3}}\,\overline{9}$	
$\overline{3}$	$16\,\overline{3}\,\overline{18}$	✓
	$16\,\overline{3}\,\overline{18}$	

$49\,\overline{6}$ divided by 3 in the Egyptian style.

Try these on your own.

PRACTICE: Divide the following by 3 using the Egyptian trick.

- 24
- 52
- $35\,\overline{9}$

ANSWERS: 8, $17\,\overline{3}$, and $11\,\overline{\overline{3}}\,\overline{36}\,\overline{108}$

Now let's figure out how to divide by 18. To do this we simply divide by 3 twice and then divide by 2. Each division by 3, of course, is the taking of $\overline{\overline{3}}$ followed by a division by 2. The following division by 18 is actually written as a multiplication by $\overline{18}$.

Most of the calculations follow the methods I've shown above; however, I've simplified a few of the answers. In the example below, we multiply $\overline{3}$ by $\overline{\overline{3}}$ and use the odd rule to get $\overline{6}\ \overline{18}$. Normally when we divide by $\overline{2}$ we get $\overline{12}\ \overline{36}$. I've simplified this to $\overline{9}$. I'll explain later how to do this, but you can think of it in the following way. Using Ahmose's table, we know that $2\times\overline{9}$ is $\overline{6}\ \overline{18}$. So we know $(\overline{6}\ \overline{18})\div2$ is $\overline{9}$.

EXAMPLE: Divide 36 by 18 using the division-by-3 trick.

	18	?	
1)	1	36	
2)	$\overline{\overline{3}}$	24	$(1)\times\overline{\overline{3}}$
3)	$\overline{3}$	12	$(2)\div2$
4)	$\overline{6}\ \overline{18}$	8	$(3)\times\overline{\overline{3}}$
5)	$\overline{9}$	4	$(4)\div2$
6)	$\overline{18}$	2 ✓	$(5)\div2$
		2	

36 ÷ 18 done as 36 × $\overline{18}$.

One nice consequence of this method is that the first column never changes no matter what number we divide by 18. Even the operations used will always remain the same.

EXAMPLE: Divide 30 $\overline{10}$ by 18 using the division-by-3 trick.

	$\overline{18}$	?	
1)	1	30 $\overline{10}$	
2)	$\overline{\overline{3}}$	20 $\overline{15}$	$(1)\times\overline{\overline{3}}$
3)	$\overline{3}$	10 $\overline{30}$	$(2)\div 2$
4)	$\overline{6}\ \overline{18}$	6 $\overline{3}$ $\overline{45}$	$(3)\times\overline{\overline{3}}$
5)	1	3 $\overline{3}$ $\overline{90}$	$(4)\div 2$
6)	$\overline{18}$	1 $\overline{3}$ $\overline{180}$ ✓	$(5)\div 2$
		1 $\overline{3}$ $\overline{180}$	

30 $\overline{10}$ ÷ 18.

PRACTICE: Divide 9 and 27 $\overline{2}$ by 18.

ANSWERS: $\overline{2}$ and 1 $\overline{2}$ $\overline{36}$

Let's finish off this section by solving the problem that motivated the division-by-3 method.

EXAMPLE: You wish to draw a figure 12 palms high. If you draw it on a grid of 18 blocks, what distance separates the each grid line?

SOLUTION: We need to divide 12 palms into 18 equal pieces.

	$\overline{18}$	?	
1)	1	12	
2)	$\overline{\overline{3}}$	8	$(1)\times\overline{\overline{3}}$
3)	$\overline{3}$	4	$(2)\div 2$
4)	$\overline{6}\ \overline{18}$	2 $\overline{\overline{3}}$	$(3)\times\overline{\overline{3}}$
5)	$\overline{9}$	1 $\overline{3}$	$(4)\div 2$
6)	$\overline{18}$	$\overline{3}$ ✓	$(5)\div 2$
		$\overline{3}$	

The grid separation is $\overline{18}\times 12$, or $\overline{3}$ palms.

MEASURING THE GRID OF ETERNITY

Constructing the 18 grid

In the previous section we saw how to divide a number by 18 in order to calculate the spacing of one grid. In the problem where the figure was to be 12 palms high, the grid lines were $\overline{3}$ palms apart. It's possible that the artist drew a baseline then drew a line $\overline{3}$ above it and then measured up another $\overline{3}$ to get the next line. He could continue until all 19 grid lines were measured out. But there's a problem with this method. The errors of the repeated measuring would accumulate, significantly altering the height of the figure.

It doesn't always make sense to do something in order. Consider, for example, the task of splitting up a 6-foot submarine sandwich between eight people. You don't know how long each piece is, but you can start by cutting the sub in half, giving two pieces.

We could now cut the sub in four by dividing each piece in two. Note that each piece is the same size.

A submarine sandwich cut in two.

We could now finally cut each fourth in half, giving us the required eight pieces.

Each half cut in half makes four pieces.

Note that since each piece is the same size, the distances from any new cut to the preceding cut to its left are all equal.

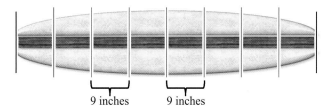

9 inches   9 inches

Each quarter cut in two makes eight pieces. Every new cut is
the same distance to the old cut to the left.

Note that we've essentially made a horizontal grid with
our sub sandwich. In the above example, we could have
made the cuts by eye, but we could have also determined
them mathematically. When we initially cut the 6-foot sub
in half, we could have performed the operation 6 ÷ 2 = 3
and made the cut 3 feet from the left. The second set of
cuts could have been determined by taking the length of
a piece and dividing it by 2, giving 3 ÷ 2 = 1.5. This means
that the first cut is 1 foot, 6 inches from the edge of the
piece. When we cut the other piece in half, we move the
same distance from the first cut, 1 foot, 6 inches, so there's
no need to recalculate. Finally, the last set of cuts is made
at intervals of 1.5 ÷ 2 = 0.75 feet, or every 9 inches. We can
use 9 inches to cut all four of the slices.

Before we start with our Egyptian example, realize that the sub has nine divider lines—seven cuts and
the two lines that demark the ends of the sub. What this
means is that a grid of eight squares will have nine cuts.
In general, there will always be one more line than the
number of grids. If you need a fence post every foot, ask
some persons how many fence posts 20 feet will require,
they'll often say 20, when in fact the answer is 21.

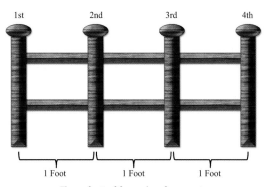

1st   2nd   3rd   4th

1 Foot   1 Foot   1 Foot

Three feet of fence has four posts.

This is called a *fence-post error*. It's easy to see the mistake in a smaller example. Consider the 3 feet of fence at
the bottom of the last column. Clearly it has four fence
posts.

We can think of this disparity mathematically in terms
of fractions. The above fence is divided by the posts into
thirds. The post labeled "2nd" is a third of the way over
from the left. So I can easily call it ⅓. Likewise, the post
marked "3rd" is two-thirds of the way over, so it can be
called ⅔. So, what should we call the post marked "1st"?
It's "none" of the way over, so it's 0⁄3, and the last is "all"
of the way over, so it is 3⁄3. The tops of our fractions are
0, 1, 2, and 3. Normally when we count to 3 we get three
numbers, but since we start from 0 we get four. Think of
the "0th" post as the baseline. You don't measure where it
is; you measure from it.

Let's now draw an Egyptian woman 10 fingers high. A
finger is a fourth of a palm. You can see this by looking at
your own hand.

4 Fingers   1 Palm

4 fingers = 1 palm.

You will need a blank piece of paper and a copy of the
Egyptian-style ruler located in the Computation Tables.
The drawing is loosely based on a real Egyptian measuring stick found today in the Louvre Museum. It can be
used to measure fingers, palms, and some fractions of a
finger. Like the original, the ruler I've provided has two
sets of numeric labels. The top row of numbers measure
the length in fingers from the left edge. In the illustration
at the top of the next page, I've marked out 5 fingers. The
bottom set of numeric labels give us fractions of a finger.
If we look between the 10 and 11 finger mark, we see $\overline{3}$.
The region between the two marks is divided into three
pieces. Hence, each is $\overline{3}$ of a finger.

Remember that our goal is to create a grid of 18 vertical regions in the space of 10 fingers. Begin by drawing a baseline across the bottom of your paper. We'll be

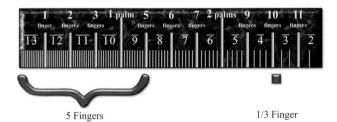

5 Fingers   1/3 Finger

Five fingers and 1/3 finger on the ruler.

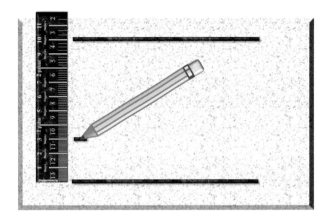

To go 3 $\overline{3}$ up, start by going up 3 fingers from the bottom.

creating a grid 18 squares high, so this line represents the $\frac{9}{18}$ place. Next, use the ruler to measure up 10 fingers to get the top line.

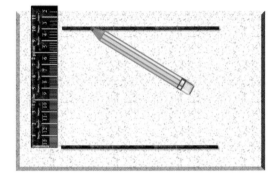

The top and base lines are 10 fingers apart.

If you recall, when dividing by 18, we started by taking a third. So we need to calculate $\overline{3}$ of 10.

	$\overline{18}$	?	
1)	1	10	
2)	$\overline{3}$	6 $\overline{\overline{3}}$	$(1) \times \overline{3}$
3)	$\overline{\overline{3}}$	3 $\overline{3}$	$(2) \div 2$

10 ÷ 3 on its way to division by 18.

We have to go up 3 $\overline{3}$ fingers from the baseline. Now take your ruler and measure 3 fingers up from the bottom. Mark this location.

To get the $\overline{3}$, look on the ruler in between the 10th and 11th finger. You will see a $\overline{3}$ over a finger divided in 3.

From the mark you just made, go up one of those small notches, representing a third of a finger. Draw a line here. This line is 3 $\overline{3}$ from the baseline. Note that it is a third of the way up the total height of 10 fingers.

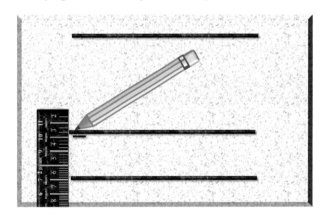

Draw a line $\overline{3}$ above the mark to make a line 3 $\overline{3}$ above the baseline.

From this line, go up another 3 $\overline{3}$ to get the two-thirds line. Alternatively we can go up 6 $\overline{\overline{3}}$ from the baseline. Note that the 6 $\overline{\overline{3}}$ is the second row of our computation, $\overline{3}$ of the way up.

Continuing, we need to divide each of the three resulting regions in thirds again. Go back to the original computation and take $\overline{\overline{3}}$ of 3 $\overline{3}$ and then take half of it. Note that in the following computation, we know that $\overline{6}$ $\overline{18} \div 2$ is $\overline{9}$ since Ahmose's table says that $2 \times \overline{9}$ is $\overline{6}$ $\overline{18}$.

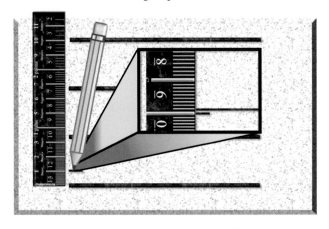

	$\overline{18}$	?	
1)	1	10	
2)	$\overline{3}$	$6\,\overline{\overline{3}}$	$(1)\times\overline{3}$
3)	$\overline{3}$	$3\,\overline{3}$	$(2)\div 2$
4)	$\overline{6}\,\overline{18}$	$2\,\overline{6}\,\overline{18}$	$(3)\times\overline{\overline{3}}$
5)	$\overline{9}$	$1\,\overline{9}$	$(4)\div 2$

$10 \div 9$ is $1\,\overline{9}$.

We now have to go up $1\,\overline{9}$ from the bottom of each of the three regions. Use the ruler to mark off one finger and then find the $\overline{9}$ marks to go up an additional ninth.

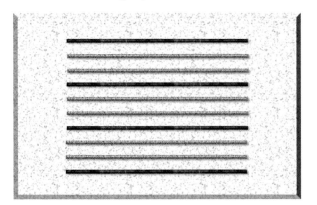

Make a mark 1 finger up and draw a line $\overline{9}$ above it.

The 10 fingers split into nine regions.

Go up $1\,\overline{9}$ in each of the three regions and draw a horizontal line. Then go up an additional $1\,\overline{9}$ above each of the new lines. You should now have 10 lines that break the 10 fingers into 9 regions.

We need 18 regions, so we'll divide each region in two. We compute this by taking half of our last answer, giving $1\,\overline{9}\div 2$, which is $\overline{2}\ \overline{18}$.

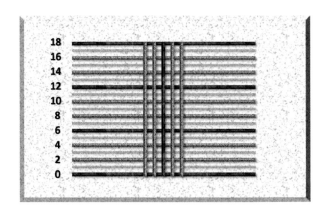

	$\overline{18}$	?	
1)	1	10	
2)	$\overline{3}$	$6\,\overline{3}$	$(1)\times\overline{3}$
3)	$\overline{3}$	$3\,\overline{3}$	$(2)\div 2$
4)	$\overline{6}\,\overline{18}$	$2\,\overline{6}\,\overline{18}$	$(3)\times\overline{\overline{3}}$
5)	$\overline{9}$	$1\,\overline{9}$	$(4)\div 2$
6)	$\overline{18}$	$\overline{2}\,\overline{18}$	$(5)\div 2$

The finished division.

So in each of the nine regions we need to draw one last line $\overline{2}\ \overline{18}$ above each of the existing lines. Since $\overline{18}$ is so small, you can choose to ignore it or use the $\overline{9}$ marks and go up half a mark. You now should have 18 grid regions. Label each of the horizontal lines 0 to 18, starting at the bottom. Now draw five vertical lines with the same spacing as the horizontal lines, $\overline{2}\ \overline{18}$ fingers apart.

The horizontal grid labeled 0 to 18 with five vertical lines.

Now that we have a grid, let's draw the perfect Egyptian woman. I'm loosely following proportions Egyptians used in the Middle Kingdom (about 2000–1700 BCE). Even in the ancient world, artists did not strictly follow rules the way a modern mathematician would, so consider the following an overly precise set of instructions.

We have a grid 4 blocks across and 18 down. Let's start with the head in the top two blocks. The ear attaches to the vertical center line in the top row as shown below. Put the eyeball on the next vertical line, going slightly above center. The nose and lips grow out to the right of this line.

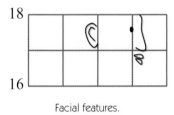

Facial features.

Now draw an almond-shaped eye around the eyeball and an eyebrow above. For the lip, draw a short line between the nose and mouth. Draw the chin. Draw the skull starting from the top of the nose. The skull actually goes off the grid almost one grid line higher. It peaks around the center line and hits the grid just to the left of the second vertical line. Curve it around, going through the grid intersection in line 17 and end it in the middle of the square.

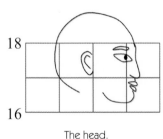

The head.

The face carries the personality of a figure, so we'll shade parts to give it more depth. Shade the side of the nose, the curl of the nostril, and the bottom of the chin.

Go about half a grid from the center line in each direction to draw the neck. Draw down to line 16 and curve in almost horizontally for the shoulders. When you go off the grid, curve the shoulder line down to begin the arms.

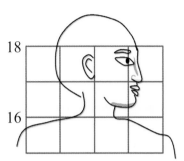

The neck, shoulders, and facial shading.

The torso consists of two curves that roughly follow the second and fourth vertical lines. On the left, the back slopes gently inward, starting about line 15 and curving outward at the small of the back on line 12. It then curves around for the buttocks, ending on line 9. The front side starts with the breast, which peaks on line 14 and ends halfway between lines 13 and 14. The rest of the front is fairly straight, although it curves slightly inward at the navel at 12 and bulges at the abdomen at 11. The front ends at line 10.

It may seem strange that the shoulders are square to the view yet the breasts are on the side. While this is not realistic, we must remember that this is a blueprint for the afterlife, not a realistic depiction. The arms need to be visible, hence the position of the shoulders, and the breasts, which are needed to make the woman fertile, must also show.

Next we draw the arms. They are both relatively straight, but they bulge ever so slightly in both the upper and forearms. The bulges thin at the elbow, around line 12, and the arms end slightly below line 10. The bulge in the forearm is closer to the elbow, making the wrist area slender.

The Egyptians drew many types of hands. In particular they often drew their subjects holding different items,

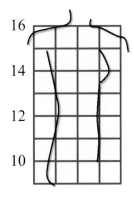

The torso.

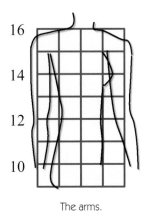

The arms.

each of which had symbolic value. I've chosen a side view of an empty hand because it's fairly easy to draw. The total length of each hand is a little more than one grid.

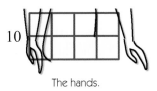

The hands.

The legs of an Egyptian woman are drawn fairly curvy. The back of the legs start at line 9 and curve inward to the small of the knee at 5. They then curve outward to the calf, which peaks around line 4 and goes in to a slender

ankle that ends at line 1. The front of the legs curves to the knee at line 5 and then goes fairly straight but slightly inward to the ankle at 1.

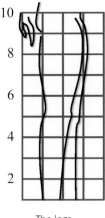

The legs.

The feet fill the bottom grid box and are about 3 grids long. There are rounded parts on the bottom at the toe, the ball of the foot, and the heel.

The feet.

You've now completed the basic outline of an idealized Egyptian woman. If yours looks less than perfect, don't worry. Egyptian artists made mistakes, too. After they finished their drawings, the chief artist would come in and correct their sketches with red paint. After this was done, specialists would come and carve and/or paint them. Of course the drawings would be fitted with hair, clothes, jewelry, and other accessories. It had to look good because someone's soul would be spending an eternity in the drawing. At some point a priest would perform the symbolic "opening of the mouth ceremony," a ritual to enable the deceased soul to enter the painting. Here's my drawing. I hope yours looks good.

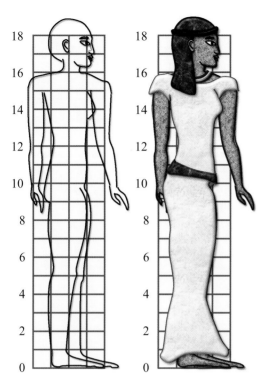

The "perfect" woman's outline and finished painting.

THE DOCK ENLISTEES

Division by 10

Almost four thousand years ago, the enlistees of the dock-yard of Naga El-Deir did something in the "brick-clay fields." What, exactly, their task was is hard to say. They did something for the "builders of the tower," and they did more things for the "eastern chapel of the glorious chambers." While we don't exactly know what they did, we do know how many enlistees it took to do each job. It's all described rather vaguely in a papyrus that archeologist George Reisner found about a hundred years ago.

The scroll is composed of many parts, ranging from lists of workers to accounts of cargo. Although there are many administrative records from the excessively bureaucratic

Egyptians, the Reisner Papyrus was a particularly lucky find. The scroll doesn't just have ancient records, it also contains the results of their computations. The section we're concerned with involves the computation of the number of workers required to do some (unknown) task.

Each entry has a vague description followed by a list of six numbers. The first three are dimensions, usually listed in cubits. One line starts with the measurements 3, 2, and 2. These numbers are believed to represent the length, width, and depth of some object. It is well known today, as it was in ancient times, that you multiply these together to obtain the volume. This is a fairly obvious relation if you think about it in concrete terms. Consider a 3-by-2-by-2 box made out of 1-by-1-by-1 blocks. You can make the base of your box using three rows of two blocks. This has

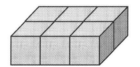

A 3-by-2-by-1 box has a volume of 3×2.

3×2 blocks, and since each block has a volume of 1, it has a total volume of 6.

To make the height of your box equal to 2, we need to put two of these 3-by-2 groupings together, one on top

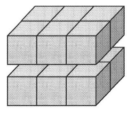

Two groups of 3-by-2 blocks make a 3-by-2-by-2 box.

of the other. This would take twice as many, or $6 \times 2 = 12$, blocks. We could have obtained 12 just by multiplying the three dimensions 3, 2, and 2.

It should come as no surprise that after the numbers 3, 2, and 2 comes the number 12, two columns later, giving the volume of the group. Another entry gives the dimensions as 8 by 5 by $\overline{4}$ and follows it with a 10, indicating that 10 is $8 \times 5 \times \overline{4}$. Although the papyrus has no calculations, we can imagine that the scribe multiplied them in two steps as follows.

EXAMPLE: Find the volume of an 8-by-5-by-$\overline{4}$ block.

8	?
1	5
2	10
4	20
8	40 ✓
	40

$\overline{4}$	?
1	40
$\overline{2}$	20
$\overline{4}$	10 ✓
	10

8×5 is 40, which in turn is multiplied by $\overline{4}$, to get 10.

PRACTICE: Find the volumes of a 3-by-4-by-5 block and a 7-by-3-by-1 $\overline{2}$ block.

ANSWERS: 60 and 31 $\overline{2}$

The number after the three dimensions but before the volume has been interpreted as "units." It's possible this means the number of blocks needed. This interpretation just increases the number of multiplications by one. For example, if I need three groups of 2-by-2-by-5 blocks, the total volume is just three groups of $2 \times 2 \times 5$, or $3 \times 2 \times 2 \times 5$.

The final column of the papyrus, "Enlistees," is the subject of this section. It is believed to show the number of workers required to complete the unstated job. The larger the volume calculated for each row, the more enlistees are required to do the work. This, of course, makes perfect sense. No matter if the job was to move, make, count, or load the material in question, more volume requires more work to accomplish the task. Because the entries are listed by day, more work in a fixed amount of time requires more workers.

It's interesting to note that most jobs are assigned to a fractional number of people. For example, in the row with the block with a volume of 12, it requires 1 $\overline{5}$ workers.

While this may seem strange to have "partial" enlistees doing a job, it's quite natural if you're trying to calculate the total number of workers for the day. For example, if you have one job that requires 4 $\overline{2}$ workers and another that requires 3 $\overline{2}$, what you really need is eight workers, four on the first job, three on the second, and one who splits his time equally between the two, hence becoming the "half" worker.

It's natural to expect that the volume and the number of workers would be connected by what mathematicians call a *linear relation*. Simply put, if one number doubles, the other number doubles, too. This is common sense because if you had twice as much to do, you would usually need twice as many people. A simple example of a linear relationship is the relation between feet and yards. We know that 5 yards is 15 feet. Similarly 10 yards is 30 feet. Note that both numbers doubled. It's also clear that 15 yards is 45 feet. In this case both numbers tripled.

In a linear relationship, there is always a *multiplier* between the two. This multiplier can be found by dividing one by the other. For example, if we divided the 15 feet by the 5 yards, we get $15 \div 5 = 3$. We could have just as easily gotten the same answer by dividing the 45 by the 15 or the 30 by the 10. The 3 in our solution represents the 3 feet that occur in every yard. Let's find the multiplier by dividing the volume by the number of enlistees. We know that a volume of 10 needs 1 worker. Clearly the multiplier is 10. We can check to see if the same multiplier works on the other entries. We know a volume of 12 needs 1 $\overline{5}$ workers, so if we're right, $12 \div 10$ should be 1 $\overline{5}$. Let's test this by doing an Egyptian division. As we can see below, our guess is in fact correct.

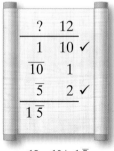

?	12
1	10 ✓
$\overline{10}$	1
$\overline{5}$	2 ✓
1 $\overline{5}$	

$12 \div 10$ is 1 $\overline{5}$.

Egyptians were perfectly capable of dividing by 10 in the above manner, but they didn't need to. They wrote their numbers in base-ten symbols just as we do. This simply means that ten of one unit makes the next. So ten 1s is 10, ten 10s is 100, ten 100s is 1,000, and so on. In hieroglyphics this is particularly simple. Examine the values of the base-ten symbols below.

| = 1   ∫ = 10,000

∩ = 10   ❧ = 100,000

℮ = 100   ☥ = 1,000,000

𝕀 = 1,000

The Egyptian units.

Just as it's clear to us that 10×30 is 300, it would be clear to an Egyptian that $10 \times ∩∩∩$ is ℮℮℮. Multiplication by 10 did nothing but change the symbols. Similarly $10 \times ∥∥$ is ❧❧ and so on. The trick works for mixed combinations of these. For example, we know that 100×32 is 320, and an Egyptian would know that $100 \times ∩∩∩∥∥$ is ℮℮℮∩∩. Division by 10 is the same process in reverse. The units simply change to the next smallest one.

EXAMPLE: What are $10 \times ❧℮℮∩∩∩$ and $☥℮℮℮∩ \div 10$?

SOLUTION:

$$10 \times ❧℮℮∩∩∩ = ☥𝕀𝕀∥℮℮℮$$
$$☥℮℮℮∩ \div 10 = ❧∩∩∩∥$$

PRACTICE: What is $10 \times ∥∥℮∩∩∩$ and $❧∥∥∥∥∩∩ \div 10$?

ANSWERS: $❧❧𝕀℮℮℮$ and $∥∥∥∥∥∥$

Dividing fractions by 10 is also very easy. We have to remember that Egyptian fractions work in the opposite way as whole numbers. So dividing a fraction by 10 works just like multiplying a whole number by 10. In hieroglyphics an Egyptian would write ¹⁄₃₀ as $\overline{∩∩∩}$. When we divide this by 10, we write $\overline{℮℮℮}$, since $10 \times ∩∩∩$ is ℮℮℮.

EXAMPLE: What is $\overline{∥∥∥} \; \overline{℮∩∩} \div 10$?

SOLUTION: $\overline{∩∩∩} \; \overline{𝕀℮℮}$

Try this one.

PRACTICE: What is $\overline{℮∥} \; \overline{𝕀℮℮} \div 10$?

ANSWER: $\overline{𝕀∩} \; \overline{❧𝕀𝕀}$

Since we don't work in hieroglyphs, let's do the same type of problems using the bar notation. The problems in this notation get even easier. Instead of replacing the symbol under the bar with another that is 10 times greater, we simply multiply them by 10.

EXAMPLE: What is $470 \; \overline{2} \; \overline{33} \div 10$?

SOLUTION: $47 \; \overline{20} \; \overline{330}$

There's only one problem when we divide by 10. We don't yet know how to divide the numbers from 1 to 9. Consider the division in hieroglyphs of ∥∥∥. We could replace these with symbols worth a tenth of the value of a |. This would mean that $∥∥∥ \div 10$ is $\overline{∩} \; \overline{∩} \; \overline{∩}$. While this is perfectly valid, an Egyptian would never write this because it fails to be a decent approximation. Here three-tenths would be interpreted as a tenth and a little more. We're off by a factor of three.

Egyptians like to have options when doing math. Division by 10 is easy for many whole numbers and all fractions; hence they would want a way to easily divide all whole numbers by 10. In order to accomplish this they used a small table. It appears in Ahmose's papyrus right after the table that tells how to double odd fractions. It simply tells how to divide the numbers 1 to 9 by 10.

$1 \div 10 = \overline{10}$	$6 \div 10 = \overline{2} \; \overline{10}$
$2 \div 10 = \overline{5}$	$7 \div 10 = \overline{\overline{3}} \; \overline{30}$
$3 \div 10 = \overline{5} \; \overline{10}$	$8 \div 10 = \overline{\overline{3}} \; \overline{10} \; \overline{30}$
$4 \div 10 = \overline{3} \; \overline{15}$	$9 \div 10 = \overline{\overline{3}} \; \overline{5} \; \overline{30}$
$5 \div 10 = \overline{2}$	

Ahmose's division-by-10 table.

The table is remarkably easy to use. We need to remember that we can divide a sum by 10 by working on the individual parts. So if we divided $70+60$ by 10 we could divide 70 and 60 by 10 and add the answers, $7+6$. So if I need to divide 437 by 10, I can rewrite the number 437 as $430+7$. I know how to divide 430 by 10, and the table tells me how to divide 7 by 10. So the answer is $43+\overline{\overline{3}}\ \overline{30}$, or simply $43\ \overline{\overline{3}}\ \overline{30}$.

EXAMPLE: What is $836\ \overline{20} \div 10$?

SOLUTION:

$$836\ \overline{20} \div 10 = (830+6+\overline{20}) \div 10$$
$$= 83+\overline{2}\ \overline{10}+\overline{200}$$
$$= 83\ \overline{2}\ \overline{10}\ \overline{200}$$

PRACTICE: What is $53\ \overline{2}\ \overline{31}\ \div 10$?

ANSWER: $5\ \overline{5}\ \overline{10}\ \overline{20}\ \overline{310}$

Now let's get back to the problem that motivated this section. Remember that to find the number of workers needed, we simply divide the volume of a box by 10. I'm going to assume the work done is loading bricks on a ship, because the records involved are from a dock and there are a number of oblique references in the document to bricks.

EXAMPLE: Bricks are piled in a 4-by-7-by-3 $\overline{8}$ cubic cubit block. How many enlistees are needed to load the bricks?

SOLUTION: Volume $= 4 \times 7 \times 3\ \overline{8}$

$$87\ \overline{2} \div 10 = (80+7+\overline{2}) \div 10 = 8\ \overline{\overline{3}}\ \overline{30}\ \overline{20} = 8\ \overline{\overline{3}}\ \overline{20}\ \overline{30}$$

Number of enlistees = volume ÷ 10

PRACTICE: Bricks are piled in a 3-by-2 $\overline{2}$-by-7 cubic cubit block. How many enlistees are needed to load the bricks?

ANSWER: $5\ \overline{5}\ \overline{20}$ or $5\ \overline{4}$

We now know all the basic Egyptian operations. We'll next learn how to simplify them.

4
SIMPLIFICATION

PEDAGOGY AND PIZZA

Adding Fractions Using Parts

Students simply do not appreciate how much time teachers put into their classes. Little do they realize that the hours they spend in class are matched by hours outside. Good teachers labor over every problem they explain in class. The order, difficulty, and complexity of each exercise is carefully planned. Apparently little has changed in almost four thousand years.

We can learn a lot about ancient mathematics by looking at texts from a pedagogical point of view. Far too often historians of mathematics select one interesting problem and analyze it to death. By taking these problems out of context, they lose valuable insights into the purpose of the example. Which problems come before it, after it, what numbers are selected, and what is said and not said—all reflect on the role of the exercise in the text.

What I want to focus on is the beginning of Ahmose's text. It starts with the times-2 table for odd fractions and the division-by-10 table. Notice that there are no references to whole-number operations. From this we can infer that the students who used this book already knew whole-number multiplication and division. The initial focus on fraction operations, the basic tools required to work with fractions, suggests that this is in fact a fraction textbook. Much of the rest of the papyrus bears this out. After the tables, it begins with examples of fraction multiplication, followed by subtraction and division. The text then continues with word problems, most of which require fractions to solve. This is exactly the way most modern texts on fractions would be organized.

What comes at the start of Ahmose's text tells us much about the fundamental ideas that will be utilized in the rest of the text. Let's examine the first real problem in the papyrus, which appears immediately after the odd-fraction and division-by-10 tables in Ahmose's text. It's a multiplication of $\bar{4}\ \overline{28}$ by $1\ \bar{2}\ \bar{4}$. The computation itself goes like this.

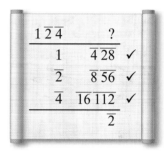

$$1\ \bar{2}\ \bar{4} \times \bar{4}\ \overline{28} = \bar{2}.$$

The first thing we should notice is that constructing $1\ \bar{2}\ \bar{4}$ on the left-hand side is trivial. This is made easy because the textbook writer doesn't want us to focus on that aspect of the multiplication. Also notice that the calculation of each line is trivial. Taking half of $\bar{4}\ \overline{28}$ to get $\bar{8}\ \overline{56}$ is obvious to anyone familiar with the doubling of whole numbers. They know that twice 4 is 8, and their teacher makes it clear that this means half of $\bar{4}$ is $\bar{8}$. Once again, this is not the focus of the problem, because it relies heavily on knowledge the student already has.

The only nontrivial part of this problem is on how the numbers in the right column, $\bar{4}\ \overline{28}$, $\bar{8}\ \overline{56}$, and $\overline{16}\ \overline{112}$, become $\bar{2}$. Recall that I said that the book starts with the multiplication and subtraction and then the division of fractions. A sharp reader may have noticed that I left out addition. The simple truth is, this is not the first problem in multiplication, since it differs little from whole-number multiplication, but rather, it is the first problem in the addition of fractions.

In the multiplication above, I left out an important part. There are comments to the side of the computation included below to explain how the addition was done. Here's a more complete version of the previous multiplication. I'll explain the added comments on the right shortly.

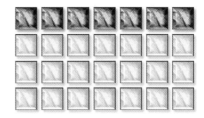

$1\,\overline{2}\,\overline{4} \times \overline{4}\,\overline{28}$ with "parts" comments.

In order to understand what's going on here, we need to go back to the pizza analogy we used to realize that $\overline{2}\,\overline{6}$ is $\overline{\overline{3}}$. We could phrase this as follows. As parts of a six-slice pie, $\overline{2}$ is three slices and $\overline{6}$ is one slice. This is four slices, which as part of a pie, are expressed as $\overline{\overline{3}}$.

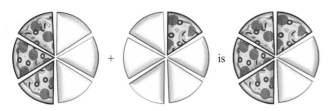

$\overline{2} + \overline{6} = \overline{\overline{3}}$ as parts of a six-slice pizza.

If we now phrase this in the same way as it appears in Ahmose's text, it might appear as follows.

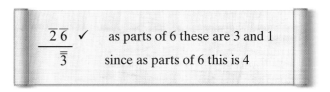

$\overline{2}\,\overline{6} = \overline{\overline{3}}$ expressed in parts.

The problem from Ahmose's papyrus uses parts of 28. Clearly $\overline{28}$ of 28 is 1. Calculating $\overline{4}$ of 28 would be relatively easy for an Egyptian who would know that half of half of 28 is 7.

A fourth of a 28-slice Sicilian pizza is 7 slices.

I want to look at an easier example than the one Ahmose picked; one that doesn't involve fractional parts. Consider the sum of $\overline{2}$, $\overline{3}$, $\overline{4}$, and $\overline{6}$. We can view these as parts of 12, since all of these numbers divide into 12 easily. Let's think of these as fractions of a foot. Since half a foot is 6 inches, $\overline{2}$ is 6 as parts of 12. Similarly a third of a foot is 4 inches, a fourth is 3 inches, and a sixth is 2 inches. So we get:

$$\overline{2}\,\overline{3}\,\overline{4}\,\overline{6} \text{ feet} = 6 + 4 + 3 + 2 \text{ inches} = 15 \text{ inches}$$

Our original problem was expressed in feet, and this means we need to convert our answer back into feet. I know 15 is 12 + 3, both of which can be easily expressed as parts of 12. As parts of 12, 12 and 3 are 1 and $\overline{4}$, respectively. I can now finish my problem.

$$15 \text{ inches} = 12 + 3 \text{ inches} = 1\,\overline{4} \text{ feet}$$

Hence, we now know that $\overline{2}\,\overline{3}\,\overline{4}\,\overline{6} = 1\,\overline{4}$. Here's another example worked out using the inches-and-feet analogy. Remember that to an Egyptian, $\overline{\overline{3}}$ of a foot is obviously 8 inches, since 8 and its half is 12.

EXAMPLE: Use inches and feet to simplify $\overline{\overline{3}}\,\overline{2}\,\overline{6}\,\overline{12}$ feet.

$$\overline{\overline{3}}\,\overline{2}\,\overline{6}\,\overline{12} \text{ feet} = 8+6+2+1 \text{ inches}$$
$$= 17 \text{ inches}$$
$$= 12+4+1 \text{ inches}$$
$$= 1\,\overline{3}\,\overline{12} \text{ feet}$$

Notice that we could not break up the 17 inches into $12+5$ because 5 cannot be made into a whole part of 12.

PRACTICE: Use inches and feet to simplify $\overline{3}\,\overline{4}\,\overline{12}$ feet.

ANSWER: $\overline{\overline{3}}$ feet

For the following problem, realize that a twentieth of a dollar is 5 pennies, a twenty-fifth is 4, and a fiftieth is 2. Realize that 10 pennies is $\overline{10}$ of a dollar.

PRACTICE: Use dollars and pennies to simplify $\overline{20}\,\overline{25}\,\overline{50}$ dollars.

ANSWER: $\overline{10}\,\overline{100}$

The above method works well only because we are familiar with feet, inches, dollars, and pennies. I want to generalize this so we can work in parts of any size. We can only guess how the Egyptians worked with parts because we see only the finished result. Since our ignorance prohibits us from working these out as an Egyptian, I'm going to modernize the method. This will make the method easier at the expense of authenticity.

We need to understand that when we convert from feet to inches, all we're doing is multiplying by 12. Similarly, when we go back into feet, we divide by 12. Consider the sum of $\overline{4}$, $\overline{6}$, and $\overline{12}$. We can think of this as feet to be converted into inches by a multiplication of 12 as follows.

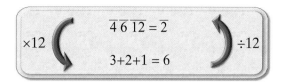

We've now computed that $\overline{4}\,\overline{6}\,\overline{12}$ as parts of 12 are $3+2+1$, or 6. We can now convert back into feet by division by 12, giving $\overline{2}$.

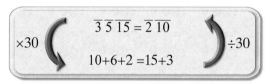

We can now do any problem. The only difficulty is finding a good way to break up a sum into easy parts of the multiplier. Below I'm going to work in parts of 30 and note that 18 can be broken into 15 and 3. Notice that we start with $\overline{3}, \overline{5},$ and $\overline{15}$ and multiply by 30, putting our answer below the three fractions. Then we add the multiplied numbers and break them up into parts of 30, putting the answer on the bottom right. Finally we divide the parts by 30, putting our answer to the right of the equal sign. All of these problems have this U-shaped solution of "down, right, up."

EXAMPLE: Add $\overline{3}, \overline{5},$ and $\overline{15}$ as parts of 30.

SOLUTION:

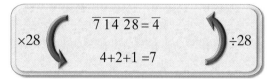

EXAMPLE: Add $\overline{7}$, $\overline{14}$, and $\overline{28}$ as parts of 28.

SOLUTION:

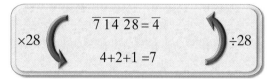

Now try these.

PRACTICE: Add $\overline{\overline{3}}, \overline{5}, \overline{6},$ and $\overline{30}$ as parts of 30.

ANSWER: $1\,\overline{15}$

PRACTICE: Add $\overline{4}, \overline{5}, \overline{10},$ and $\overline{20}$ as parts of 20.

ANSWER: $\overline{2}\,\overline{10}$

You might wonder where I get the number of parts for each problem. The answer is not easy, and I'll elaborate on this choice later in the book. However, for now there is a relatively simple method for those familiar with the modern addition of fractions. Note that in the above problem, the parts $\overline{4}$, $\overline{5}$, $\overline{10}$, and $\overline{20}$ all have nice, whole-numbered parts as part of 20 since 4, 5, 10, and 20 all go into 20 itself. This means that 20 is a common multiple of all the numbers that comprise the fractions. Since small numbers are generally easier to work with than large numbers, the least common multiple (LCM) is a nice number to use as the total number of parts. This is the same LCM that we used in grade school to find common denominators.

There is actually a close connection between our method of adding fractions and the Egyptian parts. Consider the following solution to the above problem where we added $\overline{7}$, $\overline{14}$, and $\overline{28}$ to get $\overline{4}$. The numbers were taken as parts of 28, giving $4+2+1$, or 7, which is $\overline{4}$ of 28. If we add them as fractions in the modern way, the solution looks like this:

$$\tfrac{1}{7}+\tfrac{1}{14}+\tfrac{1}{28} = \tfrac{4}{28}+\tfrac{2}{28}+\tfrac{1}{28} = \tfrac{4+2+1}{28} = \tfrac{7}{28} = \tfrac{1}{4}$$

Note that $\tfrac{1}{7}$ turned into $\tfrac{4}{28}$. This is essentially saying that $\tfrac{1}{7}$ as parts of 28 is 4. Also note the $4+2+1$ over the 28. This is the same sum we did in the parts problem. Finally, when we reduced the fraction $\tfrac{7}{28}$ to $\tfrac{1}{4}$, we essentially stated that $\tfrac{1}{4}$ is 7 parts of 28.

The methods are not exactly equivalent. We are familiar with LCMs from our method of adding fractions and our knowledge of the multiplication tables. An Egyptian would not find this as easy, but they had other methods. Consider the original problem Ahmose gave.

In order to solve this problem, Ahmose needed to add $\overline{4}$, $\overline{28}$, $\overline{8}$, $\overline{56}$, $\overline{16}$, and $\overline{112}$. The least common denominator

is 112, which wouldn't even be obvious to most modern math students. In fact, Ahmose doesn't use 112, but rather 28. This may seem strange since it gives awkward answers. For example, as parts of 28, $\overline{16}$ is $1\ \overline{2}\ \overline{4}$, something a seasoned mathematician would not see as trivial.

So why did Ahmose pick 28? The answer is remarkably simple. It's easy to determine $\overline{4}$ and $\overline{28}$ as parts of 28. They are 7 and 1, respectively. When he goes to figure out how many parts $\overline{8}$ and $\overline{56}$ are, he simply notes that these are half of $\overline{4}$ and $\overline{28}$. This is obvious because he got these numbers by taking half of $\overline{4}$ and $\overline{28}$. So if $\overline{4}$ is 7 parts of 28, then $\overline{8}$ is half of 7 parts, or $3\ \overline{2}$ parts. Similarly $\overline{16}$ is half of $\overline{8}$ so $\overline{16}$ is half of $3\ \overline{2}$ parts, or equivalently, $1\ \overline{2}\ \overline{4}$.

This is easier than it looks. When going from one line of a multiplication to the next, the scribe performs the same operation on both columns. When he deals with parts, he simply applies the same operation to the parts. The beautiful aspect of this is that the Egyptian mathematician wouldn't even pause momentarily to determine the number of parts, whereas a modern mathematician would have to contemplate for a while to determine a common multiple of $\overline{4}$, $\overline{28}$, $\overline{8}$, $\overline{56}$, $\overline{16}$, and $\overline{112}$.

	$1\ \overline{2}\ \overline{4}$	?	Parts	
1)	1	$\overline{4}\ \overline{28}$ ✓	7 1	
2)	$\overline{2}$	$\overline{8}\ \overline{56}$ ✓	$3\ \overline{2}\ \overline{2}$	(1)÷2
3)	$\overline{4}$	$\overline{16}\ \overline{112}$ ✓	$1\ \overline{2}\ \overline{4}\ \overline{4}$	(2)÷2
		$\overline{2}$		

Operations on the rows are the same for the parts.

You might argue that the Egyptian method would generate fractions that would make the problem more difficult. While this is true, the fractions they obtain are the easiest fractions for an Egyptian to work with. All the fractions are powers of 2 and, as a result, simple to add. Consider the addition Ahmose did to obtain his total number of parts. Recall that $\overline{4}$ and $\overline{28}$ were 7 and 1 parts of 28. Similarly $\overline{8}$, $\overline{56}$, $\overline{16}$, and $\overline{112}$ were $3\ \overline{2}$, 2, $1\ \overline{2}\ \overline{4}$, and $\overline{4}$ parts, respectively. When adding, you only need to know

	$1\ \overline{2}\ \overline{4}$	?	
	1	$\overline{4}\ \overline{28}$	✓
	$\overline{2}$	$\overline{8}\ \overline{56}$	✓
	$\overline{4}$	$\overline{16}\ \overline{112}$	✓
		$\overline{2}$	

$\overline{2}\ \overline{2}$ is 1, $\overline{4}\ \overline{4}$ is $\overline{2}$, $\overline{8}\ \overline{8}$ is $\overline{4}$, and so on. The numbers in the problem add up as follows:

$$(7 + 1) + (3\,\overline{2} + \overline{2}) + (1\,\overline{2}\,\overline{4} + \overline{4})$$
$$= 12 + (\overline{2}\,\overline{2}) + \overline{2} + (\overline{4}\,\overline{4})$$
$$= 13 + \overline{2}\,\overline{2}$$
$$= 14$$

So the total number of parts is 14. Remember what I said about the author of the text carefully selecting the numbers to make the method as transparent as possible? He wants his students to focus on the method, not on overly burdensome computations. As a result, this problem has the answer 14 parts. The author knows that his students can double in their sleep and would immediately be aware that 14 of 28 parts is $\overline{2}$, the answer to the first exercise in the text.

THE EASY LIFE

The G Rule

"I have seen beatings.… I have witnessed a man seized for his labor." So begins the advice of the renowned scribe Duau Khety to his son, Pepy. The warning is taking place as they sail up the Nile toward the Residence, the palace of the pharaoh. There Pepy is to be educated side by side with the sons of the foremost officials. Khety wants his son to take his studies seriously, so just like any other well-meaning father, Khety begins a long and serious lecture, which, assuming family life has changed little in the intervening millennia, would be completely ignored.

Khety spends very little of the lecture on the value of being a scribe. Rather, he warns Pepy of his fate should he fail. Khety describes the toil and hardship of the other professions of ancient Egypt one by one. He tells of the smiths whose fingers are worn as rough as the skin of a crocodile and who smell worse than fish eggs. He portrays craftsmen, bead makers, and bricklayers as overworked and whose arms are destroyed by the tiring labor. He describes reed cutters as laid low by mosquitoes and gnats. He compares potters to the buried dead, whose skin is constantly under a layer of mud. He warns his son of the oozing blisters on a gardener and the sores on fingers of the field laborers. He tells how carpenters can work a month on a roof and not be paid enough to feed their children. Khety relates how bird catchers and washermen live in constant fear of crocodiles and how the traders fear lions and barbarians. None of these professions, Khety tells his son Pepy, are free from abusive bosses. The one exception, of course, is the scribe, for he is the boss.

So learning to write is the only way to escape the horrors of life. As Khety puts it, a day in school "is more useful than an eternity of toil in the mountains." He wants Pepy to "love writing more than his mother" and "recognize its beauty." Khety ends his lecture by telling his son that no scribe ever goes hungry or even goes without the objects found in the house of a king. Pepy should be grateful for the opportunity he's been given and to be sure to pass it on to his own children.

Scribes do seem to have lived a relatively easy life. Paintings often show them reclining under the shade of a tree, recording the occasional number as the workers toil in the hot Egyptian sun. A scribe didn't go shirtless like the other workers, apparently because he was never forced to perform chores that would make him sweat. If we go by the idealistic portrayals on tomb walls, the scribes of Egypt did in fact have one of the easiest lives of the ancient world.

However, if you've ever worked an hour or more on a math problem without success, you might disagree. In the modern world, mathematics has a reputation for being anything but easy. Scribes needed to do math for their record keeping, easy or not.

As we've seen, in the previous section, adding fractions can be a bit tedious. First you're confronted with a choice in number of parts, then you need to express all your fractions in those parts, and finally you need to break up your answer efficiently into numbers that have nice parts. Although this gets easier with experience, it would still take a while if you had to do this for every problem. Are there ways to sometimes take a shortcut and bypass this process?

Consider the following simplifications of Egyptian fractions.

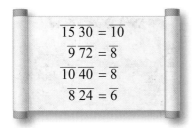

Each of these can be verified using the parts method. The number of parts needed is just the value of the second fraction on the left-hand side of each equation. For example, we can examine $\overline{15}\ \overline{30} = \overline{10}$ as parts of 30. A fifteenth of 30 is 2, a thirtieth of 30 is 1, and a tenth of 30 is 3. So the equation becomes $2 + 1 = 3$, which is, of course, true.

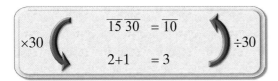

There's a pattern in the above equations. It was noticed by Richard J. Gillings and described in his book *Mathematics in the Time of the Pharaohs*, which he "humbly" named the *G rule*.

Consider the numbers by which you need to multiply or divide the successive terms of the equations to get the next term. In $\overline{15}\ \overline{30} = \overline{10}$ we need to multiply 15 by 2 to get 30, and then we need to divide 30 by 3 to get 10. Similarly, for $\overline{10}\ \overline{40} = \overline{8}$ we need to multiply 10 by 4 to get 40 and then divide it by 5 to get 8.

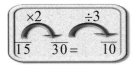 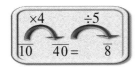

Look at rest of the relations below and see if you can find the pattern.

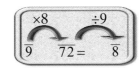 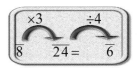

Gillings noted that in each case the number you divide by is 1 more than the number you multiplied by. This trick always works when the numbers divide the quantity evenly. Assume we have the fractions $\overline{30}\ \overline{150}$. The first thing you should notice is that we can get the 150 by multiplying the 30 by 5. At this point you should check to see if 1 more than 5 goes into 150 evenly. It does, since $150 \div 6 = 25$. This tells us that $\overline{30}\ \overline{150}$ is, in fact, $\overline{25}$.

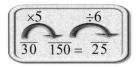

EXAMPLE: Simplify the following.

• $\overline{4}\ \overline{12}$
• $\overline{33}\ \overline{330}$

SOLUTION:

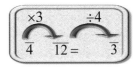 

PRACTICE: Simplify the following:

• $\overline{12}\ \overline{36}$
• $\overline{20}\ \overline{180}$

ANSWERS:

• $\overline{9}$
• $\overline{18}$

What do we do if we get something like $\overline{5}\ \overline{15}$? We can see that 15 is 5×3, but we don't get a whole number when we divide 15 by 4. The answer is simple: don't do anything. The number $\overline{5}\ \overline{15}$ is a perfectly legitimate solution in ancient Egypt. Likewise, when we see something like $\overline{4}$ $\overline{9}$, where no whole number multiplied by 4 produces 9, we also generally do nothing.

At this point you might be skeptical as to the practicality of the G rule. If we picked two fractions at random, the chance that the G rule applied to them would be miniscule. What's the point in knowing a shortcut that can rarely be used? What we need to know is that Egyptian fractions are not random numbers. Consider the following multiplication of $1\,\overline{24}$ by 6.

6	?	
1	$1\ \overline{24}$	
2	$2\ \overline{12}$	✓
4	$4\ \overline{6}$	✓
	$6\ \overline{6}\ \overline{12}$	

$6 \times 1\,\overline{24} = 6\,\overline{6}\,\overline{12}$, which is $6\,\overline{4}$.

Is it a coincidence that we ended up with two fractions, $\overline{6}$ and $\overline{12}$, where one number is twice another? Of course it's not, because we obtained $\overline{6}$ by taking half of 12. So it's not a coincidence that 2×6 is 12. The only "coincidence" is that 12 is divisible by 3. Numbers divisible by 3 are hardly rare. If the last number were not divisible by 3, such as in the pair $\overline{7}\ \overline{14}$, the fraction could not have been simplified anyway.

Here's another multiplication. On the papyrus, only the final simplified answer occurs. The simplification occurs below on the *ostracon*, a piece of broken pottery the Egyptians used for informal notes. I've placed number pairs about to be "G ruled" together in parentheses. Note the natural occurrence of double and quadruple numbers.

7	?	
1	$\overline{12}\ \overline{40}$	✓
2	$\overline{6}\ \overline{20}$	✓
4	$\overline{3}\ \overline{10}$	✓
	$\overline{2}\ \overline{8}\ \overline{12}\ \overline{20}$	

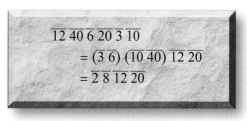

$$\overline{12}\ \overline{40}\ \overline{6}\ \overline{20}\ \overline{3}\ \overline{10}$$
$$= (\overline{3}\ \overline{6})\ (\overline{10}\ \overline{40})\ \overline{12}\ \overline{20}$$
$$= \overline{2}\ \overline{8}\ \overline{12}\ \overline{20}$$

$7 \times \overline{12}\ \overline{40} = \overline{2}\ \overline{8}\ \overline{12}\ \overline{20}$.

PRACTICE: Multiply and use the G rule to simplify your answer.

• $5 \times \overline{20}$
• $7 \times \overline{24}$

ANSWERS:

• $\overline{4}$
• $\overline{4}\ \overline{24}$ or $\overline{6}\ \overline{8}$

A number of relationships we have already encountered are merely cases of the G rule. Recall that we have used $\overline{3}\ \overline{6} = \overline{2}$, a simple example of the G rule. Also note that our rule for adding even fractions to themselves, like $\overline{6}\ \overline{6} = \overline{3}$, is a fairly trivial case of the G rule.

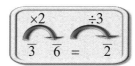
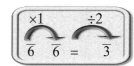

In fact, it's possible to show using modern algebra that any two Egyptian fractions that simplify to one fraction must obey the G rule. For example, the identity $\overline{2}\ \overline{6} = \overline{\overline{3}}$ doesn't seem like the G rule until you realize that $\overline{\overline{3}}$ is $\overline{1.5}$ and that $6 \div 4$ is 1.5. Another example of a difficult-to-notice G rule is $\overline{10}\ \overline{15} = \overline{6}$.

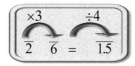
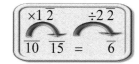

We can't be sure the ancient Egyptians explicitly used the G rule. It may be that this is just an arithmetic relationship noticed by modern mathematicians who can easily spot such patterns; however, if it were something that obvious to an ancient Egyptian scholar, it's likely that they recognized and used this mathematical tool.

In fact, the G rule is completely obvious when seen in the right context. Consider the relationship $\overline{6}\ \overline{12} = \overline{4}$. We can see that 2×6 is 12, so the answer is the fraction made by taking 12 and dividing by 3. Let's think about what these multiplicative relationships tell us about the fractions. Remember that Egyptian fractions work in the opposite way as whole numbers do. The fact that the 12 is twice as big as the 6 tells us that $\overline{6}$ is twice as big as $\overline{12}$. Interpret this literally. We can think of $\overline{6}$ as being composed of two $\overline{12}$s. This relationship becomes clear when we think of the fractions as parts of 12, where the $\overline{12}$ becomes 1 and the $\overline{6}$ becomes 2.

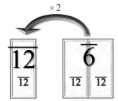

$\overline{6}$ is composed of two $\overline{12}$s.

Similarly because $12 \div 4$ is 3, we can think of $\overline{3}$ as being composed of four $\overline{12}$s.

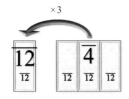

$\overline{4}$ is composed of three $\overline{12}$s.

Now let's interpret what the G rule says for this example. We start with $\overline{6}$, which is two $\overline{12}$s. We add one $\overline{12}$ to the two we already have. We're told by the G rule that this is $\overline{4}$, which is three $\overline{12}$s. Summing up, the G rule says

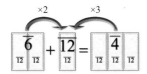

Two $\overline{12}$s and one more is three $\overline{12}$s.

that two $\overline{12}$s and one more is three $\overline{12}$s, something anyone would find as obvious.

In order to see the general pattern, let's look at another example. Consider $\overline{4}\ \overline{12} = \overline{3}$. When considering their relative size when compared to $\overline{12}$, we see that $\overline{4}$ and $\overline{3}$ are three and four pieces of size $\overline{12}$. So the G rule says that three $\overline{12}$s and one more is four $\overline{12}$s.

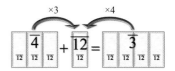

Three of something and one more is four.

When we apply the G rule, we must determine what you need to multiply the first term by to get the second term. This number represents how many pieces you start with. Adding the second term increases the total by 1. When we divide by one more, we're merely acknowledging that we have one more piece. So the G rule essentially says that if you add one, then you have one more. Clearly, some ancient mathematician could have recognized this, and I'd be surprised if no one did.

THE SUAVE SCRIBE

The Egyptian Mathematical Leather Roll

It's always an exciting day to find an ancient Egyptian scroll made of leather. The excessive cost of leather at the time meant that anything written on it was extremely important to its owner. You would expect something like a royal proclamation to appear on a roll of leather and not, as in the case of the Egyptian Mathematical Leather Roll, a table of fraction identities.

As Gillings points out in his discussion on the leather roll, the Egyptians wrote on three main types of surfaces. The cheapest was ostraca, broken pieces of pottery and similar things. Pottery is made of mud, and mud was cheap in the river valley of the Nile. Clay pots were used to hold just about everything in Egypt. Wood, glass, and similar materials were just too rare or expensive. But pots get old, chip, and break, so there was no shortage of pottery shards. Although few people could write, the middle class had to keep some records. These were generally written on economical ostraca.

The next class of writing surface was papyrus. The name comes from a plant once found in the marshes of Egypt. It was so important in the ancient world that our modern word "paper" derives from it. To make a sheet of papyrus, someone went to the mosquito-infested swamps to gather the plant. The long thin leaves were then laid out side by side. On top of this, another layer was laid perpendicular to the first layer. The two layers were then beaten until the juice of the plant came out of the leaves. The liquid acted like a glue to hold the leaves together. The sheets formed could then be joined end to end and pounded together to make long strips of paper for use in a scroll. After the leaves dried out, the writing surface was smoothed by sanding it down. Although papyrus itself probably wasn't expensive, the untold hours required to make a single sheet of paper meant it was only used for important purposes, such as an official record or textbook.

Yet papyrus wasn't the most expensive writing material; leather was. The scarcity of leather was a direct consequence of the scarcity of meat, which in the ancient world was a luxury. It's only in the latter half of the twentieth century that meat became a substantial portion of the diet, and even then only in the most prosperous countries. Cows and other livestock consume a lot of food, and a number of families can be fed on the amount of grain it takes to make enough meat for one person. In the ancient world where food was relatively scarce, arable land couldn't be wasted for animal feed. Hence, livestock were generally confined to grazing on areas unsuitable for farming, so there wasn't a lot of meat.

People still ate some meat since protein is an important part of the diet. But for commoners it was reserved for special occasions like religious festivals. Since few cows were slaughtered for food, there simply wasn't that much leather, and the resulting scarcity made it extremely expensive. So if leather was so costly, why would anyone use it for a math table? The answer is, because they could.

We can imagine a suave scribe strutting through some ancient Egyptian town in his fine linen clothes. He would then discreetly adjust his shirt so his pot belly would just barely show. In the ancient world where wages were paid in food, doing so would be the equivalent of flashing a platinum-class credit card. When he caught the eyes of some attractive young impressionable girl, he would pause, pretending to be thinking about some math problem required by his high-powered career. Slowly he would pull out his math table from under his arm. The girl would undoubtedly swoon as she eyed the leather scroll worth more than all the objects in her father's house combined. He gives her a wink and thinks to himself that life is good.

As ridiculous as the above description may sound, it's essentially correct. The use of leather for the scroll was probably more of an indicator of social status than of serving some practical value. It is interesting that some scribe used a mathematical table as symbol of class, much in the same way as a modern CEO might wear an expensive watch. If the practice of using leather for mathematical tables was widespread, it would suggest that the scribes of Egypt had wealth and social standing. If nothing else, the existence of the leather scroll demonstrates that mathematics was greatly respected in the ancient world.

I've tried to loosely mimic the use of ostraca, papyrus, and leather in the backgrounds of the computations in this book. Most of the examples are done on the image of a papyrus scroll. You can tell if a scroll is papyrus by looking carefully at the texture in the image. You should be able to see subtle lines going vertically and horizontally. This is a vestige of the reeds that were placed down in these directions in the manufacture of the papyrus. For tables, I've used a leather scroll background. You can spot these by the lack of lines in the texture and the thicker scroll rolls. For "scratch work," I've often used an ostraca background. And occasionally, I've used a "modern" looking background for operations not done in a traditional Egyptian style, such as when we compute parts through multiplication.

On the Egyptian Mathematical Leather Roll there are dozens of fraction identities. Many of these are examples of the G rule discussed earlier. For example, the relation $\overline{10}\,\overline{40} = \overline{8}$ appears near the beginning. You might be wondering that if the Egyptians knew the G rule, why would they need this table? Couldn't they just see that 40 is 4 times as large as 10 and then divide 40 by 5? I can give a couple of possible explanations.

$$\overline{10}\ \overline{40} = \overline{8} \qquad \overline{24}\ \overline{48} = \overline{16}$$
$$\overline{5}\ \overline{20} = \overline{4} \qquad \overline{18}\ \overline{36} = \overline{12}$$
$$\overline{4}\ \overline{12} = \overline{3} \qquad \overline{21}\ \overline{42} = \overline{14}$$
$$\overline{10}\ \overline{10} = \overline{5} \qquad \overline{45}\ \overline{90} = \overline{30}$$
$$\overline{6}\ \overline{6} = \overline{3} \qquad \overline{30}\ \overline{60} = \overline{20}$$
$$\overline{3}\ \overline{3} = \overline{\overline{3}} \qquad \overline{15}\ \overline{30} = \overline{10}$$
$$\overline{9}\ \overline{18} = \overline{6} \qquad \overline{48}\ \overline{96} = \overline{32}$$
$$\overline{12}\ \overline{24} = \overline{8} \qquad \overline{96}\ \overline{192} = \overline{64}$$

G-rule expressions in the Egyptian Mathematical Leather Roll.

It's relatively easy for us to simplify $\overline{10}\ \overline{40}$ because at some point in our lives, we were forced to memorize $4 \times 10 = 40$ and $5 \times 8 = 40$. The Egyptians performed their math through doubling and so never needed to learn multiplication tables. As a result, these relationships wouldn't be immediately obvious. Hence the table might serve the same purpose as our memory when we apply the G rule.

We also need to know that the leather roll has only some twenty-odd relations. It obviously is not meant to serve as a complete fraction simplification guide; its purpose might be to simply show examples. By examining these equations, perhaps students were meant to obtain some general rules. The relation after $\overline{10}\ \overline{40} = \overline{8}$ is followed by $\overline{5}\ \overline{20} = \overline{4}$. Perhaps they were intended to note the "×4, ÷5" relation embedded in the two equations. Or perhaps the equations were intended to show that you could take a known relation like $\overline{10}\ \overline{40} = \overline{8}$ and divide it by 2 in order to obtain a new relation like $\overline{5}\ \overline{20} = \overline{4}$.

$$\overline{6}\ \overline{6}\ \overline{6} = \overline{2} \qquad \overline{14}\ \overline{21}\ \overline{42} = \overline{7}$$
$$\overline{25}\ \overline{15}\ \overline{75}\ \overline{200} = \overline{8} \qquad \overline{18}\ \overline{27}\ \overline{54} = \overline{9}$$
$$\overline{50}\ \overline{30}\ \overline{150}\ \overline{400} = \overline{16} \qquad \overline{22}\ \overline{33}\ \overline{66} = \overline{11}$$
$$\overline{25}\ \overline{50}\ \overline{150} = \overline{15} \qquad \overline{28}\ \overline{49}\ \overline{196} = \overline{13}$$
$$\overline{7}\ \overline{14}\ \overline{28} = \overline{4} \qquad \overline{30}\ \overline{45}\ \overline{90} = \overline{15}$$

Non-G-rule relations on the Egyptian Mathematical Leather Roll.

The Egyptian Mathematical Leather Roll had many equalities that do not satisfy the G rule. One of the identities on the scroll is $\overline{7}\ \overline{14}\ \overline{28} = \overline{4}$.

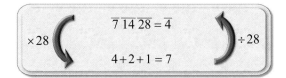

$$\overline{7}\ \overline{14}\ \overline{28} = \overline{4}$$
$$\times 28 \qquad\qquad\qquad \div 28$$
$$4 + 2 + 1 = 7$$

$\overline{7}\ \overline{14}\ \overline{28} = \overline{4}$ verified as parts of 28.

We can easily verify this relationship by considering parts of 28. In this system the identity becomes the obvious identity $4 + 2 + 1 = 7$.

This identity would be useful only for adding terms with a $\overline{7}$, a $\overline{14}$, and a $\overline{28}$. For randomly selected numbers, this would almost never happen. But as I discussed when considering the G rule, Egyptian fractions are not random. The methods involved in their use cause certain "coincidences" to happen repeatedly. Consider the following multiplication of $\overline{7}$ by $3\ \overline{2}$.

	$3\ \overline{2}$	?	
1)	1	$\overline{7}$ ✓	
2)	2	$\overline{4}\ \overline{28}$ ✓	(1) × 2
3)	$\overline{2}$	$\overline{14}$ ✓	(1) ÷ 2
		$\overline{4}\ \overline{7}\ \overline{14}\ \overline{28}$	

$3\ \overline{2} \times \overline{7} = \overline{4}\ \overline{7}\ \overline{14}\ \overline{28}$

The answer is awful. We could simplify it as parts of 28. However, we now know the identity $\overline{4} = \overline{7}\ \overline{14}\ \overline{28}$, found above, and we can see the three terms of the right-hand side within our answer. So we can simplify it as follows.

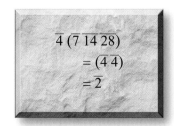

$$\overline{4}\ (\overline{7}\ \overline{14}\ \overline{28})$$
$$= (\overline{4}\ \overline{4})$$
$$= \overline{2}$$

Notice how this identity seems ready-made for my multiplication of $\overline{7}$. Try these multiplications and focus on the "coincidences" that occur. You need to be careful because these can be simplified in a number of ways. To get the answer given, be sure to use the $\overline{7}$ $\overline{14}$ $\overline{28} = \overline{4}$ identity as soon as you see it. Also be sure to put the $\overline{7}$ term in the right column.

PRACTICE: Multiply then simplify using $\overline{7}$ $\overline{14}$ $\overline{28} = \overline{4}$

- $5 \times \overline{7}$ $\overline{28}$
- 5 $\overline{4} \times \overline{7}$

ANSWERS:

- $\overline{2}$ $\overline{4}$ $\overline{7}$
- $\overline{2}$ $\overline{4}$

Now let's look at another identity on the leather roll, $\overline{18}$ $\overline{27}$ $\overline{54} = \overline{9}$. We can verify it using parts of 54 as follows.

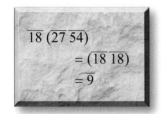

$\overline{18}$ $\overline{27}$ $\overline{54} = \overline{9}$ verified as parts of 54.

We can also confirm this identity as two applications of the G rule.

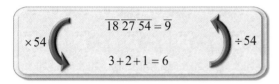

$\overline{18}$ $\overline{27}$ $\overline{54} = \overline{9}$ verified using the G rule.

Consider the multiplication of $\overline{54}$ by 6. Once again "coincidences" happen.

Notice that with the table, the answer immediately simplifies to $\overline{9}$. Also notice that while regular doubling gave us the $\overline{27}$ from the $\overline{54}$, Ahmose's double-fraction table gave us the $\overline{18}$ we needed for our identity. It should hardly come as a surprise that the pieces of Egyptian

$6 \times \overline{54} = \overline{18}$ $\overline{27}$ $\overline{54}$.

mathematics are meant to come together as a cohesive system. In the practice multiplication below, there are a number of ways it can be solved. To get an answer that matches the one given, reduce the $\overline{18}$ $\overline{27}$ $\overline{54}$ first.

PRACTICE: Multiply then simplify using $\overline{18}$ $\overline{27}$ $\overline{54} = \overline{9}$

- $\overline{36}$ $\overline{54} \times 6$

ANSWER: $\overline{6}$ $\overline{9}$

We need to be careful when dealing with ancient mathematics. Just because we find a possible use of some ancient mathematics doesn't mean that's exactly how it was used in the past. We've been using the relations of the leather roll literally. Perhaps they were used in a more abstract way. Consider the relation $\overline{14}$ $\overline{21}$ $\overline{42} = \overline{7}$. Ancient scribes may have viewed the terms on the left side proportionally, as we do when applying the G rule. They would probably have recognized that 21 is 14 and its half, 7, from their knowledge of multiplication by $\overline{\overline{3}}$. And they would have also immediately recognized that 42 is double 21. The relation could have been interpreted to mean that the solution is half of the initial term.

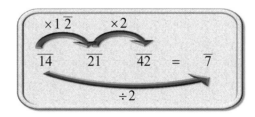

$\overline{14}$ $\overline{21}$ $\overline{42} = \overline{7}$ in relative terms.

Notice that the next term on the leather roll is in fact the same relationship. Note that in the equation $\overline{18}$ $\overline{27}$ $\overline{54}$

= $\overline{9}$, 27 is 18 and its half (9), and 54 is twice 27. Applying the same rule the solution should be generated by taking half of 18 and in fact the answer is $\overline{9}$.

EXAMPLE: Simplify $\overline{20}\ \overline{30}\ \overline{60}$.

SOLUTION: Since $20\times 1\ \overline{2} = 30$ and 30×2 is $60, \overline{20}\ \overline{30}\ \overline{60} = \overline{20\div 2} = \overline{10}$.

PRACTICE: Simplify $\overline{12}\ \overline{18}\ \overline{36}$.

ANSWER: $\overline{6}$

SELF-SUFFICIENCY

Creating Inequalities

Although the night watchman was paid in bread and beer, most professional Egyptians were paid primarily in grain. Workmen in the Valley of the Kings were paid over 500 pounds in wheat and barley each month. This is estimated to have easily fed ten people. This means that the typical Egyptian household had to be by and large self-sufficient in terms of production. They had to grind their own flour and bake their own bread. They used the barley to brew their own beer. They even had to spin their own thread out of flax and weave their own linen.

As we've seen, the people of ancient Egypt were in many ways self-reliant, making what they needed for themselves. Is it possible that Egyptian mathematicians regularly made their own tools? Ahmose's papyrus has a table for doubling odd fractions and a table for division by 10, and the Mathematical Leather Roll is a table of fractional equalities. There's no doubt the Egyptians used these, but we are confronted with the questions of how they found them and whether, if they needed expressions not found on one of the tables, they could have created them on their own.

The simple truth is that, creating Egyptian style fraction equalities is trivial, especially if we think in parts. Let's start with a number with a lot of factors, 30. The factors are 1, 2, 3, 5, 6, 10, 15, 20, and 30. The modern perspective would not consider 20 to be a factor of 30, but just

as 5 is $\overline{6}$ of 30, thus making it a factor, 20 is $\overline{\overline{3}}$ of 30. These whole numbers can be associated with fractions when thinking of them as parts of 30. Hence 2 is $\overline{15}$ of 30 and 6 is $\overline{5}$ of 30. A scribe could easily make a table deriving all of the whole parts of 30 as follows:

1)	1	30	
2)	$\overline{3}$	20	$(1)\times\overline{\overline{3}}$
3)	$\overline{2}$	15	$(1)\div 2$
4)	$\overline{3}$	10	$(2)\div 2$
5)	$\overline{10}$	3	$(4)\leftrightarrow$
6)	$\overline{5}$	6	$(5)\times 2$
7)	$\overline{6}$	5	$(6)\leftrightarrow$
8)	$\overline{15}$	2	$(3)\leftrightarrow$
9)	$\overline{30}$	1	$(1)\leftrightarrow$

The whole parts of 30.

Now pick any number in the third column of the table, say 20, and write it as the sum of other numbers on the table. For example, consider $20 = 15 + 5$. We can now write the fractional equivalents, $\overline{\overline{3}} = \overline{2} + \overline{6}$, giving us an identity. We could express this with arithmetic in the following way:

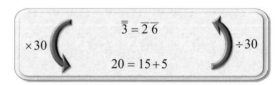

Each equation we make from the whole numbers from the above table gives us a new identity. To solve the following problem, we first convert the fractions into their corresponding parts: 10, 6, and 2. Then we add them and try to express them in numbers from the table.

EXAMPLE: Make an identity for the simplification of $\overline{3}\ \overline{5}\ \overline{15}$.

SOLUTION:

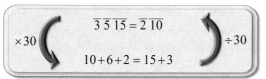

$$\overline{3}\ \overline{5}\ \overline{15} = \overline{2}\ \overline{10}$$
$$\times 30 \qquad\qquad \div 30$$
$$10 + 6 + 2 = 15 + 3$$

PRACTICE: Make an identity for the simplification of $\overline{5}$ $\overline{10}$ $\overline{30}$.

ANSWER: $\overline{3}$

We can use this method to find sums of fractions equal to a particular fraction. For example, if we want to find ways to add fractions to get to $\overline{\overline{3}}$, we simply try to add numbers from the table to get to 20. We've already seen that $15 + 5 = 20$ generates the identity $\overline{2}\ \overline{6} = \overline{\overline{3}}$. Similarly $15 + 3 + 2$, $10 + 6 + 3 + 1$, and $10 + 5 + 3 + 2$ generate $\overline{2}\ \overline{10}\ \overline{15}$, $\overline{3}\ \overline{5}\ \overline{10}\ \overline{30}$, and $\overline{3}\ \overline{6}\ \overline{10}\ \overline{15}$, respectively, all of which equal $\overline{\overline{3}}$.

PRACTICE: Using the whole parts of 30, find all identities that sum to $\overline{3}$.

ANSWERS: $\overline{5}\ \overline{10}\ \overline{30}$ and $\overline{6}\ \overline{10}\ \overline{15}$

We can now see how the division-by-10 table in Ahmose's papyrus may have been constructed. Recall that the table had values for the numbers 1 to 9 all divided by 10. As parts of 30, the fractions $\frac{1}{10}$, $\frac{2}{10}$, and $\frac{3}{10}$ become 3, 6, and 9, respectively. It's not hard to see that each number is 3 times the number that gets divided by 10. Hence as parts of 30, $7 \div 10$ becomes 7×3, or 21. We can think of this as $7 \times \overline{10}$, and the $\overline{10}$ then converts into a 3. Using the value of 7×3, we can now write 21 as $20 + 1$, which when converted back into fractions is $\overline{3}\ \overline{30}$. This is precisely the entry that appears in the table.

EXAMPLE: Using the whole parts of 30, find an expression for $4 \div 10$.

SOLUTION:

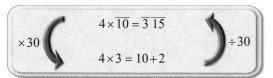

$$4 \times \overline{10} = \overline{3}\ \overline{15}$$
$$\times 30 \qquad\qquad \div 30$$
$$4 \times 3 = 10 + 2$$

PRACTICE: Using the whole parts of 30, find an expression for $3 \div 10$ and $8 \div 10$.

ANSWERS: $\overline{5}\ \overline{10}$ and $\overline{3}\ \overline{10}\ \overline{30}$

All the answers we've obtained above are the values in the table. This suggests that the division-by-10 table may have been constructed in a similar manner.

If we now choose a number different from 30, we can create a new table of parts. Consider the following table containing the whole parts of 28.

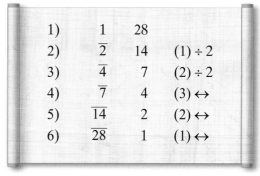

1)	1	28	
2)	$\overline{2}$	14	$(1) \div 2$
3)	$\overline{4}$	7	$(2) \div 2$
4)	$\overline{7}$	4	$(3) \leftrightarrow$
5)	$\overline{14}$	2	$(2) \leftrightarrow$
6)	$\overline{28}$	1	$(1) \leftrightarrow$

The whole parts of 28.

From this table we can notice that $7 = 4 + 2 + 1$ means $\overline{4} = \overline{7}\ \overline{14}\ \overline{28}$ and similarly form other identities. We could also create a table for the numbers from 1 to 6 when divided by 7. For example $3 \div 7$ could be determined as follows.

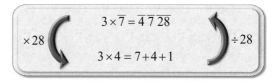

$$3 \times \overline{7} = \overline{4}\ \overline{7}\ \overline{28}$$
$$\times 28 \qquad\qquad \div 28$$
$$3 \times 4 = 7 + 4 + 1$$

$3 \div 7 = \overline{4}\ \overline{7}\ \overline{28}$.

The change from 30 to 28 has some effect on the types of identities we can create. As we see above, we can form expressions with $\overline{7}$ in them because 7 is a factor of 28. However, we lose the ability to express $\overline{3}$ and $\overline{\overline{3}}$ because 3 is not a factor of 28, although it is one of 30. The loss of $\overline{\overline{3}}$ is particularly troublesome for the expression of large fractions. It's difficult to add whole parts of 28 to numbers like 27. In fact, there is only one way to do it: $14 + 7 + 4 + 2$, which makes the particularly ugly fraction $\overline{2}\ \overline{4}\ \overline{7}\ \overline{14}$. The initial fraction $\overline{2}$ is a lousy approximation to this number, which is much closer to $\overline{\overline{3}}$, but because $\overline{\overline{3}}$ doesn't form a whole part of 28, we can't easily use it. Since there is often only one way to add the whole parts of 28 to get certain numbers, the number of identities we can find is limited.

.The problem would get even worse if we picked a number like 35. The whole parts—1, 5, 7 and 35, given below—will only sum up to a handful of numbers. For example, we can't get any of the numbers from 2 to 4 or any from 14 to 34.

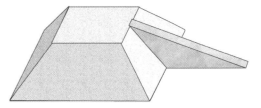

1)	1	35	
2)	2	70	$(1) \times 2$
3)	$\overline{5}$	7	$(2) \div 10$
4)	$\overline{7}$	5	$(3) \leftrightarrow$
5)	$\overline{35}$	1	$(1) \leftrightarrow$

The whole parts of 35.

The difference between 30 and 35 is due to their prime factorizations. When completely factored, 30 is $2 \times 3 \times 5$, while 35 is 5×7. The number of factors and their size determines how many whole parts a number will have. The first number has three prime factors and the second only has two. This means 30 will have more parts than 35. Since you need more, smaller numbers to multiply to a large number, the smaller the factors, the more parts. The number 28 is $2 \times 2 \times 7$. Although it has three factors, the repeated 2 reduces the number of parts slightly. In general, for the number of parts to be useful, you want it to be the product of many small, different factors.

TWO CHOICES

Creating the Times-2 Table

Ancient history consists of a few scattered remains that force speculation to complete it and create a cohesive story. We're often confronted with two or more choices with little to go on except common sense and intuition. A good example of this is in the construction of the pyramids. No one is quite sure how they were built. The stones used were often massive, but they somehow were lifted to the top of what were the tallest man-made structures until the construction of the Eiffel Tower.

It's assumed the Egyptians used a ramp, because the slopes of the pyramids were too steep to safely push the massive stones up the sides safely. The question considered is, what form did these ramps take? The simplest answer is that perhaps they were straight.

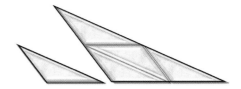

A partially built pyramid with a straight ramp.

There's a problem with this solution. The ramps would get large quickly as the pyramids got bigger. Assuming the ramps had a uniform width, doubling the height of a ramp would quadruple the amount of material required to build it. This is easily seen with an example. Consider the triangles pictured below. The triangle twice as large in length is composed of four of the smaller triangles.

A triangle twice as high is made of four smaller ones.

This is called *quadratic growth*. The size grows with the square of the length. So if you double the length, you multiply the size by 2^2, or 4. Similarly if you triple the length the size grows by 3^2, which is 9 times larger. This relation is easiest to see on the growth of square. If you take a square, you need three rows of three to make a new square 3 times larger. Clearly, three by three is 3×3, or 3^2. As the pyramid grows, the size grows rapidly. If the size

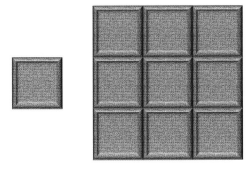

Three times as long requires three by three.

goes up 20 times, the size of the ramp grows 400 times, and so on.

So some have suggested that the ramps got too large for the Egyptians to have used this method. Instead they suggest that the ramps spiraled around the pyramids.

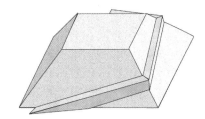

A partially built pyramid with a spiral ramp.

Both the width and height of the ramp remain fixed since it only needed to extend to the pyramid, not the ground. Hence the growth in the size of the ramp is linear. This means that if it gets twice as long, the ramp volume grows by two as well.

There's a problem with this method. The analysis assumes that the only consideration is the quantity of material required to make the ramp. There are many others. For example, which ramps are structurally sound? Remember that pyramids with seked of 5 or less eventually collapse. Similar physics holds for ramps.

We also need to consider how the shape of the ramp affects its efficiency. Image how difficult it must be to turn a stone block that weighs $2\,\overline{2}$ tons. If we assume that the block is being pushed on rollers, they need to be turned along with the block. However, they're under thousands of pounds of stone. Hence the straight ramp may be

ultimately more efficient even though it takes more work to build.

This problem gets worse when you realize that the size of the pyramid grows cubicly. In other words, when you double the height of the pyramid, the number of blocks you need grows by a factor of 8. When you need to move millions of blocks, 8 times as many difficult turns is nothing to ignore. It's not hard to imagine that a bigger ramp would require far less work. The cubic relation means that you cube the growth in length to get the growth in volume. So 5 times larger means 5^3, which is $5 \times 5 \times 5$, or 125, times as large. This is easily seen in terms of a cube. The following diagram shows that to make a cube twice as large, you need eight—$2 \times 2 \times 2$—blocks.

A block twice as large is made of eight smaller blocks.

When considering how the Egyptians created identities we also are confronted by two competing theories, neither of which can be proved or disproved. In the previous section we more or less accepted the point of view that the Egyptians started off with a set number of parts and from that generated identities. We are now going to speculate on methodologies that start with the fractions themselves rather than with the parts.

Consider the identities of Ahmose's table for multiplying odd numbers by two. In that table we see entries such as $2 \times \overline{7} = \overline{4}\,\overline{28}$. Notice the 4, 7, and 28 are all whole parts of the 28 mentioned in the last section. We can view this identity in the same way as when we created the division-by-10 table, thinking of this as $2 \div 7$. Since $\overline{7}$ is four parts of 28, $2 \times \overline{7}$ is eight parts of 28. Since $8 = 7 + 1$, all of which are whole parts of 28, we know that $2 \times \overline{7} = \overline{4}\,\overline{28}$. We can represent this algebraically as follows:

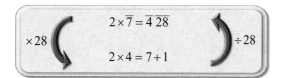

This trick always obtains the table entry if you pick the "appropriate" number of parts. However, the choice of the number of parts is not always obvious. There are rules of thumb to use when selecting the parts. One obvious rule is that you want the initial fraction to be a whole number of parts, so since we started with $\overline{7}$, we would want the number of parts to be a multiple of 7.

Instead of selecting the number of parts, I'm going to show you a method that selects the number of parts for you, and it works well for generating two term identities. Let's find an identity for $2 \times \overline{15}$. Pick any two factors of 15. I'm going to pick 5 and 1. Now put them in the bottom right of our parts computation as shown and also put the $2 \times \overline{15}$ on the top left.

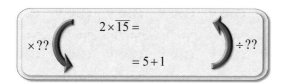

We know $5 + 1$ is 6 and 6 is 2×3, so we can put 2×3 on the bottom left.

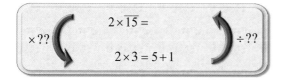

Now we need to think backward. The $\overline{15}$ corresponds to the 3, so a fifteenth of some number of parts is 3. What number of parts is this? We know a fifteenth of 15 is 1, a fifteenth of 30 is 2, so a fifteenth of 45 is 3. Hence there are 45 parts. The 45 is just 3 fifteens so we could have obtained this solution by multiplying 3×15. Now that we know the number of parts is 45, we can determine what

fractions the 5 and the 1 correspond to. When divided by 45, these become $\overline{9}$ and $\overline{45}$, respectively. Hence we get the following final answer:

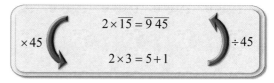

$2 \times \overline{15} = \overline{9}\ \overline{45}$ as parts of 45.

We chose the factors 5 and 1, but this method still works if we choose any two factors of 15. Let's now try 5 and 3.

EXAMPLE: Find an expression for $2 \times \overline{15}$ using the factors 5 and 3.

Start with $2 \times \overline{15}$ on the top left and put our two factors, 5 and 3, as a sum on the bottom right. We know $5 + 3$ is 8, which is 2×4. So 2×4 goes on the bottom left. We now calculate the number of parts $\overline{15}$ would be of 4, which is 15×4, or 60. Finally we determine what fractions 5 and 3 are of 60. We do the divisions $5 \div 60 = \overline{12}$ and $3 \div 60 = \overline{20}$ and put these as our final answer on the top right. We now know $2 \times \overline{15} = \overline{12}\ \overline{20}$.

SOLUTION:

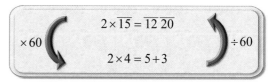

$2 \times \overline{15} = \overline{12}\ \overline{20}$ as parts of 60.

PRACTICE: Find an expression for $2 \times \overline{15}$ using the factors 3 and 1.

ANSWER: $\overline{10}\ \overline{30}$

Notice that we've finally found the entry on Ahmose's table. Before we finish with finding expressions for $2 \times \overline{15}$, try creating one using the factors 15 and 5. You will see that we get the same answer as using 3 and 1. This is because 3 and 1 can be scaled to 15 and 5 by multiplying by

5. In general we can ignore factor pairs that have common factors since they produce the same solutions as simpler pairs.

PRACTICE: Find an expression for $2 \times \overline{15}$ using the factors 15 and 5.

ANSWER: $\overline{10}\,\overline{30}$

GOOD VERSUS EVIL

What Makes a Good Identity?

The battle between the two strongest gods, Horus and Seth, disrupted the heavens and the earth. The moon itself was destroyed in the conflict. The gods had to put an end to the violence and hence formed a tribunal to settle the dispute peaceably and decide which of the two would become the king of the gods.

It had all started many years ago when the god Osiris ruled the land of Egypt. At the time he was the god of the Nile valley and embodied the fertility of the land. His skin was the green of the vegetation that flourished in his domain. Osiris was a good king, and Egypt prospered under his dominion.

Seth, god of chaos.

Seth, the brother of Osiris, was the god of the red desert. He embodied the destructive power of the harsh lands that surrounded Egypt. As a result, he was the fiercest warrior of the gods and was greatly feared. Seth was also a jealous god who watched his passive brother control all the glory of Egypt.

Seth devised a plan. He threw a party and invited the unwitting Osiris. At the end he brought out a coffin he had made especially to fit Osiris. Seth announced that whoever fits perfectly into the coffin will win a prize. One by one the guests tried it out but no one fit exactly until Osiris. At this point Seth and the other guests, who were in on the conspiracy, shut the lid, nailed it down, sealed it with lead, and threw the coffin in the Nile.

Isis, the wife of Osiris, found the body and revived it long enough to conceive a son, Horus. Ma'at, the goddess of order, declared that the dead Osiris could not remain in the land of the living, so he descended into the underworld to become its ruler. As Horus grew up, Seth repeatedly tried to kill the boy, but Horus was hidden in the papyrus plants and protected by many powerful gods.

When Horus reached manhood, he set out to avenge his father. When he found Seth, the two battled. Horus's left eye (the moon) was ripped out and torn to pieces. Seth lost a body part of his own. Let's just say that my male readers will agree that Horus won because he lost only an eye. However, the fight proved inconclusive. Thoth, the god of wisdom, found the pieces of the moon and repaired it. Horus gave the moon to his father, Osiris, and Seth healed as well.

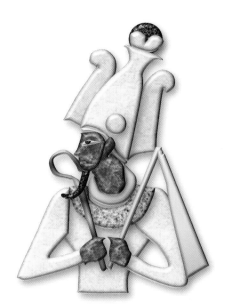

Osiris, lord of the underworld.

Horus, the son of Osiris.

they were not so shallow. Ra was not wrong for supporting Seth. Seth was needed each day to protect the sun from destruction. But neither was Seth entirely evil. Isis tricked Seth by magically disguising herself. She related a tale wherein her son was robbed of his deceased father's cattle by a stranger, to which Seth replied that the stranger should be beaten in the face with a rod so that justice could be done. Isis at this point revealed her true form to Seth and he became ashamed by his own actions. Although Seth didn't withdraw his claim, he himself reported the incident to the tribunal so he could be judged.

On the other hand, we also can't assume that Horus was purely good. He was driven by anger and the desire for revenge. When his mother showed Seth some mercy, Horus became enraged and struck off her head. While the act was not lethal to a goddess, it was still considered a great offense, and it was Seth who sought out Horus in order to punish him.

The wisest voices did not seek a victor in the conflict but, rather, reconciliation. They knew that strength was needed as well as fertility and justice. The end of the story is telling. After it was decided that Horus should be made king, the most ancient god, Atum, asked Seth why he even allowed himself to be judged when he could just have used force to take the crown for himself. Seth simply replied that they should give the kingship to Horus. For his acquiescence, Seth was taken to the heavens, where he roamed the sky as thunder.

As with Seth and Horus, when judging what is best, even in mathematics you can't be too simplistic. As we've seen, Ahmose had many choices for the value of twice a fraction such as $2 \times \overline{15}$. Let's consider all the two-term identities. These can be generated by considering factor pairs of 15.

Before they could battle anew, the gods formed a tribunal to mediate the dispute. It consisted of the Ennead, a group of nine gods, and was led by Ra, the sun god. Most of the Ennead seemed to favor Horus's claim, but Ra, one of the oldest and most venerable of the gods and also the sun god, sided with Seth. The tribunal repeatedly sent letters to the elder gods seeking advice. The first letter was sent to Neith, mother of all things. She sent a reply that justice should be done and the rightful heir should take the throne. This did nothing but enrage the sun god, who declared that the kingship was too much for someone so young.

Letters were exchanged between Atum, the god of the beginning and end of time, and Osiris, both of whom supported Horus. Each time, Seth flew into a rage and challenged Horus to some competition to decide the matter. Horus won, often with the aid of his mother, who used trickery to offset Seth's brute force. After eighty years, Atum ordered that Seth be placed in shackles and awarded the crown to Horus.

On the surface, the above tale might seem a simplistic story of good triumphing over evil. Although the Egyptians did have a strong sense of right and wrong,

Pair	Identity		
(15,1)	$2 \times \overline{15}$ =	$\overline{8}$	$\overline{120}$
(5,1)	$2 \times \overline{15}$ =	$\overline{9}$	$\overline{45}$
(3,1)	$2 \times \overline{15}$ =	$\overline{10}$	$\overline{30}$
(5,3)	$2 \times \overline{15}$ =	$\overline{12}$	$\overline{20}$

All possible two-term identities for $2 \times \overline{15}$.

The original author of Ahmose's papyrus needed one identity for his table but had at least four reasonable identities to choose from. Although we can't be sure why he selected the identities found at the beginning of the scroll, we can look at his choices and try to backward engineer the process through which they were selected. Below are four criteria he seems to have used, and you will see why they are very practical.

1. Good Approximation

When we first examined Egyptian fractions, I pointed out that we can think of the multiple fractions as approximations that get refined with each term. For example we can think of $\overline{8}\ \overline{120}$ as roughly an eighth. It differs from one-eighth by a hundred and twentieth. Since 120 is much larger than 8, the error is extremely small and the approximation of the value being an eighth is quite good.

The fraction $\overline{12}\ \overline{20}$ offers a lousy approximation. The expression makes the claim that the value is roughly a twelfth, but the error, a twentieth, is fairly large compared to a twelfth. Half of $\overline{12}$ is $\overline{24}$, which is smaller than the error of $\overline{20}$, so we're off by more than half. This is far too large for $\overline{12}$ to be a good approximation.

Below is a visual representation of the four approximations. The length of each bar represents $2\times\overline{15}$. The light bar on the left of each represents the first fraction. This can be thought of as the approximation. The closer it is to the length of the whole bar, the better the approximation. The dark bar to the right is the error represented by the second fraction. The smaller the error, the better the approximation.

It's easy to see that $\overline{8}$ is the best approximation in the lot. If this were the only criterion, that's what we would select. However, there's more. At this point let's

eliminate $\overline{12}\ \overline{20}$ from contention because it is such a bad approximation.

2. Even Fractions

Egyptians had to double fractions constantly in their computations. When they had to double odd fractions like $\overline{15}$, they had to consult a table or their memory. Imagine the frustration of Egyptian students multiplying $\overline{15}$ by 4. In the very first step they would need to consult a table to determine the value of $2\times\overline{15}$. If the table used the value $\overline{9}\ \overline{45}$, the problem of odd fractions would rear its ugly head once again. The student would have to do two look-ups: one for $\overline{9}$ and one for $\overline{45}$. The result would be an untidy collection of four fractions. While they could be reduced to two fractions using the G rule, it's still more work than a simple multiplication should entail.

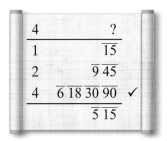

$4\times\overline{15}=\overline{5}\ \overline{15}$ using $2\times\overline{15}=\overline{9}\ \overline{45}$.

On the other hand, if the table had the value $\overline{8}\ \overline{120}$ or $\overline{10}\ \overline{30}$, the problem of odd fractions would go away, and the next doubling would be trivial. There would also only be two fractions on the third line, eliminating the need to simplify the expression.

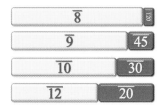

The approximations of $2\times\overline{15}$.

$4\times\overline{15}=\overline{4}\ \overline{60}$ using $2\times\overline{15}=\overline{8}\ \overline{120}$.

Out of the fifty entries in Ahmose's table, only six start with odd fractions. Two of the six, $\overline{3}$ and $\overline{\overline{3}}$, are easy to double. The Egyptians clearly know that the table should contain as many even fractions as possible, and anyone who has spent any time multiplying the Egyptians' way knows this, too. So, we will ignore the possibility that $2 \times \overline{15} = \overline{9}\,\overline{45}$.

3. Avoid Large Numbers

Which numbers would you rather multiply: 7×5 or 83249×2734? If you're like most humans, you won't need me to explain why Egyptians didn't like large numbers, but let me give you a few other reasons that might not be immediately obvious. The first thing we must remember is that their table only doubles odd fractions from $\overline{3}$ to $\overline{101}$. If they needed to double $\overline{103}$, they would have had a slight problem.

There's also the concern about fair division. Recall how we first examined Egyptian fractions by slicing loaves of bread. If we wanted to split up two loaves of bread between 15 workers, each would get $2 \times \overline{15}$. We have only two solutions left to consider, $\overline{10}\,\overline{30}$ and $\overline{8}\,\overline{120}$. For the first, we would cut the two loaves into tenths. This would make 20 slices, of which we would then hand out 15, one to each worker, leaving 5.

Two loaves cut in tenths with 15 slices handed out.

We would then cut each of the remaining 5 in thirds, giving us 15 smaller slices, exactly one for each. Since these slices are a third of a tenth, they have size $\overline{30}$, so each worker received $\overline{10}\,\overline{30}$.

The five remaining are cut in thirds and handed out.

Let's see what would happen if we tried to do this with the answer $\overline{8}\,\overline{120}$. Initially there is no problem. We cut each loaf in 8 pieces, making a total of 16. Of these we hand out 15, one to each worker.

Two loaves cut in tenths with 15 slices handed out.

The one slice that remains is then cut in 15 slices. A fifteenth of an eighth is one-twentieth. These "slices" are nothing but crumbs, having been cut far too thin. In fact, the one slice would probably be too thin to cut easily. The workers would gladly accept the $\overline{8}$-size slice but would scoff at their second piece. As a practical matter, physical items cut in size $\overline{120}$ are just not workable.

The last slice, cut in 15 parts, is too small.

So, as we have seen, large numbers are difficult to work with both in fractions and in the physical world. Hence it would be natural for the Egyptians to avoid expressions like $\overline{8}\,\overline{120}$. After we eliminate this choice, we are only left with the value of $\overline{10}\,\overline{30}$ for $2 \times \overline{15}$. Perhaps it should come as no surprise that this is precisely the entry that the author of Ahmose's papyrus selected.

4. As Few Fractions As Possible

When we considered $2 \times \overline{15}$ above, we obtained a number of two-fraction expressions from which to choose. However when we consider $2 \times \overline{17}$, we're not so lucky. This is because 17 is prime: it only has two factors, 1 and itself. If we were to limit ourselves to the factor method of generating identities, then we would only get the following equation.

The following expression is wrong in oh so many ways. Both terms are odd and 153 is too large. Even worse, when

Pair	Identity
(17,1)	$2 \times \overline{17} = \overline{9}\ \overline{153}$

There is only one two-term identity for $2 \times \overline{17}$.

we would have to double $\overline{153}$, we would be out of luck since our table goes only to $\overline{101}$. The Egyptians realized that it's better, in general, to work with as few fractions as possible—but there are exceptions. For this problem they chose the identity $2 \times \overline{17} = \overline{12}\ \overline{51}\ \overline{68}$. It's a decent approximation: all the numbers are relatively small and two of the three are even. It's not difficult to see why they might have selected this identity over the only two-term expression $\overline{9}\ \overline{153}$.

When I use the above four criteria to rate fractional identities, I frequently choose exactly the same entries that appear on Ahmose's table. For most of the ones with which I differ, I can still see why the other value was selected. However, there are a few whose values seem poorly chosen. At this point we have to realize that it's all but impossible to understand exactly how the Egyptians viewed these identities and the criteria by which they selected them. In the end, we must accept that this is all speculation. When we encounter something that does not precisely fit our theories, we're forced to keep an open mind.

POINTS OF VIEW

Modern Equations and Egyptian Identities

Years ago I was told the following anecdote about Paul Erdös, arguably the greatest mathematician in the last half century. While still a student, he demonstrated a solution to a problem using what would later be called the *probabilistic method*. By viewing certain mathematical structures as the outcome of random choices, he was able to assign probabilities to them and use these values to answer questions.

The ideas he considered were often trivial. Consider rolling two dice. You're told that the probability of rolling a 13 is 0. This tells you that it is impossible to do. On the other hand, if you're told the probability of rolling a 10 is 1 in 12, then you know there must be a way of rolling a 10. This is mathematically certain even if you don't know anything about dice. This is because $\frac{1}{12}$ is greater than 0. Erdös also used the notion that if two events are likely to happen, then it must be possible for them to happen at the same time. For example, I am almost certain to roll more than a 3. I am also almost certain not to roll an 11. I can now conclude that there must be a way to roll a number more than 3 that's not 11. All you need to know to prove this is that the probability of each is more than half. Intuitively this is trivial. If we've both read more than half the articles of a magazine, then some article must have been read by both of us. Erdös's ideas were simple but their use was inspired.

Upon hearing of the method, an elder mathematician declared that it was pointless. He realized that every problem that could be solved with the probabilistic method could be solved by combinatorial methods, that is, mathematical methods that count complex objects. It's essentially the difference between saying "I have 3 of 4" and "There is a $\frac{3}{4}$ths chance that any one of them is mine." The reason there is a probability of $\frac{5}{18}$ of rolling a 9 is that there are 10 ways out of 36 of obtaining this number. The fact that the 10 is not 0 means there is a way to roll a 9.

On the surface, the grumpy mathematician was right. In some logical sense, the two methods are redundant. Yet after a few years, the probabilistic method was used to answer many problems that had remained unsolved by people using combinatorial methods. What the objecting professor missed was that how you view things changes how you think about them. Mathematicians think about probabilities in different ways than we visualize quantities. Erdös didn't change the math; the numbers in essence stayed the same. Rather, he changed our conception of the numbers. New thoughts meant new ideas, which in turn meant new solutions.

The way we do mathematics today is radically different than the way it was done for most of human history. Everything is thought of in terms of algebra. Open any undergraduate or graduate text and you will see lines and lines of algebra, even if the subject has nothing to do with numbers. Human reasoning itself has been reduced to algebra. Consider the following expression (don't worry if you don't understand it):

$$\forall(a,b)\in S^2, (a \rightarrow b) \rightarrow (\sim b \rightarrow \sim a)$$

What if I were to tell you the above symbols are obvious? We know that if it is raining, then the ground is wet. If you noticed the ground was not wet, what would you conclude? Hopefully you would realize that it's not raining. That's essentially what the above expression says. The idea embodied in the above expression is simple. Realizing that it is simple is the hard part.

The *distributive law* can be represented algebraically as $(a+b)c = ac+bc$. When I get an expression like $(1+3) \times 2$, the parenthesis tell me to add the 1 and 3 first giving 4×2, which in turn is 8. However the distributive law says we can multiply the 2 by the individual numbers first, getting 1×2 and 3×2, or equivalently, 2 and 6. Then we can add them to get 8. How did we know we would get the same answer? The distributive law promised us we would.

$$(1+3) \times 2 = 4 \times 2 = 8$$
$$(1+3) \times 2 = 1 \times 2 + 3 \times 2 = 2 + 6 = 8$$

The two ways produce the same solution.

As a mathematician, I feel compelled to prove anything I state. So how do I prove the distributive law? The simple truth is I can't. A *law* in mathematics is something that can't be proven. Often called *axioms*, these statements are fundamental truths that must be accepted on faith.

Open a copy of Euclid's *Elements*, a textbook that's more than two thousand years old, and you will find a proof of the distributive law. How could they have proved something that has no proof? The simple answer is that they didn't look at it the way we do. What we assume, what we require justification for, and the methods we accept as valid are shaped by our view that math is algebra.

Take, for example, how the Greeks looked at math in terms of geometry.

Let's now try to view the distributive law from the Greek perspective. In geometry there are only lines and shapes. There is no multiplication. So how can we express 3×2 in geometry? Recall that the area of a rectangle is the product of the base and the height. We can think of 3×2 as the area of a rectangle with a base of 3 and a height of 2.

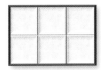

3×2 is just a 3-by-2 rectangle.

Now the distributive law tells us that $(3+1) \times 2$ is the same as $(3 \times 2) + (1 \times 2)$. Thinking in terms of rectangles, $(3+1) \times 2$ is just a 4-by-2 rectangle where the 4 is viewed as the combination of a piece of 3 and 1. The sum $(3 \times 2) + (1 \times 2)$ is just two rectangles, a 3-by-2 and a 1-by-2, combined. So the theorem says that the big rectangle, $(3+1)$ by 2, has the same area as the two small rectangles 3 by 2 and 1 by 2. Look at the diagram below. It shows that the large rectangle has simply been cut in two pieces and that the area of the whole is just the sum of the pieces it has been divided into.

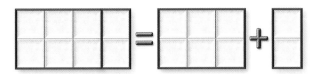

$(3 + 1) \times 2 = (3 \times 2) + (1 \times 2)$ expressed geometrically.

So to a Greek, the distributive law simply says that if you combine two rectangles of the same height, you don't change the area. The area of the pieces is the same as the area of the whole. In fact, this is the axiom the Greeks used: a whole is the sum of its parts. When we call this the distributive law, we're trying to impose our algebraic understanding on the Greeks. Algebra has changed the way we think and has distorted the way we think about the mathematics of others.

The rise of algebra grew hand in hand with the rise of written mathematics. Before the printing press was invented, paper was rare and hence expensive. Mathematics was placed on paper, primarily as an archive. Calculation was mainly done with an abacus or with other physical representations. Reasoning was done orally and transferred that way from master to pupil. When paper became more common, more and more computations were done on paper.

Imagine I asked you to add 2 and 3 on your fingers. You would first hold up two fingers and then raise three more. In the end you would have five fingers raised, representing the fact that $2+3 = 5$. However, if someone walked in the room after the computation, they would see only five fingers and not know how they came to be raised. They wouldn't be able to tell if you made a mistake adding two plus two or if you added one and four. A paper computation is different. Someone looking at the sum $2+3 = 5$ will see the elements of the computation on the paper after the work is completed.

The paper leaves a record, and the work can then be verified by someone familiar with the methods of computation. It in essence becomes a proof. Over the course of the last few centuries, mathematics gradually became more and more algebraic. As mathematicians became used to seeing their computations, they began to associate these ink marks with mathematics itself. It grew into the process of justifying a conclusion.

There are a number of benefits to algebra. For one thing, it's short. It's much easier to write $2+3 = 5$ than "two things when combined with three other things become a total of five things." As you can see, the rhetorical mathematics of the Greeks can appear quite cumbersome. But it's also precise. The diagrams of the Greeks can be misleading. Consider two crossing lines and the angle "between them." As you look at the diagram below realize that there are two perfectly valid angles between the lines and it's unclear which one I'm referring to.

Is the angle between the lines the top or the left angle?

Algebra is viewed as infallible. Words can be ambiguous and diagrams misleading, but no one will question if $2+3$ is 5. The more aspects of mathematics that could be boiled down to symbolic manipulation, the more reliable it appeared to be.

This transformation changed the way we looked at mathematics. The symbols were no longer just a shorthand language for expressing ideas but also became the focus of mathematical thinking. Problems, relations, and methods rooted in the real world became replaced with the infamous x and the equations it inhabited. Rote manipulations according to preassigned rules replaced intuitive thought and reasoning. Many persons simply know the steps required to solve a problem but don't really understand what they're actually doing or why they're doing it.

Consider the following analogy. My word processor in some sense understands the rules of English. If I type a singular noun followed by a plural verb, it will not only point out the mistake but even offer a suggestion on how to rephrase it. However, it would be foolish to suggest that my computer actually understands English. It has no concept of the ideas I'm trying to express with the words and often gives me bad advice precisely because it doesn't understand the context. When a student says that they like math but they don't get word problems, it's almost exactly the same thing. They know rules for manipulating symbols, but they can't use them to answer questions about the world since they don't understand what they mean.

Let's now look at the algebraic treatment of Egyptian mathematics. If you don't know any algebra or find it particularly abhorrent, it won't hurt you to skip what I'm about to do. Technically, what follows isn't even Egyptian mathematics—it's modern tools mimicking Egyptian forms. However, if you can tolerate some algebra, it will help you understand the difference between modern and ancient methods and perhaps give you some valuable insight into both.

Let's first consider the Egyptian rule for doubling an even fraction. We know that $2 \times \overline{6}$ is $\overline{3}$. We got this, of course, by taking half of 6. Here's an algebraic version. Note that the first and last steps convert the expression back and forth between Egyptian and modern notations.

EXAMPLE: Using algebra, prove that $2 \times \overline{6} = \overline{3}$.

PROOF: $2 \times \overline{6} = 2 \times \frac{1}{6} = \frac{2}{1} \times \frac{1}{6} = \frac{2 \times 1}{1 \times 6} = \frac{2}{6} = \frac{1}{3} = \overline{3}$

PRACTICE: Using algebra, prove that $3 \times \overline{12} = \overline{4}$.

When I first showed you the doubling-even rule, I tried to convince you of it by showing you simple examples. Hopefully examining them enabled you to understand why it worked; however, it may not have. Examples only work well if you already have some intuitive model with which to compare them. If you don't, then you need to examine many examples to build the intuition from scratch. So "proof" by example isn't really a proof at all but simply an attempt to relate the current situation to something you already accept as true. The above algebra, in some sense, is proof. If you accept the rules of algebra, you have to accept the conclusion that $2 \times \overline{6} = \overline{3}$.

We still haven't gone completely modern. Today we like to generalize. We don't write something like a tree + a tree is two trees. It's somehow redundant with a dog + a dog is two dogs. We want an expression that encompasses this idea in all its forms so we say $x + x = 2x$. The variable x, which we encounter so often in algebra, is nothing but a pronoun. Just like the word "it," x appears in many mathematical sentences, taking on many different meanings. In general, it just means "some number."

The rule $2 \times \overline{6} = \overline{3}$ is no different than the rule $2 \times \overline{18} = \overline{9}$. This suggests that just as we replace "dogs" and "trees" with x, we need to do the same with 6 and 18. However, they're both even, and the rule won't work properly if they're odd. Nothing about x suggests that it is even. We now need to ask ourselves, what's special about even numbers? The answer is that they're all twice a whole number. The numbers 6 and 18 are 2×3 and 2×9, respectively. This means that an even number can be expressed as $2x$ and hence even fractions can be written as $\overline{2x}$, where x is assumed to be a whole positive number.

When we replace 6 and 18 with 2×3 and 2×9, the equations form a nice pattern. Consider that $2 \times \overline{6} = \overline{3}$ becomes $2 \times \overline{2 \times 3} = \overline{3}$ and $2 \times \overline{18} = \overline{9}$ becomes $2 \times \overline{2 \times 9} = \overline{9}$. If you notice the pattern, you are a good modern mathematician. In general $2 \times \overline{2x} = \overline{x}$. Here's the proof below.

THEOREM: $2 \times \overline{2x} = \overline{x}$

PROOF: $2 \times \overline{2x} = 2 \times \frac{1}{2x} = \frac{2}{1} \times \frac{1}{2x} = \frac{2}{2x} = \frac{1}{x} = \overline{x}$

Let's give the G rule the algebraic treatment. Consider $\overline{15}\ \overline{30} = \overline{10}$. This works because 2×15 is 30 and $30 \div 3$ is 10. I can write $\overline{15}\ \overline{30} = \overline{10}$ as

$$\overline{3 \times 5}\ \overline{2 \times 3 \times 5} = \overline{2 \times 5}$$

Note the extra "$2\times$" in the second term. This makes it twice the first term. This algebraically is representing the fact that $15 \times 2 = 30$. The lack of a "$\times 3$" in the last term means you get rid of it by dividing the middle term by 3. This represents $30 \div 3 = 10$. The 5 has no particular meaning. The "$\times 2, \div 3$" relationship would hold if we changed the 5 to 7 or any other number. The 5 above is our x in the equation. So we can write and prove the following:

THEOREM: $\overline{3x}\ \overline{2 \times 3 \times x} = \overline{2x}$

PROOF: $LHS = \frac{1}{3x} + \frac{1}{6x} = \frac{2}{6x} + \frac{1}{6x} = \frac{3}{6x} = \frac{1}{2x} = RHS$

PRACTICE: Prove $\overline{5x}\ \overline{4 \times 5 \times x} = \overline{4x}$

Above we proved the "$\times 2, \div 3$" and the "$\times 4, \div 5$" forms of the G rule. For the G rule to work, all that's important is that the number you divide by be 1 more than the number you multiply by. Let's now generalize this concept. If we multiply by y, we need to divide by 1 more, or $y + 1$. Hence our generalized G rule becomes the following:

THEOREM: $\overline{(y+1)x}\ \overline{y \times (y+1) \times x} = \overline{yx}$

PROOF: $LHS = \frac{1}{(y+1)x} + \frac{1}{y(y+1)x} = \frac{y}{y(y+1)x} + \frac{1}{y(y+1)x} = \frac{y+1}{y(y+1)x}$
$= \frac{1}{yx} = RHS$

As we look at the above algebraic representation of the G rule, we need to ask ourselves a few questions. While the above symbols precisely define and verify the relation, do they give us any real insight into the G rule? If I initially explained the rule in these symbolic terms, would you have learned it as fast or would you have even understood it at all? Do you remember when I expressed the G rule as "if you add 1, then you have 1 more"? Such a simple understanding of the rule seems obscured by the above symbolic representation. It's still there. In the second line of the above proof there are two fractions: one with a numerator of y and the other with 1. The addition of the second fraction somehow embodies the idea of "add 1." The next line in the proof has a single fraction with a numerator of $y+1$, which embodies the idea of "1 more." However, when most people apply the rules of algebra, they are performing rote operations and not trying to comprehend the ideas that the symbols embody.

The purpose of this essay is not to belittle algebra. Its precision is often necessary to firm up loose, intuitive ideas. The powers of computation can't be denied. Rhetoric and visual models may increase understanding, but they are often cumbersome when seeking a specific answer to a question. It's important to keep a balance between formal symbolic methods and their intuitive meaning. But this book is about Egyptian mathematics, and I doubt few people could argue that the above equations give any real insight into the mind of an ancient mathematician.

5
TECHNIQUES AND STRATEGIES

PRECISION, PYRAMIDS, AND PESU

The Basics of Simplification

The four cardinal points of the compass were extremely important to the Egyptians. The human world ran north-south along the Nile. Trade, raw materials, and people continuously moved up and down the river. The life-bringing flood came from the south and worked its way to the north. Throughout their entire lives most Egyptians rarely moved significantly in any other direction.

Death was another matter entirely. At this time, Egyptians hoped to join their gods, who moved east to west when embodied as the sun, moon, and stars. The entrance into the netherworld was in the land of the setting sun, and hence all tombs were built on the western side of the Nile. Many had a false door painted on or carved into the western inner wall of the tomb. Appearing as mere decoration to the living, this door was accessible only by the spirits of the deceased. Shortly after death, an Egyptian's soul would begin the journey to the Hall of Osiris.

The pharaohs made a much grander entrance into the afterlife, and the pyramid was instrumental to this end. The shape of the pyramid is usually thought of as mimicking the primordial mound from which the Egyptian universe was formed. It is probably more accurate to think of it as symbolic of a ray of light. The *benbenet* was the sacred icon of Heliopolis, the Egyptian city of the sun. It was shaped exactly like a pyramid but smaller. Benbenet were often placed on top of obelisks, forming the pointy tips, such as the one on the top of the Washington Monument. The cult of the sun chose the benbenet as the shape of their icon presumably because it looked like a ray of sunshine. This shape was selected for the tomb of a pharaoh because the dead king walked up to the stars

on a ray of sunlight; hence, the pyramid can literally be thought of as a representation of a ramp to heaven.

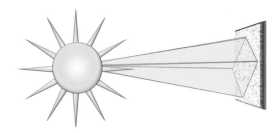

A ray of sunshine is similar to a pyramid.

For the ramp to work, it needs to be pointed to the west; as a result, pyramids were aligned with their sides running east-west and north-south. The precision of the alignment is remarkably accurate. The angular alignment of the Great Pyramid of Khufu is off by only 3 minutes and 26 seconds of arc, which is about a seventeenth of 1 degree.

In order to get a rough estimate of the magnitude of the error, we need to remember that objects look bigger when they make a bigger angle with your eye. The farther away something is, the smaller it looks because it forms a smaller angle.

If you hold out your thumb at arm's length, it has an angular width of about 2 degrees. This means that the error of the Egyptians' north-south line is about a thirty-fourth of this. This is equivalent to an error of roughly 1 inch every 100 feet.

They probably found the north-south orientation using the stars, which move in a complete circle about once every day. The center of these circles falls on the earth's polar axis, which by definition points north-south. One way the ancient Egyptians could have found true north

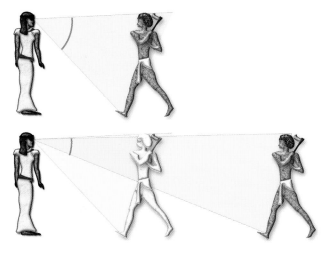

Farther objects make smaller angles with the eye.

The thick blue line represents a thirty-fourth of a thumb
width at arm's length.

is by seeing where a star in the northern hemisphere rises
and sets. Halfway between these two points is true north.

It's easier to measure objects on the ground than in
the sky, and in measuring the former, the Egyptians were
even more accurate. The seked of the Great Pyramid of
Kufu was $5\,\overline{2}$, which is 51 degrees, 50 minutes, and 34 seconds of arc. The actual angular slopes of the above two

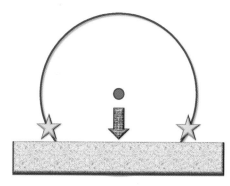

True north is halfway between where a star rises and sets.

pyramids are both 51 degrees and 50 minutes. Including
possible rounding errors, this means that the slope is off
from $5\,\overline{2}$ seked by at most 1 minute of arc. This extraordi-
narily accurate slope is off by less than one-hundredth of
a thumb width.

Such accurate measurements were presumably accompanied by similarly accurate computations. To attain this
kind of precision, the numbers used would have many
fractional components. As we've seen, in Egyptian mathematics fractions beget fractions. For example, doubling
an odd fraction usually increases the number of fractions
in an expression by one or two. The scribes would have
had to reduce these numbers quickly and efficiently. To
do so, they would have applied the identities we've mastered to transform the resulting clutter of fractions into
simple, useable numbers.

In the last chapter we learned of many different types
of identities. Our goal in this section is to learn how to put
them all together. In Ahmose's papyrus there's a problem
of calculating the pesu of 80 loaves made of $3\,\overline{2}$ hekat of
grain. Since the pesu is the number of loaves per hekat,
the former is calculated by dividing 80 by $3\,\overline{2}$. The division
is done as follows. I've added line 7 to include a step I assume the scribe did in his head.

		80	
1)	1	$3\,\overline{2}$	
2)	10	35	$(1)\times 10$
3)	20	70✔	$(2)\times 2$
4)	2	7✔	$(3)\div 10$
5)	$\overline{3}$	$2\,\overline{3}$✔	$(1)\times\overline{3}$
6)	$\overline{7}$	$\overline{2}$✔	$(4)\leftrightarrow$
7)	$\overline{14}$	$\overline{4}$	$(6)\div 2$
8)	$\overline{21}$	$\overline{6}$✔	$(7)\times\overline{3}$
	$22\,\overline{3}\,\overline{7}\,\overline{21}$		

$80\div 3\,\overline{2}$ is $22\,\overline{3}\,\overline{7}\,\overline{21}$.

The multiplication itself is interesting and far from obvious. I'd love to talk about it, but we need to focus first
on simplification. It's not immediately clear that the right
column adds to 80. When confronted with the sum of 70,

7, 2 $\overline{3}$, $\overline{2}$, and $\overline{6}$, we first add the integer parts, getting 79 $\overline{2}$ $\overline{3}$ $\overline{6}$. When trying to simplify the fractional part, we first add equal fractions, of which there are none. Then we look for applications of the G rule. Start with $\overline{2}$ and see if it goes into $\overline{3}$ or $\overline{6}$ evenly. Clearly, it doesn't go into 3, but it does go into 6 three times, although 6 divided by 1 more than 3, which is 4, is not a whole number. We then go on to $\overline{3}$ and notice that if it is doubled, we get 6, and when we divide by 3, we get 2. This means that $\overline{3}$ $\overline{6}$ is just $\overline{2}$.

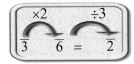

This means that we have 79 $\overline{2}$ $\overline{2}$. Of course $\overline{2}$ $\overline{2}$ is just 1, bringing the total to 80. The Egyptians always did such simplifications without writing anything down; however, if they were to write out these computations, it might appear as follows:

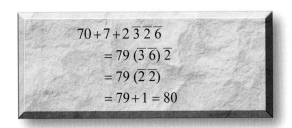

$$70 + 7 + 2\,\overline{3}\,\overline{2}\,\overline{6}$$
$$= 79\,(\overline{3}\,\overline{6})\,\overline{2}$$
$$= 79\,(\overline{2}\,\overline{2})$$
$$= 79 + 1 = 80$$

After Ahmose performed this division, he did a check. If his solution of $80 \div 3\,\overline{2}$ is 22 $\overline{\overline{3}}$ $\overline{7}$ $\overline{21}$ was correct, then 22 $\overline{\overline{3}}$ $\overline{7}$ $\overline{21}$ $\times 3\,\overline{2}$ should be 80. Ahmose performed the following straightforward multiplication.

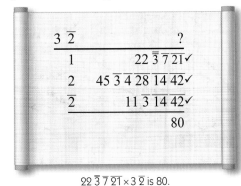

3 $\overline{2}$	?
1	22 $\overline{\overline{3}}$ $\overline{7}$ $\overline{21}$ ✓
2	45 $\overline{3}$ $\overline{4}$ $\overline{28}$ $\overline{14}$ $\overline{42}$ ✓
$\overline{2}$	11 $\overline{3}$ $\overline{14}$ $\overline{42}$ ✓
	80

22 $\overline{\overline{3}}$ $\overline{7}$ $\overline{21}$ $\times 3\,\overline{2}$ is 80.

Once again it's not clear that the sum of the fractions is 80. The mass of fractions seems overwhelming to a beginner. As you gain experience it does get easier. The strategy I use is broken down into the following steps:

- Add integers.
- Add $\overline{\overline{3}}$ to $\overline{\overline{3}}$ or $\overline{3}$.
- Add equal fractions, such as $\overline{4}$ $\overline{4}$ or $\overline{7}$ $\overline{7}$, using doubling.
- Apply the G rule, for example, $\overline{5}$ $\overline{20} = \overline{4}$.
- Look for special relations found on the Egyptian Mathematical Leather Roll.
- Use parts if necessary.

Our job is to add the ugly collection of numbers found in the above multiplication that supposedly add up to 80:

$$22\,\overline{\overline{3}}\,\overline{7}\,\overline{21},\ 45\,\overline{3}\,\overline{4}\,\overline{28}\,\overline{14}\,\overline{42},\ 11\,\overline{3}\,\overline{14}\,\overline{42}$$

Start by applying the first three steps at once, grouping whole numbers, $\overline{\overline{3}}$ and $\overline{3}$, and equal fractions. Doing this we get

$$(22\ 45\ 11)\,(\overline{\overline{3}}\ \overline{3})\,(\overline{14}\ \overline{14})\,(\overline{42}\ \overline{42})\,\overline{3}\,\overline{4}\,\overline{7}\,\overline{21}\,\overline{28}$$

The integers add up to 78. The $\overline{\overline{3}}$ and $\overline{3}$ sum to 1, a fact that every Egyptian mathematician had to have known. The two $\overline{14}$s and $\overline{42}$s are easy to add because they are even, giving $\overline{7}$ and $\overline{21}$, respectively. This process results in

$$78\ 1\ \overline{7}\ \overline{21}\ \overline{3}\ \overline{4}\ \overline{7}\ \overline{21}\ \overline{28}$$

Notice that once again we have multiple whole numbers and repeated fractions, so we repeat some of the steps we just took. First we group them:

$$(78\ 1)\,(\overline{7}\ \overline{7})\,(\overline{21}\ \overline{21})\,\overline{3}\,\overline{4}\,\overline{28}$$

Then we add. Note: to add the $\overline{7}$s and the $\overline{21}$s we need to consult Ahmose's doubling table. Doing so we get

$$79\ \overline{4}\ \overline{28}\ \overline{14}\ \overline{42}\ \overline{3}\ \overline{4}\ \overline{28}$$

Once again we group and add.

$$79\ \overline{3}\,(\overline{4}\ \overline{4})\,\overline{14}\,(\overline{28}\ \overline{28})\,\overline{42}$$
$$79\ \overline{3}\ \overline{2}\,(\overline{14}\ \overline{14})\,\overline{42}$$
$$79\ \overline{2}\ \overline{3}\ \overline{7}\ \overline{42}$$

At this point, the first three rules are played out. We now consider the fourth rule. In order to apply the G rule, we check for which numbers go into which evenly and then see if when we divide by 1 more than the result, we get a whole number. For the $\overline{2}$, we see that 2 goes into 42 twenty-one times, but $42 \div 22$ is not whole. However, 7 goes into 42 six times, and $42 \div (6+1)$ is 6, so $\overline{7\,42}$ is just $\overline{6}$. We now get

$$79\,\overline{2}\,\overline{3}\,\overline{6}$$

Once again the G rule applies since 3 goes into 6 two times, and the result of $6 \div (2+1)$ is a whole number; hence, $\overline{3\,6}$ is just $\overline{2}$. Finally we are left with two $\overline{2}$s, which add up to $\overline{1}$, which is just 1. We get 80, just as we knew we should.

$$79\,\overline{2}\,(\overline{3\,6})$$
$$79\,(\overline{2\,2})$$
$$80$$

It seems like a coincidence that the fractions simplified so well, but as I've often said, "coincidences" are common in Egyptian mathematics. Here's another example:

EXAMPLE: Add $4\,\overline{\overline{3}}\,\overline{46} + 5\,\overline{46}\,\overline{60} + 3\,\overline{\overline{3}}\,\overline{23}$.

$$4\,\overline{\overline{3}}\,\overline{46} + 5\,\overline{46}\,\overline{60} + 3\,\overline{\overline{3}}\,\overline{23}$$
$$= (4 + 5 + 3) + (\overline{\overline{3}}\,\overline{\overline{3}}) + (\overline{46}\,\overline{46}) + \overline{23}\,\overline{60}$$
$$= (12 + 1) + \overline{3} + (\overline{23}\,\overline{23})\,\overline{60}$$
$$= 13\,\overline{3}\,(\overline{12}\,\overline{60})\,\overline{276}$$
$$= 13\,\overline{3}\,\overline{10}\,\overline{276}$$

Now try this one:

PRACTICE: Add $8\,\overline{\overline{3}}\,\overline{15}\,\overline{120} + 2\,\overline{10}\,\overline{120} + 9\,\overline{3}\,\overline{15}$.

ANSWER: $20\,\overline{4}$.

Accurate measurements beget fractions. Operating on these numbers magnifies the problem, but as we see, the proper application of identities can help contain the resulting mess. You may have noticed that I didn't use the last one of the six steps to simplify fractions. The step "use parts if necessary" requires a bit more sophistication and will be discussed in the next section.

THE SOLAR EYE

Building Egyptian Fractions

Thoth, the god of wisdom and scribes, had two tasks to perform. He had no choice because it was a direct command from Ra. The first task was completed when he found Ra's eye, the Solar Eye, in the land of Nubia. Thoth knew that the second task, bringing the eye home, would be both difficult and dangerous.

According to ancient Egyptian myths, the eyes of the gods were beings in their own right, having their own personalities and some degree of free will. They embodied the very power of the god and could be sent out independently of the rest of the divine body to accomplish difficult tasks. Ra's eye had refused to come home because she was angry. Even worse, she had transformed herself into a lioness, the form female goddesses take when they want to commit acts of violence. Hence Thoth found himself face to face with the full power of Ra at a time when she was uncontrolled, angry, and looking for a fight.

The eye had already killed everyone in human form who had approached her. Thoth only avoided destruction by transforming himself into a baboon. Now that Thoth

The Solar Eye in lioness form.

had found the eye of Ra, he needed to convince her that she belonged with Ra. Thoth tried pleas and taunts, but none of these tactics seemed to work. Eventually he won her over with amusing but pointed tales. One of the stories, about a lion and a mouse, goes like this. Once upon a time, a powerful lion of the mountains noticed that the other animals were in great distress. They were each wounded, chained, or tortured. When the lion asked who had done this, each animal had the same reply: man. The lion had never seen a man but vowed that he would avenge the animals of his land.

As the lion searched for man, he accidentally stepped on a mouse. The mouse begged mercy and promised that if the lion spared his life, he would save the lion one day. The thought of a tiny mouse saving the great lion amused him so much that he let the mouse go. Although the lion never found man, he did find one of his traps. The lion fell in to a covered pit and became ensnared in the net below. For hours the lion tried to break free, but to no avail. Finally the mouse came and freed the lion by chewing through his bonds, repaying his debt.

The point of the story, of course, is that even the mighty, like the Solar Eye, need the aid of the meek and that acts of goodness are repaid by the universe. So the eye returned to Ra, where she was met with music and dancing. These helped pacify Ra's eye, transforming her from her deadly lioness form.

My point in telling this story is to emphasize that there is a big difference between finding something and justifying that what you've found is what you want. Thoth found Ra's eye, but his task was not finished until he could convince her that she belonged to Ra. Thoth essentially had to prove what he found was the Solar Eye. Consider the following mathematical equation:

$$\sqrt{x^3 - 12x - 3} + \log_2(2x^2 + 30) = 23$$

One solution is $x = 7$. Any math student familiar with the above functions and the use of a calculator can replace the x's above with 7's on the left side and get 23, verifying the solution. Once presented with the solution, it's easy to check. However, I dare anyone to try to solve this equation and arrive at $x = 7$ as a solution. Finding and proving are very different operations.

As we've seen, the ancient Egyptians frequently checked their answers, at least in their texts, if not in practice. We've also seen that they were unable to prove many of their statements, at least in the modern sense, because they were unable to abstract, as we do through algebra. This does not mean they ignored proof. The Egyptians proved much of what they did, but they reserved proof for specific as opposed to general statements. One place

we see this is in Ahmose's justification of the table of doubling odd fractions.

I've portrayed Ahmose's table in a modern way: as two columns, the first as the fraction to be doubled and the second as its value when doubled. But this is not how it appears in Ahmose's papyrus. For example, I've given the entry $\overline{7}, \overline{4}\,\overline{28}$ to mean that $2 \times \overline{7} = \overline{4}\,\overline{28}$. But in fact, the beginning of the entry appears as $\overline{4}\ 1\ \overline{2}\ \overline{4},\ \overline{28}\ \overline{4}$. The $\overline{4}$ and the $\overline{28}$ are in red in the scroll, and oddly enough, the 7 doesn't even appear except as part of a multiplication next to the entry, which looks roughly as follows:

?	2
1	7
$\overline{2}$	$3\ \overline{2}$
$\overline{4}$	$1\ \overline{2}\ \overline{4}$ ✓
2	14
4	28
$\overline{28}$	$\overline{4}$ ✓
$\overline{4}\ \overline{28}$	

2 ÷ 7 is $\overline{4}\ \overline{28}$.

The key to understanding what's happening here is to realize that the calculation is determining the value of $2 \div 7$, which is in fact equivalent to $2 \times \overline{7}$. Notice that the checked rows have $1\ \overline{2}\ \overline{4}$ and $\overline{4}$, which are precisely the black numbers that appear in the table entry, $\overline{4}\ 1\ \overline{2}\ \overline{4},\ \overline{28}\ \overline{4}$. It's possible that the line of numbers should be read as "a fourth of 7 is $1\ \overline{2}\ \overline{4}$ and a twenty-eighth of 7 is $\overline{4}$." The important mathematical property of the $1\ \overline{2}\ \overline{4}$ and the $\overline{4}$ is that they add up to 2, and hence the checked entries of the right column of the division will add up to 2.

There's a lot going on here and it's mildly confusing, especially because there are at least two valid interpretations. One is that they are deriving the value for the table, and the other is that they are proving it. Mathematicians often use the two terms interchangeably, but they really are different. A derivation is how you find something, and a proof is how you verify that what you found is what you claim it to be. Most modern texts greatly favor the latter, so students see mathematical ideas being verified, but they are woefully ignorant of the process of mathematical discovery and creation.

If we view the above calculation as a derivation, then we are dividing 2 by 7 and get the lines $\overline{4}$, $1\,\overline{2}\,\overline{4}$, and $\overline{28}$, $\overline{4}$. Note that $1\,\overline{2}\,\overline{4}$ and $\overline{4}$ add up to 2, as is demonstrated by the following:

$$1\,\overline{2}\,\overline{4} + \overline{4}$$
$$= 1\,\overline{2}\,(\overline{4}\,\overline{4}) = 1\,(\overline{2}\,\overline{2}) = 2$$

We now know that $2 \div 7$ is $\overline{4}\,\overline{28}$. I personally disagree with this interpretation because the divisions for the other elements of the table make no sense. Consider the division in the calculation of $2 \div 65$.

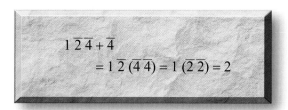

?	2
1	65
$\overline{39}$	$1\,\overline{\overline{3}}$ ✓
3	195
$\overline{195}$	$\overline{3}$ ✓
$\overline{39}\,\overline{195}$	

Interpreting the above calculation as a division makes no sense. The first line follows from the second by a "division" of 39. Never have we encountered a one-step division by 39 in any Egyptian problem, and I have no idea how it could be done. Even if steps were skipped, it's impossible to see how or why the problem was done this way.

Let's instead view these calculations as proofs. In other words, the scribe knows the answer before he starts. Now when we consider the first calculation of $2 \div 7 = \overline{4}\,\overline{28}$, we assume the scribe is already aware of this and instead wants to verify it.

Division is the opposite of multiplication. If I divide 15 by 3 to get 5, I can verify this solution by multiplying 3 by 5 to get 15. Similarly if a scribe wants to verify $2 \div 7 = \overline{4}\,\overline{28}$, he can perform the multiplication of $7 \times \overline{4}\,\overline{28}$ in order to get 2. Note the following multiplication of 7 by $\overline{4}\,\overline{28}$. Except for the column headings and sums, it looks precisely like the division of 2 by 7.

$\overline{4}\,\overline{28}$	?
1	7
$\overline{2}$	$3\,\overline{2}$
$\overline{4}$	$1\,\overline{2}\,\overline{4}$ ✓
2	14
4	28
$\overline{28}$	$\overline{4}$ ✓
	2

$7 \times \overline{4}\,\overline{28} = 2$ looks just like $7 \div 2 = \overline{4}\,\overline{28}$.

You should also realize that these column headings and sums don't actually appear in the Egyptian papyrus; rather, I put them in for clarity. So it's unclear if this is $7 \times \overline{4}\,\overline{28}$ or $2 \div 7$; hence the two interpretations. I'm not even sure whether an Egyptian would greatly distinguish between the two. In fact, they often would phrase something like $2 \div 7$ as something that loosely translates to "multiply by 7 to get 2." Now the $\overline{4}$ and $\overline{28}$ are not byproducts of the calculation but rather numbers that are consciously constructed by the scribe. When the second column adds to 2, the proof that $2 \div 7$ is $\overline{4}\,\overline{28}$ is provided by the fact that $7 \times \overline{4}\,\overline{28}$ is 2.

Let's accept this interpretation and try to provide our own proofs. Consider the following identity from Ahmose's table: $2 \times \overline{15} = \overline{10}\,\overline{30}$. We need to realize that this means that $2 \div 15$ is $\overline{10}\,\overline{30}$, and to prove that this identity is true, we will show that $15 \times \overline{10}\,\overline{30}$ is 2. Starting with 1 and 15 in the first row, we will need to create a $\overline{10}$ and a $\overline{30}$ in the first column. When we check off the corresponding rows, we should get a total of 2.

We have four basic tools at our disposal. They are doubling, taking $\overline{\overline{3}}$, dividing by 10, and switching. Creating the

$\overline{10}$ is trivial. We have 1 initially in the first column, and dividing by 10 will immediately give us $\overline{10}$. The $\overline{30}$ has an additional factor of 3 in it. Taking $\overline{\overline{3}}$ of the $\overline{10}$ puts the 3 in, and dividing by 2 finishes the job. Here's the calculation.

	$\overline{10}$	$\overline{30}$	?	
1)		1	15	
2)		$\overline{10}$	$1\,\overline{2}\,$✓	$(1)\div 10$
3)		$\overline{15}$	1	$(2)\times\overline{\overline{3}}$
4)		$\overline{30}$	$\overline{2}\,$✓	$(3)\div 2$
			2	

Constructing $\overline{10}\ \overline{30}$ to multiply by 15.

As you can see, the $1\,\overline{2}$ and the $\overline{2}$ add up to 2, proving that $2\div 15$ is $\overline{10}\ \overline{30}$.

Now let's look at the table entry for $2\div 11$ is $\overline{6}\ \overline{66}$. We'll need to show that $11\times\overline{6}\ \overline{66}$ is 2. Constructing the $\overline{6}$ is easy. We start with 1, divide it by 2 and then by 3. The division by 3 is performed by multiplying by and then dividing by 2. So the proof starts as follows:

	$\overline{6}$	$\overline{66}$	?	
1)		1	11	
2)		$\overline{2}$	$5\,\overline{2}$	$(1)\div 2$
3)		$\overline{3}$	$3\,\overline{3}$	$(2)\times\overline{\overline{3}}$
4)		$\overline{6}$	$1\,\overline{2}\,\overline{3}\,$✓	$(3)\div 2$

Now that we've got the $\overline{6}$, we need to construct the $\overline{66}$. This number can be more difficult to construct because it has an 11 as a factor. There is no natural way I know to construct an 11, but we always need to remember that Egyptian numbers are not random. The $\overline{66}$ is there precisely because we are dividing by 11. If we switch the first row, we get $\overline{11}$ in the first column. In order to get $\overline{66}$, we need to divide by 2 and then by 3, giving

	$\overline{6}$	$\overline{66}$	?	
1)		1	11	
2)		$\overline{2}$	$5\,\overline{2}$	$(1)\div 2$
3)		$\overline{3}$	$3\,\overline{3}$	$(2)\times\overline{\overline{3}}$
4)		$\overline{6}$	$1\,\overline{2}\,\overline{3}\,$✓	$(3)\div 2$
5)		$\overline{11}$	1	$(1)\leftrightarrow$
6)		$\overline{22}$	$\overline{2}$	$(5)\div 2$
7)		$\overline{33}$	$\overline{3}$	$(6)\times\overline{\overline{3}}$
8)		$\overline{66}$	$\overline{6}\,$✓	$(7)\div 2$
			2	

We can verify that the sum is 2 with the following calculation:

$$1\,\overline{2}\,\overline{3} + \overline{6} = 1\,\overline{2}\,(\overline{3}\ \overline{6}) = 1\,(\overline{2}\ \overline{2}) = 2$$

A sharp reader will notice that we could have built the 66 on the right and then switched, but I wanted to provide more practice in constructing fractions.

In general we give our fractions the factors of 2, 3, and 5 by using division by 2, multiplication by $\overline{\overline{3}}$, and division by 10, respectively. The first is the most obvious. If I have a $\overline{7}$ and I want a $\overline{14}$, I simply divide the $\overline{7}$ by 2.

EXAMPLE: Build a $\overline{20}$ in the first column when the first row is $\overline{5}$, 9.

Note that since we start with a $\overline{5}$ and need a $\overline{20}$, we're off by a factor of 4, which is two 2s. Hence we divide by 2 twice.

SOLUTION:

	$\overline{5}$	9	
1)			
1)	$\overline{5}$	9	
2)	$\overline{10}$	$4\,\overline{2}$	$(1)\div 2$
3)	$\overline{20}$	$2\,\overline{4}$	$(2)\div 2$

PRACTICE: Build a $\overline{56}$ in the first column when the first row is $\overline{7}$, 12 $\overline{5}$.

ANSWER: The fourth row should be $\overline{56}$, 1 $\overline{2}$ $\overline{40}$.

Sometimes we need a factor of 3 in our fraction, such as if we had a $\overline{7}$ and wanted a $\overline{21}$. This can be accomplished by a division by 3, which is the combination of dividing by 2 and multiplying by $\overline{\overline{3}}$.

EXAMPLE: Build a $\overline{21}$ in the first column when the first row is $\overline{7}$, 8.

SOLUTION:

1)	$\overline{7}$	8	
2)	$\overline{14}$	4	$(1) \div 2$
3)	$\overline{21}$	2 $\overline{3}$	$(2) \times \overline{\overline{3}}$

PRACTICE: Build a $\overline{15}$ in the first column when the first row is $\overline{5}$, 11 $\overline{2}$.

ANSWER: The third row should be $\overline{15}$, 3 $\overline{\overline{3}}$ $\overline{6}$.

Now when we need a factor of 5 in a fraction, we divide by 10. This also adds a factor of 2 that often needs to be eliminated by doubling.

EXAMPLE: Build a $\overline{30}$ in the first column when the first row is $\overline{6}$, 11.

Note that since we start with a $\overline{6}$ and need a $\overline{30}$, we're off by a factor of 5.

SOLUTION:

1)	$\overline{6}$	11	
2)	$\overline{60}$	1 $\overline{10}$	$(1) \div 10$
3)	$\overline{30}$	2 $\overline{5}$	$(2) \times 2$

PRACTICE: Build a $\overline{25}$ in the first column when the first row is $\overline{5}$, 7 $\overline{4}$.

ANSWER: The third row should be $\overline{25}$, 1 $\overline{3}$ $\overline{15}$ $\overline{20}$.

We simply combine the methods to obtain factors of 2, 3, and/or 5.

EXAMPLE: Build a $\overline{60}$ in the first column when the first row is $\overline{2}$, 10 $\overline{5}$.

Note that since 60 has a 5 and a 3 in it that is not found in the 2, we start by inserting them.

SOLUTION:

1)	$\overline{2}$	10 $\overline{5}$	
2)	$\overline{20}$	1 $\overline{50}$	$(1) \div 10$
3)	$\overline{30}$	$\overline{3}$ $\overline{75}$	$(2) \times \overline{\overline{3}}$
4)	$\overline{60}$	3 $\overline{150}$	$(3) \div 2$

PRACTICE: Build a $\overline{300}$ in the first column when the first row is $\overline{20}$, 15 $\overline{3}$.

ANSWER: The third row should be $\overline{300}$, 1 $\overline{45}$.

The last skill we need to master is switching. This moves missing factors from the right column to the left column.

EXAMPLE: Build a $\overline{35}$ in the first column when the first row is $\overline{4}$, 14.

Note that we need a 7 in the 35, but we know how to build only 2, 3, and 5. However, there is a 7 in the 14 that can be moved to the left by switching. Once there, we will need to get the 5 by dividing by 10 and then getting rid of extraneous 2s.

SOLUTION:

1)	$\overline{4}$	14	
2)	$\overline{14}$	4	(1)↔
3)	$\overline{140}$	$\overline{3}\ 15$	(2)÷10
4)	$\overline{70}$	$\overline{3}\ \overline{10}\ \overline{30}$	(3)×2
5)	$\overline{35}$	$1\ \overline{3}\ \overline{5}\ \overline{15}$	(4)×2

PRACTICE: Build a $\overline{33}$ in the first column when the first row is $\overline{8}$, 11.

ANSWER: The fourth row should be $\overline{33}$, $2\ \overline{\overline{3}}$.

Now we're ready to do some proofs. Below I'll show that $2 \div 79$ is $\overline{60}\ \overline{237}\ \overline{316}\ \overline{790}$. Although this looks difficult, it's important to note that 237, 316, and 790 are all multiples of 79. While this is not obvious in the proof, perhaps it was clear in the derivation. In other words, perhaps the scribe found this identity in a way in which 237, 316, and 790 were found by taking multiples of 79. I'm going to construct these in the Egyptian manner of creating them fully formed in the right column before switching.

EXAMPLE: Prove that $2 \div 79$ is $\overline{60}\ \overline{237}\ \overline{316}\ \overline{790}$.

SOLUTION: We will do this by showing that $79 \times \overline{60}\ \overline{237}\ \overline{316}\ \overline{790}$ is 2, as shown at the top of the next column on this page.

The right column sums to 2, as can be seen in the following computation:

$$1\ \overline{5}\ \overline{10}\ \overline{60} + \overline{3} + \overline{4} + \overline{10}$$
$$= 1\ \overline{3}\ \overline{4}\ \overline{5}\ (\overline{10}\ \overline{10})\ \overline{60}$$
$$= 1\ \overline{3}\ \overline{4}\ (\overline{5}\ \overline{5})\ \overline{60}$$
$$= 1\ (\overline{3}\ \overline{3})\ \overline{4}\ (\overline{15}\ \overline{60})$$
$$= 1\ \overline{3}\ (\overline{4}\ \overline{12})$$
$$= 1\ (\overline{3}\ \overline{3}) = 2$$

	$60\ 237\ 316\ 790$	?	
1)	1	79	
2)	$\overline{2}$	$39\ \overline{2}$	(1)÷2
3)	$\overline{3}$	$26\ \overline{3}$	(2)×$\overline{\overline{3}}$
4)	$\overline{6}$	$13\ \overline{6}$	(3)÷2
5)	$\overline{60}$	$1\ \overline{5}\ \overline{10}\ \overline{60}$ ✓	(4)÷10
6)	2	158	(1)×2
7)	3	237	(1)+(6)
8)	4	316	(6)×2
9)	10	790	(1)×10
10)	$\overline{237}$	$\overline{3}$ ✓	(7)↔
11)	$\overline{316}$	$\overline{4}$ ✓	(8)↔
12)	$\overline{790}$	$\overline{10}$ ✓	(9)↔
		2	

PRACTICE: Prove that $2 \div 27$ is $\overline{18}\ \overline{54}$.

PRACTICE: Prove that $2 \div 17$ is $\overline{12}\ \overline{51}\ \overline{68}$.

CHOOSE YOUR SALVATION

Choosing the Right Parts

Ancient Egypt existed more or less intact for almost three thousand years, but in this book I've misleadingly presented a unified view of the Egyptian mythos. In the first chapter I told you that Ptah, the craftsman god, created the world. If you were to ask this question of someone from Memphis, the capital of Egypt during the Old Kingdom, they'd probably agree with me. However, if you asked another Egyptian from another place or time, they could easily disagree; for example, if you asked someone from Hermopolis or Heliopolis, they might answer Thoth or Ra, respectively.

As time progressed, the cities of Egypt waxed and waned. As a city grew in importance, so did their gods. At the end of the Old Kingdom, about 2700–2200 BCE, the city of Memphis lost its primacy. After a period of disorder, Thebes, in southern Egypt, took over. Their chief god, Amun, the hidden one, became the king of the gods. They smoothed the transition by merging their gods with one of the old gods, Ra, creating Amun-Ra, the new creator of the universe. Some priests tried to work the old mythology into the new. The merging of gods and the reworking of divine family trees led to a twisted sense of history. Kings claimed to be both the son of Ra and the protector of their father Osiris and did not see any contradiction in having two dads. Perhaps they didn't think it was their place to question the gods or perhaps they didn't take such familial associations as literally as they might be interpreted today.

Views of the importance and roles of the gods varied not only by time and place, but also by profession and social class. Kings seemed to favor Ra, the god of the all-powerful sun, and Horus, the embodiment of the living god king. High priests leaned toward gods of knowledge, such as Ptah, the craftsman god; Thoth, the god of wisdom and learning; and Ma'at, the goddess of order. Scribes worshiped Imhotep, the deified architect of the first stone pyramid, god of scribes, dreams, and healing.

Competing theologies led to competing views of the afterlife. Those persons who favored Ra believed that the blessed spent eternity on the Boat of Millions. This giant ship carried Ra in the form of the sun around the earth each day. In Abydos, they worshiped Osiris, the benevolent god of the dead. Osiris ruled the underworld from the Field of Reeds, where he was joined by the souls of the dead who achieved salvation.

Although Abydos was never as politically powerful as cities like Memphis or Thebes, the cult of Osiris grew in importance among the people. While circling the universe daily in the Boat of Millions might be exciting, most Egyptians preferred the idyllic, peaceful existence in the Field of Reeds. Osiris's realm was very much like a perfect version of Egypt itself. Despite the insistence by

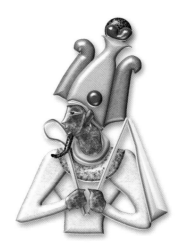

Osiris, god of the underworld.

the theocracy that the Boat of Millions would be one's final resting place, belief in the Field of Reeds persisted. It was so pervasive that it lasted until Christianity dominated Egypt, and even then it merged with that religion, forming the conceptual basis of the Christian heaven. The people chose the afterlife that most suited them.

There's an unusual parallel in Egyptian mathematics. The Egyptians had a fair amount of choice in the fractions they used. They could decide if the answers to a calculation were to their liking. If not, they would change them. As we will see, they could even choose the first terms in their identities, carefully picking the numbers they liked.

Just as every entry in Ahmose's times-2 table was checked, the same is true for most entries of the division-by-10 table. The check was done in an unusual way. After the 9-entry table, there are six word problems, each of which serves as proof that one of the table entries is correct. For example, the table tells us that $2 \div 10$ is $\overline{5}$. The second problem after the table states that two loaves of bread divided between 10 men is $\overline{5}$ of a loaf for each. The answer is clearly pulled off the table. However, it then proceeds to check the solution by multiplying 10 by $\overline{5}$, showing that the result is 2. The check appears as follows:

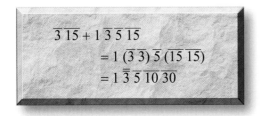

The sum of the entries in the right column can be initially calculated as follows:

$$\overline{3}\ \overline{15} + 1\ \overline{3}\ \overline{5}\ \overline{15}$$
$$= 1\ (\overline{3}\ \overline{3})\ \overline{5}\ (\overline{15}\ \overline{15})$$
$$= 1\ \overline{\overline{3}}\ \overline{5}\ \overline{10}\ \overline{30}$$

Now we know the answer should be 2, but instead we got the long sequence $1\ \overline{\overline{3}}\ \overline{5}\ \overline{10}\ \overline{30}$. As we've seen, this can be simplified using parts. Each of the fractions is easily measured as parts of 30. We know that $\overline{\overline{3}}$ of 30 is 20, $\overline{5}$ of 30 is 6, $\overline{10}$ of 30 is 3, and $\overline{30}$ of 30 is 1. So when calculating parts we get the following:

$$\times 30 \quad\curvearrowright\quad \overline{\overline{3}}\ \overline{5}\ \overline{10}\ \overline{30} = 1 \quad\curvearrowleft\quad \div 30$$
$$20 + 6 + 3 + 1 = 30$$

Since the fractions add up to 1, the answer $1\ \overline{\overline{3}}\ \overline{5}\ \overline{10}\ \overline{30}$ becomes 2, exactly the answer we expected.

In this case, we knew ahead of time that the answer was 2, so clearly the four fractions had to simplify. But what if we don't know the answer? Some answers the ancient Egyptians left as is and others they simplified. For example, in the eyes of an Egyptian, $8\ \overline{6}\ \overline{20}$ would probably be deemed acceptable, whereas $8\ \overline{6}\ \overline{10}$ might require simplification.

The answer is not black and white. It has its roots in the ideas we discussed in the section "What Makes a Good Identity," in chapter 4. Recall that a number expressed as the sum of fractions can be evaluated on a number of criteria. Perhaps the most important one is whether or not the fraction forms a *good approximation*.

Imagine we solved a problem and obtained a solution of $\overline{6}\ \overline{10}$. The pair of fractions $\overline{6}\ \overline{10}$ is not a good approximation. The error is more than 50% since 10 is less than 6×2. A tenth is just too large when compared with a sixth. I might then be tempted to simplify the fractions using parts of 30 as follows:

$$\times 30 \quad\curvearrowright\quad \overline{6}\ \overline{10} = \overline{5}\ \overline{15} \quad\curvearrowleft\quad \div 30$$
$$5 + 3 = 6 + 2$$

The answer of $\overline{5}\ \overline{15}$ is a better approximation since a fifth is three times larger than a fifteenth. Thus the error is about 33%. However, you might remember that the Egyptians favored even fractions over odd ones, so in some ways, the original solution of $\overline{6}\ \overline{10}$ is better than $\overline{5}\ \overline{15}$. Even fractions are better because you don't need to consult Ahmose's table to double them. However, if this is a final answer to a problem, we're not going to double it again. Hence $\overline{5}\ \overline{15}$ is better unless we planned on using it in another computation. There's no hard-and-fast rule, but the Egyptians seemed content with a difference of around three times as large. So, for example, $\overline{10}\ \overline{32}$ would probably be acceptable because 32 is more than 3×10, but $\overline{8}\ \overline{20}$ might not be because 20 is less than 3×8.

This rule of thumb seems to get weaker the farther down the line of fractions you get. For example, Ahmose's table tells us that $2 \times \overline{19}$ is $\overline{12}\ \overline{76}\ \overline{114}$. The fraction $\overline{76}$ is less than a sixth of $\overline{12}$, since $6 \times 12 < 76$ and hence is a good separation of fractions. However the $\overline{76}$ and the $\overline{114}$ are very close. Somehow this was not a deal breaker to the ancient scribes. I'm not sure if this was because $\overline{114}$ is so small that they felt it didn't matter or just because finding a better set of fractions led to other mathematical difficulties. For the following problems, aim to have the third fraction at least twice as large as the previous one.

EXAMPLE: Simplify $\overline{2}\ \overline{7}\ \overline{15}$ or explain why there is no need.

SOLUTION: 7 is more than 3×2, so the first pair are fine. While 15 is roughly twice 7, it's the second pair in the sequence, so it is fine.

EXAMPLE: Simplify $\overline{4}\ \overline{6}\ \overline{18}$ or explain why there is no need.

SOLUTION: 6 is much less than 3×4 so we will use parts of 36, the least common multiple.

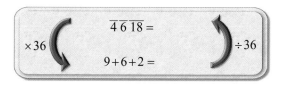

The terms $\overline{4}, \overline{6}$, and $\overline{18}$ turn into $9+6+2 = 17$. We now need to break 17 into whole parts of 36. The larger the initial part, the better the approximation will be. The largest part no more than 17 is 12, a third of 36. This leaves five parts, which can be broken into $4+1$, which represent a ninth and a thirty-sixth of 36.

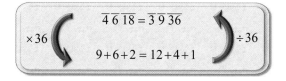

So $\overline{4}\ \overline{6}\ \overline{18}$ simplifies to $\overline{3}\ \overline{9}\ \overline{36}$.

EXAMPLE: Simplify $\overline{6}\ \overline{18}\ \overline{24}$ or explain why there is no need.

SOLUTION: While 18 is 3×6, the 18 and 24 are too close. Simplify with parts of 72.

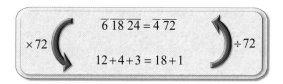

So $\overline{6}\ \overline{18}\ \overline{24}$ simplifies to $\overline{4}\ \overline{72}$.

PRACTICE: Simplify the following or explain why there is no need.

- $\overline{5}\ \overline{21}\ \overline{40}$
- $\overline{3}\ \overline{4}\ \overline{36}$
- $\overline{6}\ \overline{20}\ \overline{24}$

ANSWERS:

- It is simplified.
- $\overline{2}\ \overline{9}$
- $\overline{4}\ \overline{120}$

You should not think of the difference between fractions as isolated conditions. Each one adds or subtracts from the total suitability of a fractional expression. So we may tolerate $\overline{4}\ \overline{16}\ \overline{80}$ and $\overline{2}\ \overline{16}\ \overline{24}$, but we would not tolerate $\overline{4}\ \overline{16}\ \overline{24}$. Neither pair is bad enough to be a deal breaker but together they are unacceptable. There's a good reason for this. If we just had $\overline{4}\ \overline{16}$, the error in approximating this as $\overline{4}$ is $\overline{16}$; however, if we approximate $\overline{4}\ \overline{16}\ \overline{24}$ as $\overline{4}$, then the error is $\overline{16}\ \overline{24}$. This is a significantly large number relative to $\overline{4}$, so we might simplify it as follows:

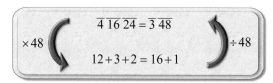

There are other considerations for when to simplify, like the simplicity of computation. For example, we would probably try to reduce $\overline{10}\ \overline{20}\ \overline{50}$ but not $\overline{11}\ \overline{21}\ \overline{51}$. The common factor of 10 would make the number of parts smaller in the first expression. The factors of 2 and 5 within the 10 would provide us with more options when breaking the total into parts. The 11, 21, and 51 of the other fractions would lead to difficult computations and extended answers if we tried to reduce it.

Occasionally using parts doesn't seem to simplify an answer. Consider $\overline{4}\ \overline{5}$. It's an obvious candidate for simplification because the two fractions are so close together. However, when we choose the obvious number of parts, 20, we get the following:

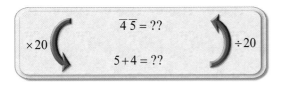

We can't break up the total of 9 into a better sum. That's because there is no factor of 20 greater than 5 but less than 9. There's a trick to dealing with this problem. It more or less requires us to estimate the answer before we start. Obviously $\overline{4}\ \overline{5}$ is a little less than $\overline{4}\ \overline{4}$, which is $\overline{2}$. Since $\overline{2}$ is already bigger than $\overline{4}\ \overline{5}$, our answer can't start with $\overline{2}$ because all we can do is add other fractions that simply make it larger. So what could our answer start with? We can't start with a $\overline{2}$, since it's too big, but if we start with $\overline{4}$, it's too small, because the error, $\overline{5}$, is too large. Hopefully a little common sense should tell us if $\overline{2}$ is too big and $\overline{4}$ is too small, then perhaps $\overline{3}$ is what we need.

So now we're confronted with the question, why didn't we get $\overline{3}$ as the start of our solution when we simplified $\overline{4}\ \overline{5}$ with parts of 20? The problem is, there just isn't any nice number of parts that is a third of 20. In order to get a $\overline{3}$, we need the number of parts to be a multiple of 3. Hence we should have used 60 parts. Let's try it and see what happens.

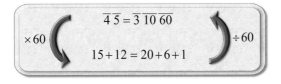

What we just did was essentially pick the first term in our simplification. We could have written the $\overline{3}$ down before multiplying by 60, getting $15 + 12 = 20 + ??$ on the bottom row. At this point we need to realize that we need 7 more, which must then be broken into nice parts of 60.

EXAMPLE: Simplify $\overline{10}\ \overline{11}$ so that the solution starts with $\overline{6}$.

For this problem, the obvious number of parts is 110. However, this is not easily split into six parts. We're missing a factor of 3, so we'll use 330 instead. Note that a sixth of 330 will be 55.

SOLUTION:

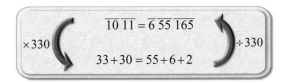

PRACTICE: Simplify $\overline{3}\ \overline{5}$ so the solution starts with $\overline{2}$.

ANSWER: $\overline{2}\ \overline{30}$

It's possible to mess up when choosing the first term. Let's try to simplify $\overline{10}\ \overline{20}$ so that it starts with $\overline{6}$ and see what happens.

EXAMPLE: Try to simplify $\overline{10}\ \overline{20}$ starting with $\overline{6}$. What goes wrong and why?

SOLUTION:

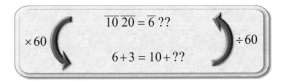

There's nothing (positive) we can add to 10 to get 9. Hence $\overline{6}$ is more than $\overline{10}\ \overline{20}$ and we can't add other fractions to $\overline{6}$ to make them equal.

PRACTICE: Try to simplify $\overline{8}\ \overline{32}$ starting with $\overline{5}$. What goes wrong and why?

Now that we know how to start a simplification with any fraction, we need to know how to choose this value. There seem to be three primary considerations: it should be a good approximation, share common factors, and be easy to break into parts.

Consider $\overline{13}\ \overline{15}$. It's obviously bad. It turns out that $\overline{7}$ is the best approximation for this sum. So we could try to start our simplification with $\overline{7}$. This will lead to unruly fractions.

$$\times 1365 \quad \overset{\displaystyle \overline{13}\ \overline{15} = \overline{7}\ \overline{1365}}{105+91 = 195+1} \quad \div 1365$$

There are problems with using the best approximations. It's unlikely that an Egyptian scribe would even be sure of what the best approximation was. If I were to describe to modern mathematicians how to find the best approximation for $\overline{13}\ \overline{15}$, I would tell them to take the harmonic mean of 13 and 15, divide by 2, and round up, giving the number to put under the fraction bar. If you don't understand what I just said, don't worry, an Egyptian scribe probably wouldn't understand it either. Even if you find the best approximation, it often leads to ugly answers because they tend to involve odd fractions and huge parts that make for tedious hand calculations.

An easy way to get a good estimate of the best approximation is to pick an even number somewhere from 13 to 15, making sure it's at least as close to the larger number as to the smaller. The closer it is to the middle, the better the approximation. Take half of this number and write it as a fraction. In our example the only even number in this range is 14, and half of it is 7, so $\overline{7}$ is our good approximation. In this case, it found the best.

EXAMPLE: Find a good approximation for $\overline{21}\ \overline{26}$.

SOLUTION: The even numbers between the two are 22 and 24. 24 is at least as close to 26 as 21, so half of 24 as a fraction, $\overline{12}$, is a decent approximation.

As you try the following practice problem, remember it's better to pick a number closer to the top than to the bottom.

PRACTICE: Find a good approximation for $\overline{33}\ \overline{37}$.

ANSWER: $\overline{18}$.

Once we find a good approximation, we need to find one that works well. Let's go back to $\overline{13}\ \overline{15}$. The fraction $\overline{7}$ makes a good approximation. If we make the number under the bar larger, the fraction gets smaller, which is fine, but the approximation gets worse and worse. We can't make the number smaller because it might get larger than the sum itself. As long as we stay above but close to the good approximation, we should be fine. So we need to consider fractions like $\overline{8}$, $\overline{9}$, and $\overline{10}$ instead of $\overline{7}$. As we will see, all of these are reasonable choices.

The choice of $\overline{8}$ is a good one because it's filled with factors that are easy to break into parts. The more 2s and 3s a number is composed of, the better, and 8 is composed of all 2s. So when we use $\overline{8}$ as our first fraction, the following fractions will be relatively easy to find as parts.

$$\times 1560 \quad \overset{\displaystyle \overline{13}\ \overline{15} = \overline{8}\ \overline{60}\ \overline{520}}{120+104 = 195+26+3} \quad \div 1560$$

It turns out that $\overline{8}$ is probably not the best choice. It does have some advantages: the 2s in the 8 make all the fractions in the identity even and hence easier to work with. But its main flaw is that the number of parts, 1560, is enormous. What if we used $\overline{10}$ instead? The reason it's a good choice is that there is already a 5 in the $\overline{15}$, so all we need is a 2 to complete the $\overline{10}$. This greatly reduces the number of parts and we still get a 2 in them.

$$\times 390 \quad \overset{\displaystyle \overline{13}\ \overline{15} = \overline{10}\ \overline{26}\ \overline{195}}{30+26 = 39+15+2} \quad \div 390$$

EXAMPLE: Consider the simplification of $\overline{20}\ \overline{23}$. Which is the best fraction to start with: $\overline{11}$, $\overline{12}$, $\overline{13}$, or $\overline{14}$? Explain and simplify with this fraction.

SOLUTION: $\overline{12}$ is the best since it has many small factors and shares a 4 with $\overline{20}$.

$$\times 1380 \quad \left(\begin{array}{c} \overline{20}\ \overline{23} = \overline{12}\ \overline{115}\ \overline{690} \\ 69 + 60 = 115 + 12 + 2 \end{array} \right) \div 1380$$

PRACTICE: Consider the simplification of $\overline{18}\ \overline{21}$. Which is the best fraction to start with: $\overline{11}$, $\overline{12}$, or $\overline{13}$? Explain and simplify with this fraction.

ANSWER: $\overline{12}\ \overline{54}\ \overline{756}$, $\overline{12}\ \overline{63}\ \overline{252}$, or $\overline{12}\ \overline{84}\ \overline{126}$

While I have to admit that simplifying fractions is perhaps the most tedious part of Egyptian mathematics, it generally needs to be done only at the end of a problem and only when the answer doesn't "feel" right. You could get by without reducing at all. Your answers would still be technically correct; however, there would be differences between your solutions and those of the ancient scribes.

EASY UNITS

Working with Parts

The units people work in tend to be fixed by practice and not by mathematicians. I'm sure at one time a foot was literally the length of a foot. Similarly a yard is the distance from your nose to the tip of your outstretched hand. The problem is that feet and arms vary in size, so sooner or later a civilization needs to standardize its units.

This is where the insight of the mathematician comes in. While it's possible to make a foot just the average foot length, it is foolish to do so because different units need to work together. My foot is about 10 inches long and my "yard" is about 36 inches. This means that I would need to multiply by 3.6 to convert from feet to yards. A good mathematician would realize that this is an overly difficult number to work with, so she would fudge the units, altering the lengths to make exactly 3 feet in a yard. Hence our foot of 12 inches is much larger than a typical human foot. Math needs to be easy to be practical, and easy numbers make for easy computations.

The Egyptians were clearly aware of this because all of their units were relatively easy to convert. We already know a palm is 4 fingers. To go from one to the other, all we need to do is double twice, or conversely, take half, and half again. A standard cubit is 6 palms. To convert palms into the standard cubit, we simply need to divide by 2 and then divide by 3. A khet is 100 cubits. As we've seen, multiplying and dividing by 10 usually requires only a change in symbol. For example, ⟍⟍ cubits is II khet.

The same ease of conversions also holds for volumes. A hekat is roughly equal to the modern gallon. The volume of a box a cubit on each side is equal to 30 hekat. Larger volumes were measured in double or quadruple hekat, which simply required doubling to convert. A khar is 20 hekat, just a doubling followed by a multiplication by 10. Converting from cubic cubits, for example, 30 hekat, into khar required multiplying by $\overline{\overline{3}}$.

Everything was easy until some pharaoh demanded that his royal cubit have one more palm than everyone else's. I imagine that he made this proclamation to two scribes, the first of whom declared that 7 palms in a royal cubit was no good since division by 7 was awkward. After the first scribe was decapitated, the second agreed that the royal cubit was a wonderful idea. I wonder how many scribes lost their heads before they agreed that there should be 5280 feet in a mile?

Remember that when performing conversions, the operations work "backward" from the table. For example, since 1 palm = 4 fingers, then 4 times the number of palms is the number of fingers.

The moral is that math needs to be easy to be useful. Any and all methods suggested by historians of mathematics should be relatively easy to perform given the knowledge and techniques of the ancients. There's a serious problem with the interpretation of Egyptian mathematics that I've put forward. In order to work with the fractions, we've needed to work in parts. This requires multiplications and divisions to convert back and forth between the original fractions and the parts. So doing a multiplication requires us to do more multiplications, and

1 palm = 4 fingers

1 common cubit = 6 palms

1 khet = 100 cubits

1 setat = 1 khet^2

1 setat = 10,000 cubits^2

1 double hekat = 2 hekat

1 quad. hekat = 4 hekat

1 cubit^3 = 30 hekat

1 khar = 20 hekat

1 khar = $\bar{\bar{3}}$ cubit^3

1 royal cubit = 7 palms

Unit conversions.

this process violates the principle that math needs to be easy to be practical. My goal in this section is to convince you that working in parts doesn't require nearly as much multiplication and division as you might think.

Let's consider the problem of adding $\bar{5}$ and $\bar{6}$. The obvious number of parts is 30, and the computation might look like this.

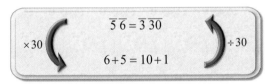

$$\bar{5}\ \bar{6} = \bar{3}\ \overline{30}$$
$$\times 30 \qquad \div 30$$
$$6 + 5 = 10 + 1$$

Notice that the first two terms of the top row, $\bar{5}$ and $\bar{6}$, and the first two of the bottom row, 6 and 5, seem interchanged. This is not a coincidence. In order to understand this, we need to realize that the 30 parts we picked is just 5×6.

When we multiplied $\bar{5}$ by the 30 parts to get 6, we really just multiplied $\bar{5}$ by 5×6. Algebraically we can think of the $\bar{5}$ and the 5 as cancelling each other. An Egyptian might think of it in the following way. When we multiply 5×6 by $\bar{5}$, we're really asking, what is a fifth of five groups of six? The answer is of course, one group of six. Similarly, when we multiply $\bar{6}$ by 5×6 to get 5, we are acknowledging that a sixth of six groups of five is one group of five.

Now consider adding $\overline{17}$ and $\overline{23}$. I don't know the value of 17×23, but it really doesn't matter. I know that $\overline{17}$ as parts of 17×23 is 23 and that $\overline{23}$ as parts of 17×23 is 17.

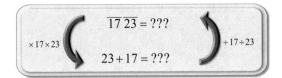

$$\overline{17}\ \overline{23} = ???$$
$$\times 17 \times 23 \qquad \div 17 \div 23$$
$$23 + 17 = ???$$

If we think in these terms, certain problems become simple. If I needed to know what $\bar{7}$ is as parts of 7×11, I'd answer 11, because the $\bar{7}$ and 7 cancel each other.

EXAMPLE: Fill in the bottom left of the following parts table.

$$\overline{15}\ \overline{21} = ???$$
$$\times 15 \times 21 \qquad \div 15 \div 21$$
$$\underline{\ \ } + \underline{\ \ } = ???$$

SOLUTION: $\overline{15}$ as parts of 15×21 is 21, and $\overline{21}$ as parts of 15×21 is 15, so we get

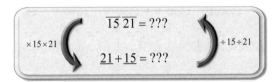

$$\overline{15}\ \overline{21} = ???$$
$$\times 15 \times 21 \qquad \div 15 \div 21$$
$$\underline{21} + \underline{15} = ???$$

PRACTICE: Fill in the bottom left of the following parts table.

$$\overline{27}\ \overline{31} = ???$$
$$\times 27 \times 31 \qquad \div 27 \div 31$$
$$\underline{\ \ } + \underline{\ \ } = ???$$

ANSWER: The bottom row is 31 + 27.

Sometimes we add extra parts. When we simplify $\overline{17}\ \overline{23}$, one strategy is to select the first fraction. A good number to select is $\overline{10}$, and this adds an extra factor of 10 into the mix. All we need to do is to multiply our parts by 10. So if $\overline{17}$ of 17×23 parts is 23, then $\overline{17}$ of $17 \times 23 \times 10$ is 23×10, or 230. While we still have to perform a multiplication, we need to remember that we selected the first fraction in part because it was an easy number to work with.

EXAMPLE: What is $\overline{7}$ in parts of $5 \times 7 \times 2$?

SOLUTION: $5 \times 2 = 10$.

PRACTICE: What is $\overline{5}$ as parts of $3 \times 5 \times 11$?

ANSWER: 33

Let's go back to our original example of adding $\overline{5}$ and $\overline{6}$. We turned them into parts of 6 and 5, respectively. The next step is to add these to get 11 and then break the 11 into parts. Here's the standard solution. The key for us to solve this problem is to break 11 into factors of 30. We know that both 10 and 1 are factors of 30 and that $11 = 10 + 1$, so we get the following solution:

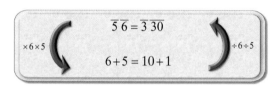

$$\times 6 \times 5 \quad \overline{5}\,\overline{6} = \overline{3}\,\overline{30} \quad \div 6 \div 5$$
$$6 + 5 = 10 + 1$$

The problem is that we know factors of many numbers because we've memorized multiplication tables. For example, since we know 4×7 is 28, we also know 4 and 7 are factors of 28. Egyptians would not realize this as readily as we do. However, I believe that the ancient scribes could easily have performed the above calculations in their head. I will now try to convince you that the above calculations are almost exactly like the Egyptians' standard multiplication method.

The trick is to "tip" the above table 90 degrees. In essence, the rows above become the columns of an Egyptian operation. Let's start with just the 6 and 5, which are trivially derived from the $\overline{5}$ and $\overline{6}$.

	?	11	
1)	$\overline{5}$	6	
2)	$\overline{6}$	5	$(1) \leftrightarrow$

Notice that the first column is just the original $\overline{5}$ and $\overline{6}$ from the problem. These numbers are found in the first row of our above parts calculation. The second column contains 5 and 6. These are just the parts from the second row above. So the first column represents the original numbers, while the second represents the corresponding parts. Note that we can think of the second row as a standard row switch. The $\overline{5}$ and 6 change places becoming $\overline{6}$ and 5.

We put 11 over the second column because it's the number of parts we need. While 5 and 6 work, it produces the bad approximation $\overline{6}\ \overline{5}$. We need to start with a number closer to 11 than 6 to get a better answer. A sharp scribe would realize that if he doubled the 5, he got 10, a number very close to 11.

	?	11	
1)	$\overline{5}$	6	
2)	$\overline{6}$	5	$(1) \leftrightarrow$
3)	$\overline{3}$	$10 \checkmark$	$(2) \times 2$

Since we now have 10 of 11 parts, there's only one more part to go. We can easily get 1 by dividing the 10 by 10. This gives us the solution of $\overline{3}\ \overline{30}$.

	?	11	
1)	$\overline{5}$	6	
2)	$\overline{6}$	5	$(1) \leftrightarrow$
3)	$\overline{3}$	$10 \checkmark$	$(2) \times 2$
4)	$\overline{30}$	$1 \checkmark$	$(3) \div 10$
	$\overline{3}\ \overline{30}$		

The significance of the above operation is clear only when you compare it to the following parts calculation. The first column above is $\overline{5}\ \overline{6}\ \overline{3}\ \overline{30}$ while the first row below is $\overline{5}\ \overline{6} = \overline{3}\ \overline{30}$, exactly the same numbers. Now compare the second column above with the second row below. Once again, they are the same.

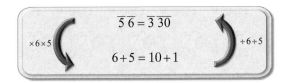

$$\overline{5}\,\overline{6} = \overline{3}\,\overline{30}$$

$\times 6 \times 5$

$$6 + 5 = 10 + 1$$

$\div 6 \div 5$

Let's look at another example. Consider adding $\overline{6}$ and $\overline{8}$. Let's use parts of 6×8. We could use the LCM, but there's really no need because the parts will be easy to compute. We know that the corresponding parts will be the terms reversed, 8 and 6, whose total is 14.

$$\overline{6}\,\overline{8} =$$

$\times 6 \times 8$

$$8 + 6 =$$

$\div 6 \div 8$

At this point, let's switch to an Egyptian-style computation. Note that the total number of parts we need is 14. We need a "nice" number of them close to but under 14. If we double 6 we get 12, which will do nicely. This leaves 2 parts. There are a number of ways to obtain 2. I'll do it by halving 8 twice.

	?	14	
1)	$\overline{6}$	8	
2)	$\overline{8}$	6	(1)$\leftrightarrow$
3)	$\overline{4}$	12 ✓	(2)×2
4)	$\overline{12}$	4	(1)÷2
5)	$\overline{24}$	2 ✓	(4)÷2
	$\overline{4}\,\overline{24}$		

Now let's put this into a parts-styled computation. Remember the columns above become the rows below. We leave some terms, like row 4 above, because they are not used in the identity.

$$\overline{6}\,\overline{8} = \overline{4}\,\overline{24}$$

$\times 6 \times 8$

$$8 + 6 = 12 + 2$$

$\div 6 \div 8$

The truth is we don't know how Egyptians did part calculations. Clearly when we use our knowledge of factors, we're taking liberties applying our modern methods. Yet it is encouraging that similar operations resemble Egyptian-styled computations. Here's one more example without all the explanation.

EXAMPLE: Add $\overline{6}\,\overline{10}$ as parts of 6×10. Use Egyptian multiplication tricks to figure out their parts and the parts of the final solution. Include a multiplication-like document to show their derivation.

$$\overline{6}\,\overline{10} = \overline{4}\,\overline{60}$$

$\times 6 \times 10$

$$10 + 6 = 15 + 1$$

$\div 6 \div 10$

	?	16	
1)	$\overline{6}$	10	
2)	$\overline{10}$	6	(1)$\leftrightarrow$
3)	$\overline{12}$	5	(1)÷2
4)	$\overline{4}$	15 ✓	(1)+(3)
5)	$\overline{60}$	1 ✓	(1)÷10
	$\overline{4}\,\overline{60}$		

Most of the operations are so simple that I doubt the Egyptians would have had the need to write such a document to compute their parts. Try this one on your own. Realize that to go from 20 parts to 4 parts you can divide by 10 to get 2 parts then double.

PRACTICE: Add $\overline{10}\,\overline{14}$ as parts of 10×14. Use Egyptian multiplication tricks to figure out their parts and the parts of the final solution. Include a multiplication-like document to show their derivation.

ANSWER: $\overline{7}\,\overline{35}$

Let's use these skills to find another way to construct Ahmose's doubling table. I'm going to focus on the doubling of odd primes because the factor method I introduced earlier in the book doesn't produce nice answers. Let's start with $2 \times \overline{7}$. We can think of this as a simplification problem by rewriting it as $\overline{7}\ \overline{7}$. There are no evens between 7 and 7, but remember that larger numbers also work, so let's choose 8. Half of 8 is 4, which is a nice number as well as a good approximation. So let's write $\overline{7}\ \overline{7}$ as $\overline{4}$ and something else.

To do this let's work in parts of 7×4. We immediately know that the corresponding parts of $\overline{7}$ and $\overline{4}$ are 4 and 7, respectively. This gives us the following partial solution.

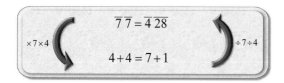

$$\overline{7}\,\overline{7} = \overline{4}\ ?$$
$$4 + 4 = 7 + ?$$
$$\times 7 \times 4 \qquad \div 7 \div 4$$

Clearly we need one more part on the right, and we can calculate it by dividing the 4 parts by 2 twice. The calculations and final answer appear as follows:

	?	8	
1)	$\overline{7}$	4	
2)	$\overline{4}$	7 ✓	(1) ↔
3)	$\overline{14}$	2	(1) ÷ 2
4)	$\overline{28}$	1 ✓	(3) ÷ 2
	$\overline{4}\ \overline{28}$		

$$\overline{7}\,\overline{7} = \overline{4}\ \overline{28}$$
$$4 + 4 = 7 + 1$$
$$\times 7 \times 4 \qquad \div 7 \div 4$$

Let's try doubling $\overline{37}$. Half of 38 is 19, a lousy number because it's odd and has no whole parts except for 1 and itself. The number 20 isn't a bad choice since it is a good approximation and has parts comprising 2s and a 5. However, the author of the scroll used 24. It's a better number because it is all 2s and 3s, although it's a slightly worse approximation. Here's the problem as an example.

EXAMPLE: Construct Ahmose's doubling of $\overline{37}$ starting with the fraction $\overline{24}$. Do all parts as a table.

SOLUTION:

	?	48	
1)	$\overline{37}$	24	
2)	$\overline{24}$	37 ✓	(1) ↔
3)	$\overline{74}$	12	(1) ÷ 2
4)	$\overline{111}$	8 ✓	(3) × $\overline{\overline{3}}$
5)	$\overline{148}$	6	(3) ÷ 2
6)	$\overline{296}$	3 ✓	(5) ÷ 2
	$\overline{24}\ \overline{111}\ \overline{296}$		

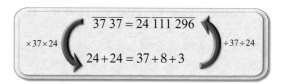

$$\overline{37}\,\overline{37} = \overline{24}\ \overline{111}\ \overline{296}$$
$$24 + 24 = 37 + 8 + 3$$
$$\times 37 \times 24 \qquad \div 37 \div 24$$

Both of the preceding answers are exactly the same as the entries in Ahmose's table. Try these two practice problems.

PRACTICE: Construct Ahmose's doubling of $\overline{11}$ starting with the fraction $\overline{6}$. Do all parts as a table.

ANSWER: $\overline{6}\ \overline{66}$

PRACTICE: Repeat for $\overline{13}$ starting with an $\overline{8}$.

ANSWER: $\overline{8}\ \overline{52}\ \overline{104}$

In the above "doubling" problems we selected $\overline{4}$, $\overline{24}$, $\overline{6}$, and $\overline{8}$ as the starting fractions of our solution. These are all extremely easy numbers to work with in Egyptian math since they all contain 2s and 3s. These choices made the breakup of the remainder into parts much easier. While it's impossible to know the exact method used to create the Ahmose's table, he seems to have followed exactly the same strategy. Let's look at the entries on the doubling table and focus on the prime fractions. Let's ignore $\overline{3}$, since doubling it is just $\overline{3}$. The next 22 prime fractions when doubled start with the fractions $\overline{3}$, $\overline{4}$, $\overline{6}$, $\overline{8}$, $\overline{12}$, $\overline{20}$, $\overline{24}$, $\overline{30}$, $\overline{36}$, $\overline{40}$, $\overline{42}$, and $\overline{60}$. Many occur a number of times, of course. With the exception of 42, which appears only once, all of the other numbers are composed of only 2s, 3s, and 5s. Even the 5s occur only if there are at least two 2s or 3s. This is probably not pure chance. It appears as if the creators of the table were intentionally selecting numbers that were easy to work with, and that should hardly come as a surprise.

REMAPPING A NATION

Four Lines or Fewer

You can see most of the stars visible at your latitude at some time in the night. While at any given moment, only half the sky is visible, the celestial sphere rotates, continually "lifting" new stars into the heavens. At a given time of year, some stars are simply too close to the sun and are hidden by the glare. However, the sun moves around the sphere of stars once a year and will, in a couple of months, move away, allowing the star to reappear in the eastern sky minutes before the dawn. The first such appearance of a star is called its *helical rising*.

Sopdet, now called Sirius, is by far, the brightest star in the sky. The Egyptian name translates into something like "she who is sharp," and her helical rising marked the beginning of the Egyptian year. It was a particularly important time for the residents of the Nile valley since by this day all the canals had to be dredged and the dikes repaired in preparation for the coming flood.

Nilometers, the device Egyptians use to measure the depth of the Nile, were continually checked. At the appropriate time, canal gates were opened and anything valuable had to be moved to higher ground. The rising waters were directed by canals into every region of the valley. When the canals were full, the gates were closed and the water was held captive by dikes that partitioned the arable land of Egypt. The fertile silt would settle to the bottom of the standing water, reinvigorating the soil. The water would slowly be absorbed into the ground, waiting to be taken up by the crops. Finally the canal gates would be reopened, draining any excess water back into the receding Nile. At this point the soil was ready for the wheat, barley, and flax that the Egyptians needed for their sustenance and clothing, but there was one problem.

The force of the water and the new layer of silt could push away or cover the markers or landmarks that distinguished one Egyptian's property from another's. Even worse, the very course of the Nile could easily change, making old boundaries obsolete. Every year the educated, math-literate class of Egyptians had to engage in a massive land survey, redrawing the property lines of an entire nation.

A farmer would want a plot equal to the one he had in previous years. It was often difficult to guarantee exactly the same stretch of land. However, if the farmer was given the same amount of land of the same quality as his old plot, it's unlikely that there would be an objection. Hence the area of a plot was its vital feature. As a result, many of the area problems that come down to us from ancient Egypt are phrased as determining the size of fields. A common unit of area used in ancient times was a *cubit strip*, which was equal to 100 square cubits. While this is equal to the area of a square 10 cubits on a side, it is more properly thought of as an area of a 1-cubit-by-100-cubit rectangle. If a row of fields were 100 cubits long, you could then find the area in cubit strips simply by measuring the width in cubits.

The area in cubit strips of a field 100 cubits deep is the width.

When we calculate areas today using familiar equations, such as "one-half base times height" for triangles, our answers come in simple units. If our lengths are measured in inches, the formulas give us areas in square inches. Similarly using measurements in feet produces solutions in square feet, and so on. A cubit strip is not such a unit. Egyptians typically measured such lengths in cubits, so standard modern equations will produce areas in square cubits not cubit strips. Hence, we need a way to convert square cubits into cubit strips, just as a modern surveyor might need a way to convert square feet into acres.

Using base times height, we can calculate the area of the above 23-cubit-strip strip field to be 100×23, or 2300 square cubits. So 23 cubit strips is equal to 2300 square cubits. Hopefully you realize that we can convert from cubit strips into square cubits by multiplying by 100, a very easy operation. Similarly square cubits can be converted into cubit strips by dividing by 100.

We've already seen how to calculate the area of a rectangle and a triangle. However, one of the field shapes discussed in Egyptian mathematics is a *truncated triangle*, known to mathematicians as a *trapezoid*. Humans like order. They want their fields to be nice rectangles all in a row. However nature doesn't always comply. The walls of the Nile valley are not straight, and the river meanders in a seemingly random path within the confines of the canyon walls. A field cut on one side by such a barrier closely resembles a trapezoid.

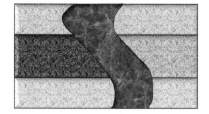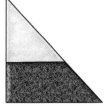

A rectangular field cut by a river forms a *truncated triangle*.

Fortunately, as the Egyptians knew, these areas are easy to compute. Instead of multiplying the height by the base, you simply multiply the height by the average base.

Consider the following trapezoid with a height of 20 and bases of 4 and 6. The average base is just the sum of the two bases divided by 2.

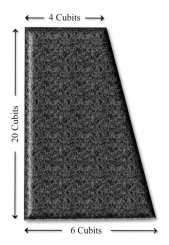

A trapezoid with a height 20 and bases of 6 and 4.

EXAMPLE: Find the area of a trapezoid with a height of 20 and bases of 6 and 4. Give the answer in square cubits and cubit strips.

SOLUTION: Add the bases to get 10. Take half to get 5, the average base. The area is average base times the height, so multiply 5 by the height of 20, to get 100:

5	?
1	20 ✓
2	40
4	80 ✓
	100

The above answer, 100, is in square cubits. Since the Egyptians measured fields in cubit strips, we need to convert. Multiply by $\overline{100}$, the number of cubit strips in a square cubit, to get the answer, 1, in cubit strips.

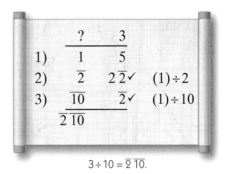

$$3 \div 10 = \overline{2}\,\overline{10}.$$

PRACTICE: Find the area of a trapezoid with a height of 9 and bases of 10 and 13. Give the answer in cubits and cubit strips.

ANSWERS: $103\,\overline{2}$ cubits and $1\,\overline{40}\,\overline{100}$ cubit strips

The area problem I want to consider is actually much simpler than finding the area of a trapezoid. The problem simply states that ten fields have an area of seven *setat*, a unit of 10,000 square cubits. What's the area of each? The problem is solved by simply calculating $7 \div 10$. The division is performed as follows:

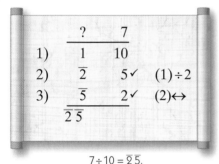

$$7 \div 10 = \overline{2}\,\overline{5}.$$

Although it's not a complicated solution, it's short and sweet and doesn't follow the usual pattern. Immediately after this problem is the same type of problem with different numbers.

EXAMPLE: Five fields have an area of 3. What's the area of each?

Both solutions get their answers in just a few lines and the computations seem unsystematic. It is the nature of Egyptian mathematics to offer surprising solutions, giving the mathematician freedom to compute as she pleases. In the hands of a skilled practitioner, this can result in unexpectedly short computations.

While it's important to know what the Egyptians knew, it's as important to think like they did. Their system of computation gave them great flexibility, and knowing how to exploit this is what gives their mathematics such power. Consider the options you have when going from one line in a multiplication to another. You can double, halve, take the $\overline{\overline{3}}$, divide by 10, sometimes multiply by 10, and switch the terms. You don't even have to do this with the previous line, but rather, you can use any line above it. When you include the option of adding any pair of lines together, the number of choices you're confronted with is enormous.

Egyptian math presents us with a series of choices, and to make choices you have to think. Modern math gives us rules and steps. Performing its operations requires rote memorization, repetition, and perseverance. It may be tedious and boring, but if you follow the instructions to the letter, eventually you will be rewarded with the correct answer. This is the exact opposite of Egyptian math. To the uninitiated it often seems impossible, but that's because they don't know which choices to make and continually make the wrong ones. An Egyptian computation can take untold pages or a handful of lines depending upon the skill of a mathematician.

Let's consider the example of $35 \div 96$. Someone only mildly familiar with the Egyptian method would set up the table and begin mindlessly taking half until they're about to hit a fraction. Here's how it might look.

?	35
1	96
$\overline{2}$	48
$\overline{4}$	24✓
$\overline{8}$	12
$\overline{16}$	6✓
$\overline{32}$	3✓

After the amateur mathematician checks off the appropriate numbers, he realizes he still needs 2 to get to 35. Mindlessly he switches the top row and then doubles it. The continuation might look like this.

?	35
…	…
$\overline{96}$	1
$\overline{48}$	2✓
$\overline{4}\ \overline{16}\ \overline{32}\ \overline{48}$	

After he gets the answer of $\overline{4}\ \overline{16}\ \overline{32}\ \overline{48}$, his teacher glares at him, and he begins to simplify, repeatedly applying the G rule.

$$\overline{4}\ \overline{32}\ (\overline{16}\ \overline{48})$$
$$= (\overline{4}\ \overline{12})\ \overline{32}$$
$$= \overline{3}\ \overline{32}$$

Now that he's done, he declares Egyptian mathematics to be awkward and pointlessly tedious. But then, to his horror he glances at the work of the student sitting next to him and he sees this.

?	35
1	96
$\overline{2}$	48
$\overline{3}$	32✓
$\overline{32}$	3✓
$\overline{3}\ \overline{32}$	

In a mere four lines with no simplification, she got the same answer as he did. How did she know to take the $\overline{\overline{3}}$ of the second line and then to switch the third line? She must have got lucky, he thinks, but he doesn't realize that she has the insight. Someone with insight sees the number 15 and immediately then gets the numbers 30, $7\ \overline{2}$, 10, 150, $\overline{15}$, and $1\ \overline{2}$ in her head, which are the double, half, $\overline{\overline{3}}$, times 10, switch, and tenth of 15. Not only that, but she will instinctively know how these numbers relate to other numbers in the problem. She saw 48 and immediately thought of its $\overline{\overline{3}}$, 32, and knew it was close to 35. In fact, it was the switch of $\overline{3}$ away.

Thinking this way is at first difficult, but the brain adapts. Just as riding a bicycle seems to go from impossible to easy overnight, so does Egyptian math. It just takes practice, and that's exactly what this section is for. In the following problems I will give you a starting line and a goal. The goal number has to appear in the right column in four lines or less. Let's look at an example.

EXAMPLE: Starting with $\overline{13}$, 20, create the number 150 on the right in four lines or less.

SOLUTION:

1)	$\overline{13}$	20	
2)	$\overline{130}$	2	$(1) \div 10$
3)	$\overline{2}$	130	$(2) \leftrightarrow$
4)	$\overline{2}\ \overline{13}$	150	$(1)+(3)$

I like the four-lines-or-less exercise. The numbers you need to create are just far enough away that the solutions are not obvious, but they are close enough that a little trial and error will usually find them. The exercises I'm providing here are the same questions I give my students in an hour-long math lab. Working in groups, they can usually find all the solutions in the time allotted. I can do it in a few minutes. I'm not trying to brag but rather convince you that anyone with practice will be able to do this easily.

Each of the problems can be solved with three applications of the following operations:

- Double
- Halve
- Multiply by $\overline{\overline{3}}$
- Divide by 10
- Switch a line
- Add any number of lines

In special circumstances, Egyptians multiplied by 10. I'm going to disallow this because it doesn't always work. Also remember that you can't switch lines that contain fractions consisting of more than one term, such as $\overline{3}\ \overline{10}$ or $2\ \overline{7}$. Good luck.

PRACTICE: Starting with the given rows, create the last number on the right in four lines or less.

- 11, 90 to 3
- 1, 45 to 31
- 11, 18 to 45
- 12, 9 to $\overline{36}$
- $\overline{4}$, 50 to 58
- 4, 150 to 5
- $\overline{6}$, 80 to 89
- 10, 36 to 48
- $\overline{6}$, 40 to 100
- 20, 12 to $\overline{45}$

By now you should realize that solving these problems is like solving puzzles. It requires more wit than repetition and patience. You're now beginning to understand the true nature of Egyptian computation.

WHAT IS AN NB.T?

Completions

The tenth problem of the Moscow Mathematical Papyrus asks the reader to determine the area of an *nb.t*. Unfortunately this word never appears anywhere else, and hence it's untranslatable. It's similar to the word for "hill," which gives us some clue. The problem also states that the nb.t has a diameter of $4\ \overline{2}$, which suggests that some part of it is circular. Further confusing matters, it also has a "[?]" followed by the Egyptian equivalent of *d*. The "[?]" means that the word is unreadable, possibly damaged. The *d* is untranslatable without some context. There are many possible shapes that loosely resemble a hill and have a radius. The math doesn't help much. Any area involving circles can only be approximated, so any shape with roughly the area calculated in the exercise is a candidate.

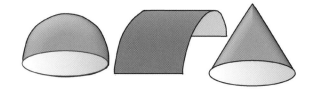

Possible shapes of an nb.t.

We may never know what an nb.t is, but we can examine the mathematical calculations. Part of the procedure of finding this unknown figure's area has the student remove a ninth from 8. Taking $\overline{9}$ of a number is relatively straightforward.

	$\overline{9}$	?	
1)	1	8	
2)	$\overline{2}$	4	$(1) \div 2$
3)	$\overline{3}$	$2\ \overline{\overline{3}}$	$(2) \times \overline{\overline{3}}$
4)	$\overline{6}$	$1\ \overline{3}$	$(3) \div 2$
5)	$\overline{9}$	$\overline{3}\ \overline{6}\ \overline{18}$ ✓	$(4) \times \overline{\overline{3}}$
		$\overline{\overline{3}}\ \overline{6}\ \overline{18}$	

We now need to remove this ninth from 8. The answer given is a simple $7\,\bar{9}$, but how did they get it? We've never subtracted fractions. Adding is easy. You put the two answers side by side and simplify if it suits you, but pulling $\bar{3}\,\bar{6}\,\bar{18}$ out of a whole number seems difficult at first. But just as division is like multiplication, Egyptian subtraction is much like addition.

Ancient civilizations thought in positive terms, not negative, and hence avoided inverse operations. When they phrased inverse operations, they usually stated them in terms of the standard operations. So you usually don't ask, "What is 8 divided by 2?" but rather, "What do I multiply 2 by to get 8?" You don't take the square root of 25, but rather ask, "What do I multiply by itself to get 25?" The ancient Egyptians would not have removed 4 from 7 in a subtraction but rather completed 4 to 7.

Completions on whole numbers were considered trivial in Ahmose's text on fractions, but the papyrus does give examples of the subtraction of fractions. True to the ways of thinking in the ancient world, it's treated as an addition problem with an unknown. Fortunately it's done with a method very similar to our computation of parts. It's simple enough that it's best to learn it by examining an example or two.

EXAMPLE: Complete $\bar{4}\,\overline{20}$ to $\bar{2}$.

To solve this we are going to literally express this as $\bar{4}\,\overline{20}$ plus what, is $\bar{2}$. I'm going to express this in the parts diagram that we're all familiar with.

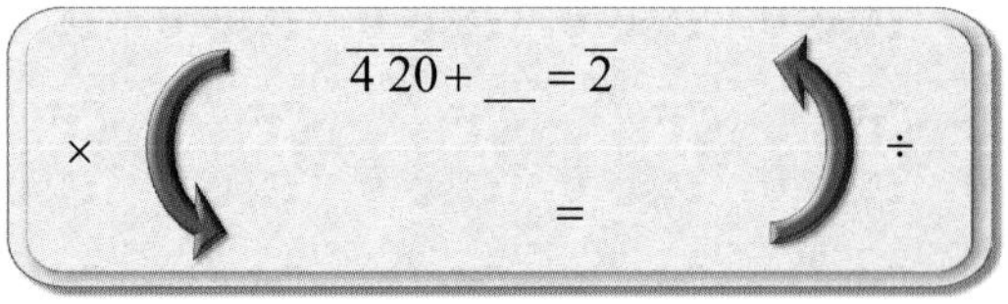

We now pick a good number of parts for these fractions. An obvious choice here is 20.

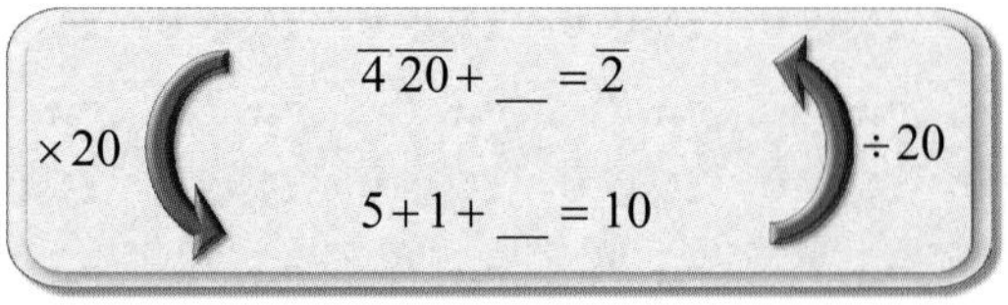

Looking at the problem on the bottom, you should realize that we're essentially asking what we should add to 6 to get 10. The answer is 4, and we fill it in.

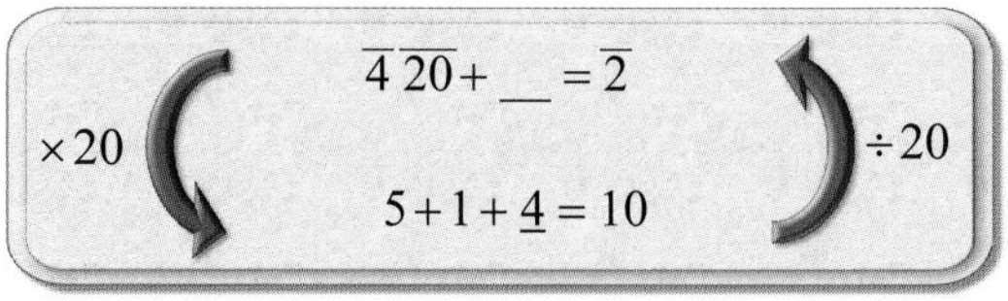

The only thing left to do is take 4 back to the top line by dividing it by 20, giving $\bar{5}$. The final solution appears as follows:

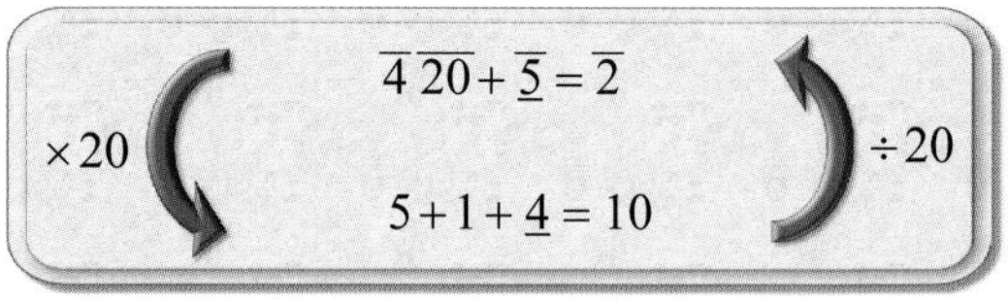

SOLUTION:

Note that we've essentially shown that $\bar{2}$ minus $\bar{4}\,\overline{20}$ is $\bar{5}$.

EXAMPLE: Complete $\bar{3}\,\bar{6}$ to $\bar{\bar{3}}\,\overline{15}$.

Once again we try to find an easy number of parts to work with. In this problem, a good number of parts is 30. Multiplying by 30 we get the following:

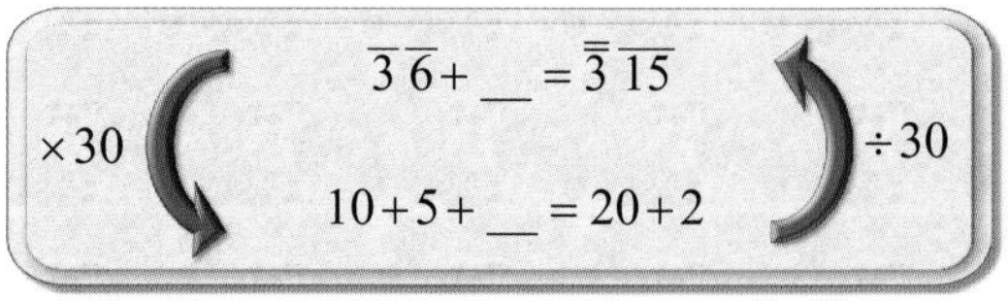

We have 15 on the left and 22 on the right, so the left needs 7 more. Unfortunately 7 is not a nice part of 30. Hence we need to break it into pieces. The obvious choice is $6+1$. When we divide these by 30 to get the fractions, the solution looks like this.

SOLUTION:

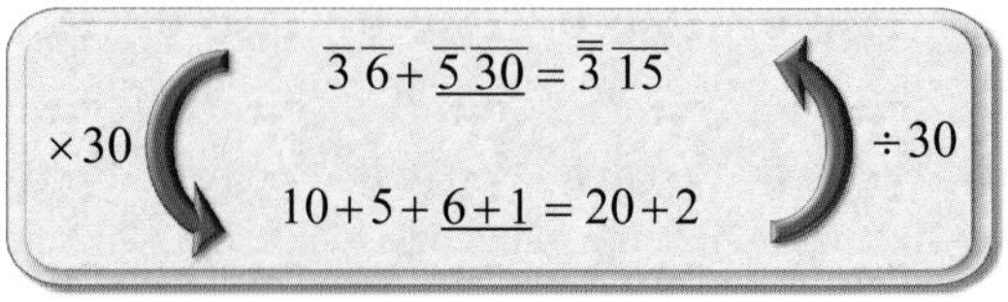

Hence $(\overline{\overline{3}}\,\overline{15}) - (\overline{3}\,\overline{6})$ is $\overline{5}\,\overline{30}$. Try these.

PRACTICE: Complete $\overline{4}\,\overline{5}$ to $\overline{2}$.

ANSWER: $\overline{20}$

PRACTICE: Complete $\overline{11}\,\overline{66}$ to $\overline{6}$.

ANSWER: $\overline{22}\,\overline{66}$

Sometimes we don't have enough parts. For example, if we were to try to complete $\overline{7}$ to 1, an obvious but wrong choice for parts is 7. We would be confronted with $1 + \underline{\quad} = 7$. Of course a 6 could fill in the blank, but the problem is that 7 has only 1 and 7 as parts, and there's no good way to make 6. A good choice for parts is 28, a number the Egyptians used frequently when dealing with sevenths. A better choice of parts is 42. In order to understand why, we need to remember that Egyptians often estimated answers before calculating them. The fraction $\overline{7}$ is near 0, and we're completing it to 1. This fraction is huge, and an Egyptian would have realized that it's going to be at least $\overline{\overline{3}}$. So we need a 3 in the number of parts. If we throw in a 2 for good measure, the number of parts is $7 \times 3 \times 2$, or 42. We can put the $\overline{\overline{3}}$ in the answer ahead of time, giving the following:

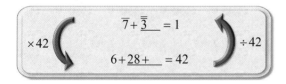

At this point we see that we're 8 shy of 42. Since 8 is easily broken into 7 and 1 we get this solution.

EXAMPLE: Complete $\overline{7}$ to 1.

When computing parts, it never hurts to throw in a 2 or two, especially when the "obvious" number of parts is odd. In the following practice problem, the obvious

number of parts is $3 \times 5 = 15$. Notice that it's odd. If we put a 2 in it, this will give us more flexibility in selecting parts. So, instead of 15, use $3 \times 5 \times 2 = 30$ parts.

PRACTICE: Complete $\overline{3}\,\overline{5}$ to 1.

ANSWER: $\overline{3}\,\overline{10}\,\overline{30}$

Sometimes you will know that the answer will be a large fraction very close to 1. The only single fraction near 1 is $\overline{\overline{3}}$. In these cases it helps to throw a 3 into the parts if it is not already there. Otherwise you will never get thirds.

EXAMPLE: Complete $\overline{14}$ to 1.

Here the obvious choice for parts is 14. However $\overline{14}$ is tiny, and hence the distance to 1 is large. Therefore we should use $3 \times 14 = 42$ for the parts. Normally we think of the "nice" parts of 42 as its factors. But remember that 28 is $\overline{\overline{3}} \times 42$, so 28 also is a part of 42. The solution is unpleasantly long, but this often happens to fractions near 1.

SOLUTION:

$$\overline{14} + \overline{\overline{3}} + \overline{6} + \overline{14} + \overline{42} = 1$$
$$3 + 28 + 7 + 3 + 1 = 42$$

The following problem uses the same trick. You need to add a 3 to get $\overline{\overline{3}}$ into your answer. Since 3×11 is odd, adding a 2 for good measure gives $3 \times 11 \times 2 = 66$ parts. Remember that 44 is $\overline{\overline{3}}$ of 66 and hence is a valid part.

PRACTICE: Complete $\overline{11}$ to 1.

ANSWER: $\overline{\overline{3}}\,\overline{6}\,\overline{22}\,\overline{33}$

Here are a few final tricks. If we want to complete $4\,\overline{3}\,\overline{8}$ to $5\,\overline{8}\,\overline{20}$, we can cancel from both sides. Hence we could remove 4 from both sides and completely eliminate the $\overline{8}$. Hence the solution would be equivalent to the answer to "complete $\overline{3}$ to $1\,\overline{20}$. Finally, there's no need to use this method to complete whole numbers. So if we were asked to complete $\overline{3}\,\overline{10}$ to 8, we should complete it to 1. Then we add 7 to our fractional answer to go from 1 to 8.

PRACTICE: Complete $8\,\overline{2}\,\overline{10}$ to $12\,\overline{2}\,\overline{5}$.

ANSWER: 4 $\overline{10}$

DON'T DISTURB MY CIRCLES

Finding the Area of a Circle

More than two thousand years ago the mathematician and engineer Archimedes was contemplating a diagram of circles he had drawn on the floor. The aging Sicilian was apparently unaware of the commotion outside. Archimedes' king had made the mistake of talking with the Carthaginians, the mortal enemies of the Romans. For this offense, the vengeful and violent leaders of Rome decided that the inhabitants of Syracuse must die. For three years the war machines invented by Archimedes held off the Roman legions, who simply camped outside the city walls waiting. Finally, after a festival filled with drunken revelry, Roman spies opened the city gates and the slaughter began. When a Roman soldier entered Archimedes' room, the mathematician told the warrior to not disturb his diagrams. For this, Archimedes received a spear plunged through his chest. Thus died a great, perhaps the greatest, mathematician.

Archimedes was a favored son of Syracuse. As the father of mechanics, the numerization of the physical laws of the universe, he created machines of unprecedented power. Those for war had kept the most powerful army in the world at bay for years. At the same time, he paralleled his mechanical creations with pure mathematics. He used math to justify his new laws, and hypothetical machines as examples to discover new laws in math. He was a mathematician literally thousands of years ahead of his time.

Archimedes is credited for finding, with proof, the area and volumes of many shapes. In particular, he found the area of a circle and the volume and surface area of a sphere, all of the equations involving the elusive constant π. The proofs he provided were done with a method that the Greeks called *exhaustion*. It is a complicated but rote method of proving the measure of a figure. The main problem with the method of exhaustion is that it's preferable to know the answer before you start. In other words, it's better at verifying a guess rather than finding a solution.

In the ancient world, mathematicians didn't know that the π involved in the circumference of a circle was related to the constant involved in other aspects of circles and spheres. Archimedes had to know the relation between the two constants before he applied exhaustion, and he explained how he knew the answer in his masterful book, *The Method*. This book had been long forgotten but was rediscovered at the dawn of the twentieth century. A copy of the text had survived to the modern day. Unfortunately, it had been erased and overwritten with a medieval religious text. For a long time no one knew of the treasure hidden under the ink of the holy tome. Upon its discovery, scholars were able to read the remnants of most of the barely visible, original text.

The argument Archimedes gives for the area of a circle is both imaginative and beautiful in its simplicity. Imagine a circular rug made of concentric strands in loops. Cut the

The circular rug cut to the middle.

rug in a straight line from somewhere on the outside to the center.

Each circular loop is now a length of rope that can be straightened out. The resulting figure is basically a triangle. This is a result of the linear relation of curves, something Archimedes would have known well. In simple terms, if you go twice as far from the center of the circle, the loop gets twice as long. Similarly, if you go twice as far

When straightened out, the loops form a triangle.

down from the peak of a triangle, the corresponding line gets twice as long.

The height of the triangle is just the radius of the old circle. If you don't see this, just count the bands in both diagrams. Both the height of the triangle and the radius of the circle in the above diagrams consist of six bands of equal width. The base of the triangle is just the outside loop, whose length is the circumference of the circle. The area of a triangle is just ½ base × height. If we replace "base" with the modern equation for the circumference, $2\pi r$, and r for the height, where r is just a name for the radius, then the area of the triangle becomes $\frac{1}{2}\,2\pi r \times r$. This simplifies to the equation that is familiar to many math students, πr^2.

This abstract cutting of a figure into thin pieces is often called *slicing*. Archimedes used slicing arguments to find many things, like the volume and surface area of a sphere and the volume of a paraboloid. He used such arguments to find centers of mass. His understanding of center of mass led him to the law of the lever and pulley, which in turn aided his construction of the war machines that defended Syracuse from the Romans. He also used the notion of the center of mass to develop his theory of hydrostatics, with which he determined whether an object would float, how far it would sink in the water, and whether it would be stable or tip over. Perhaps his understanding of the center of mass led him to conclude, almost two thousand years before Copernicus, that the sun, not the earth, was the center of the solar system.

Just like Leibniz, Newton, and others from the Age of Reason would do, Archimedes thought of shapes as consisting of the sum of infinitely thin slices. Just like they would, he applied these ideas to find areas, volumes, slopes, and centers of mass. Just like they would, he applied his methods and conclusions to the physical world. Archimedes died a few steps away from the invention of calculus The only real difference is that he determined his numerical solutions by comparative arguments rather than by algebraic rules such as Leibnitz's $\mathrm{d}x^2 = 2x\mathrm{d}x$. In my opinion, at the time of his death, Archimedes had brought mathematics within a generation of the invention of calculus.

So why did it take almost two thousand years for these ideas to reappear? Some blame the Romans, who had no time for the intellectual musings of those of Greek descent. While partly true, the great intellectual center of Alexandria in Egypt would remain free of Roman control for another two hundred years, which was plenty of time to develop calculus. Some blame the Greek preference for the theoretical over the practical. This explanation again is partly true, but there are many exceptions, such as their laws of reflection and refraction. It also assumes there is no theoretical appeal to the notions of calculus.

I, being completely biased and never missing a good opportunity for self-righteous moralizing, have my own opinion: Archimedes didn't follow the rules. The method of exhaustion had clearly defined steps, each of which had clearly defined justifications. Greek mathematicians probably balked at the notion of adding infinitely thin items, a method that at the time had no precise theoretical justification. By supplying exhaustion proofs as well as slicing proofs, Archimedes allowed the intellectually fettered to stay within their comfort zone.

The newborn calculus was presented to the mathematical establishment to receive its judgment. They looked it over and applied the standards for what used to be mathematics, thereby hindering the growth of the new mathematics. They examined the baby calculus and in essence declared, "It can't hold a job, drive a car, or even use a toilet!" But people like Archimedes, Newton, and Leibniz would look at the child and could see it for what it could become. They had the imagination to see its potential worth and not be chained by the existing standards and procedures of the so-called intellectual world. The old rules simply didn't apply, and those predisposed to following rules just couldn't understand. Even in the Age of Reason, two thousand years later, the ideas of calculus had many detractors. Fortunately Newton and Leibnitz were able to win the debate that Archimedes had lost.

Because Archimedes was able to relate the area of a circle to the area of a triangle whose base was the circumference of the circle, he was able to connect the constant of π used in the circumference equation with constants used to compute area. Unfortunately for the author of Ahmose's papyrus, this understanding was well over one thousand years into his future.

As we've seen, some in the ancient world knew that if they multiplied the diameter of a circle by $3\,\overline{8}$, they would obtain a number close in length to the circumference of a circle. Moderns would interpret this as the equation Circ. = π × Diam. We would also obtain the area of a circle by computing π × Radius². Note the Radius² in the equation. It has been long known that areas are related to the squares of numbers. For the actual shape of a square, a simple square is taken. For other shapes, the square needs to be multiplied by some constant, like ½ in the formula for the area of a triangle or π in the formula for the area of a circle. In theory, it doesn't matter if you multiply before or after you square the number; however, which you choose will change the value of the constant.

Consider the following example. Imagine the zigzag shape pictured below where the two long sides are twice as long as the short sides. You could find the area of this figure by multiplying the short side by 2 and then squaring it. So if the short edges were 5 inches long, we can find

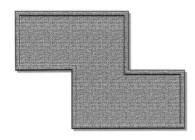

The zigzag with the top and bottom lengths exactly twice as long as the other six.

the area by doubling it to get 10 and then squaring it to get 100.

However, we could just as easily have squared 5 first to get 25 and then multiplied this result by the new constant, 4, to get 100. Notice when we multiplied before squaring the constant was 2 and when we multiplied after, it was 4. Both methods will always give the same answer. When computing the area of a circle, we're confronted with the same choice.

We moderns use the equation π*r*². In other words, we square the *r* first and then multiply by the constant π. The Egyptians chose the other route, multiplying first and then squaring. As a result, their constant is not as recognizably similar to our π as is their 3 8̄. They choose the constant 8/9, used this to multiply the diameter, and then

squared the result. You can obtain an approximation for our π by taking 8/9, doubling it, and then squaring it to get 3 9̄ 27̄ 81̄. It's easy to see how anyone without Archimedes' insight would miss the relation between the approximations 8/9 and 3 8̄. Would anyone even realize that 8/9 is actually an approximation for half of the square root of π? Archimedes' discovery that these numbers are in fact related is far from obvious.

There have been many attempts to determine where the approximation of 8/9 came from. I suspect it came from empirical measurements. In other words, they got the value from practical experience. For example, some scribe may have kept a record of how many tiles it took to cover some circular floor. After a minute or two of computation, it would become apparent that it was about the same number of tiles required to cover a square floor of with a side of eight-ninths of the diameter. It turns out that other fractions close to the actual value have denominators that are difficult to work with in Egyptian math or are excessively large. So even though this is just an estimate, it's not surprising that they settled on this value.

The Egyptians inadvertently made a huge mistake by choosing to multiply before squaring. The computations are just extremely complex and lengthy. The reason for this lies in the fraction 8/9 itself. Recall that in chapter 2 we added to the standard list of fractions because we needed reasonable approximations to numbers between 2̄ and 1. So we added one tick mark to our ruler in this region. For small fractions, the ruler is dense with tick marks. It's easy to approximate fractions in this range. However, near 1 the gaps are relatively huge, even with the addition of 3̄̄. The closer you are to 1, the worse the problem is. Unfortunately 8/9 is very close to 1. From personal experience, I can tell you that Egyptian math just becomes more difficult in this region.

This section, "Finding the Area of a Circle," is somehow out of place. I'm not going to teach you a new technique, but rather give you problems to solve that incorporate many of the ideas from this chapter in a few good problems. (By the way, you should always run in fear when a mathematician refers to an exercise as "a good one.") These problems will give you practice and in the name of fairness to modern math show you that Egyptian math doesn't always work out easily.

On its surface, the Egyptian method for finding the area of a circle is trivial. You first remove a ninth of the diameter using completion and then square the result. Let's start with a problem that won't give us any trouble. There are trickier, faster ways to solve it, but I want to do it in a way that won't change significantly when we pick harder numbers. Just so the method is clear, here are the steps the Egyptians used to estimate the area of a circle:

- Multiply $\overline{9}$ by the diameter.
- Subtract the answer from the diameter.
- Square the new answer.

EXAMPLE: Use the steps above to estimate the area of a circle with a diameter of 18.

SOLUTION: First calculate $\overline{9}$ of 18 to remove $\overline{9}$ from the diameter.

$\overline{9}$	?
1	18
$\overline{2}$	9
$\overline{3}$	6
$\overline{6}$	3
$\overline{9}$	2 ✔
	2

Now remove the ninth; in other words, subtract 2 from 18. The Egyptians would do this by completing 2 to 18.
The answer is 16.
All that's left to do is to square this value, basically turning it into an area. So we need to square 16.
256 is the area of the circle.
That wasn't so bad, but that's just because we started with a number whose $\overline{9}$ was trivial. It should come as no surprise that this is the very example in Ahmose's papyrus that first shows how to find the area of a circle. He used simple numbers to keep the focus on the method and not on the computation. We can test this approximation by using the modern equation πr^2. Since the diameter is 18, the radius is 9 and $\pi \times 9^2$ is roughly 254.47, which is

16	?
1	16
2	32
4	64
8	128
16	256 ✔
	256

a little over a half of 1 percent away from the Egyptian's value of 256.

PRACTICE: Use the steps above to estimate the area of a circle of diameter 27.

ANSWER: 576

Now let's try one whose solution is far from trivial.

EXAMPLE: Use the steps above to estimate the area of a circle of diameter 10.

SOLUTION: First calculate $\overline{9}$ of 10.

$\overline{9}$	?
1	10
$\overline{2}$	5
$\overline{3}$	3 $\overline{3}$
$\overline{6}$	1 $\overline{2}$ $\overline{6}$
$\overline{9}$	$\overline{\overline{3}}$ $\overline{3}$ $\overline{9}$ ✔
	1 $\overline{9}$

Complete 1 $\overline{9}$ to get 10.
When we complete $\overline{9}$ to get 1 we get

$$\overline{9} + \overline{\overline{3}}\,\overline{6}\,\overline{18} = 1$$
$$2 + \underline{12 + 3 + 1} = 18$$
×18   ÷18

The answer is 8 $\overline{\overline{3}}$ $\overline{6}$ $\overline{18}$.
Square 8 $\overline{\overline{3}}$ $\overline{6}$ $\overline{18}$.

	8 $\overline{\overline{3}}$ 6 18	???	
1)	1	8 $\overline{\overline{3}}$ 6 18	
2)	2	17 $\overline{\overline{3}}$ 9	(1)×2
3)	4	35 $\overline{2}$ 18	(2)×2
4)	8	71 $\overline{9}$ ✓	(3)×2
5)	$\overline{3}$	5 $\overline{3}$ 6 18 27 ✓	(1)×$\overline{3}$
6)	$\overline{3}$	2 $\overline{\overline{3}}$ 6 9 54	(5)÷2
7)	$\overline{6}$	1 3 12 18 108 ✓	(6)÷2
8)	$\overline{12}$	$\overline{\overline{3}}$ 24 36 216	(7)÷2
9)	$\overline{18}$	$\overline{3}$ 9 36 54 324 ✓	(8)×$\overline{3}$

We now have the unenviable job of adding 71 $\overline{9}$, 5 $\overline{\overline{3}}$ $\overline{6}$ $\overline{18}$ $\overline{27}$, 1 $\overline{3}$ $\overline{12}$ $\overline{18}$ $\overline{108}$, and $\overline{3}$ $\overline{9}$ $\overline{36}$ $\overline{54}$ $\overline{324}$. The whole numbers add to 77. The fractional parts can be added as follows:

$$9\ \overline{\overline{3}}\ 6\ 18\ 27\ 3\ 12\ 18\ 108\ 3\ 9\ 36\ 54\ 324$$
$$= (\overline{\overline{3}}\ \overline{3})(\overline{9}\ \overline{9})(\overline{18}\ \overline{18})(\overline{6}\ \overline{12})(\overline{27}\ \overline{54})(\overline{108}\ \overline{324})\ \overline{3}\ \overline{36}$$
$$= 1\ \overline{6}\ \overline{18}\ \overline{9}\ \overline{4}\ \overline{18}\ \overline{81}\ \overline{3}\ \overline{36}$$
$$= 1\ (\overline{18}\ \overline{18})(\overline{3}\ \overline{6})\ \overline{4}\ \overline{9}\ \overline{36}\ \overline{81}$$
$$= 1\ \overline{9}\ \overline{2}\ \overline{4}\ \overline{9}\ \overline{36}\ \overline{81}$$
$$= 1\ (\overline{9}\ \overline{9})\ \overline{2}\ \overline{4}\ \overline{36}\ \overline{81}$$
$$= 1\ (\overline{2}\ \overline{6})(\overline{18}\ \overline{36})\ \overline{4}\ \overline{81}$$
$$= 1\ \overline{3}\ (\overline{4}\ \overline{12})\ \overline{81}$$
$$= 1\ (\overline{\overline{3}}\ \overline{3})\ \overline{81}$$
$$= 2\ \overline{81}$$

So the total is 77 and 2 $\overline{81}$, or 79 $\overline{81}$. I have to admit that the above computation was painful, but it still amazes me how much it simplifies. Nothing we did above was difficult; the number of fractions involved just made the computation particularly tedious.

If the Egyptians had squared and then multiplied by the constant, things would have been much easier. They could have used $\overline{2}$ $\overline{4}$ $\overline{32}$ as their multiplier, which is essentially equivalent to using the Babylonian value of π, 3 $\overline{8}$. Now the problem would have gone like this.

EXAMPLE: Find the area of a circle of diameter 10 squaring first.

SOLUTION: Square the 10.

10	?
1	10
2	20 ✓
4	40
8	80 ✓
	100

Now multiply by the constant $\overline{2}$ $\overline{4}$ $\overline{32}$.

$\overline{2}$ $\overline{4}$ $\overline{32}$	?
1	100
$\overline{2}$	50 ✓
$\overline{4}$	25 ✓
$\overline{8}$	12 $\overline{2}$
$\overline{16}$	6 $\overline{4}$
$\overline{32}$	3 $\overline{8}$ ✓
	78 $\overline{8}$

We obtained almost the same answer in a tiny fraction of the time. We could get a closer answer by tweaking our constant a little higher. Say, perhaps $\overline{2}$ $\overline{4}$ $\overline{30}$, which would actually take fewer steps to calculate.

PRACTICE: Find the area of a circle of radius 12. Then redo using the square-first method.

ANSWERS: 113 $\overline{\overline{3}}$ $\overline{1}$ and 112 $\overline{2}$

PRACTICE: Find the area of a circle of radius 7. Then redo using the square-first method.

ANSWERS: $38 \; \overline{\overline{3}} \; \overline{27} \; \overline{81}$ and $38 \; \overline{4} \; \overline{32}$

I'VE GOT NOTHING

Effectively Using ÷10 and ×$\overline{3}$

In some chapters, the historical context of this book writes itself. However, in this chapter I've been frequently confronted with the difficulty of finding some historical connection with a subtle math technique. In this section, please forgive me because I've decided to tell you about something that I personally find amusing but that has no link to the mathematics that follows.

In popular culture, the portrayal of college life rarely centers on the classroom. Comedic movies and television shows focus on parties, drug use, and sex. How exaggerated these depictions are probably varies from school to school and student to student; however, the stereotype dominates our imagination. You might be surprised to find out that this outlook is thousands of years old.

Young Egyptian scribes learned by copying texts. In some sense it doesn't matter what they wrote as long as they learned the many symbols of hieroglyphs and hieratic. The teachers in the House of Wisdom often used these copying exercises to teach other things in the content of the texts. Sometimes the content was the format of a document like an official letter. Other times the content was a moralizing speech. Here's an example of a faux letter from a teacher to a less-than-serious student.

> I hear … that you are neglecting your writing and spending all your time dancing, going from tavern to tavern, always reeking of beer. […] If only you realized that wine is a thing of the devil. … You sit in the house with girls around you. […] You sit in front of the wench, sprinkled with perfume. … You reel and fall on your belly and are filthied with dirt.

As we've seen, multiplications and divisions can often be performed in multiple ways. Sometimes one way is far more efficient than the other. Consider the following two calculations for the division of 12 by 36. The first scribe mindlessly took half repeatedly and switched when he had gone too far. The second scribe noticed the 12 and realized she could get this using the division-by-3 trick of taking half and then taking $\overline{\overline{3}}$. By noticing a relationship between 12 and 36, she was able to shorten the computation. In some sense she was lucky that 12 was exactly one of the numbers easily reached from 36.

36 ÷ 12 done by a careless (left) and a careful (right) scribe.

Now consider the following two divisions of 13 by 36. The two numbers in the quotient are no longer precisely related. While we can't easily get 13 from 36, it often helps just to get close, and 12 is close to 13. The first division is mindless and long. The second proceeds as above, getting 12. We're now 1 away from 13, which can easily be obtained with a switch. While we didn't get 13 immediately, the smaller gap is relatively easy to make up. Also notice that the approximation $\overline{3} \; \overline{36}$ is better than $\overline{4} \; \overline{9}$. This is because "getting close" really means being more accurate.

Since 12 is closer to 13 than 9, the second method is more efficient and gets a better answer.

PRACTICE: Divide 15 by 42 in four lines.

ANSWER: $\bar{\bar{3}}\ \overline{42}$

In general, if you are close enough to the target value that taking $\bar{\bar{3}}$ will put you below it, use $\bar{\bar{3}}$ instead of $\bar{2}$. You don't always have to obey this rule, particularly when your number is not a multiple of 3 because taking the $\bar{\bar{3}}$ introduces fractions into the problem. Egyptians were not that worried about such fractions, but it can add a level of difficulty. It's your choice.

If you do decide to perform this operation on a whole number, the only fractions you'll get are $\bar{3}$ and $\bar{\bar{3}}$. Here are two divisions of 7 by 10, one avoiding and one using the trick. Below, we will get $6\,\bar{\bar{3}}$ when we're trying to get 7. Realize that we need $\bar{3}$ more and hence we must construct this answer.

?	7
1	10
$\bar{2}$	5 ✓
$\bar{5}$	2 ✓
$\bar{2}\ \bar{5}$	

?	7
1	10
$\bar{\bar{3}}$	$6\,\bar{3}$ ✓
$\overline{10}$	1
$\overline{15}$	$\bar{3}$
$\overline{30}$	$\bar{3}$ ✓
$\bar{\bar{3}}\ \overline{30}$	

The trick on a non-multiple of 3.

Notice that the time it takes to get the $\bar{3}$ we need to reach 7 takes more steps than if we didn't, particularly since we were able to construct the 2 on the left using a switch. However, the answer on the right is better than the one on the left because it is a better approximation. We probably would be expected to simplify the solution on the left using parts, eliminating any time saved.

Try this one. Note that you will need the right column to add to 16, but you will get $14\,\bar{\bar{3}}$. This means you will need 1 and $\bar{3}$ more, which can be constructed separately. Do this problem both ways. After simplifying you should get the same answer.

PRACTICE: Divide 16 by 22.

ANSWER: $\bar{\bar{3}}\ \overline{22}\ \overline{66}$

If you end up with a $\bar{3}$ in your answer, you will need to construct a $\bar{\bar{3}}$ in order to obtain the answer.

EXAMPLE: Divide 18 by 52.

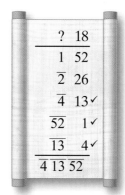

The trick on a non-multiple of 3.

As we can see again, the trick may not save time, but it generally gives better answers. Now that we're no longer terrified by fractions, we can take $\bar{\bar{3}}$ instead of half if it gets us closer to the desired value. Compare these two divisions. The first is the standard procedure, and the second takes advantage of the fact that $5\,\bar{2}$ is close to 6. Of course we need $\bar{2}$ more to make 6.

EXAMPLE: Divide 6 by 22.

SOLUTIONS:

?	6
1	22
$\bar{2}$	11
$\overline{11}$	2 ✓
$\bar{6}\ \overline{66}$	4 ✓
$\bar{6}\ \overline{11}\ \overline{66}$	

?	6
1	22
$\bar{2}$	11
$\bar{4}$	$5\,\bar{2}$ ✓
$\overline{22}$	1
$\overline{44}$	$\bar{2}$ ✓
$\bar{4}\ \overline{44}$	

Getting closer with an extra division by 2.

PRACTICE: Divide 11 by 21.

ANSWER: $\overline{2}\ \overline{42}$

The last trick of this section is, in a sense, almost the exact opposite of the $\overline{\overline{3}}$ trick. Above we multiplied by $\overline{\overline{3}}$ instead of $\overline{2}$ because we didn't want the number to shrink that much. That's because we were close to the target number. However, if you're far from the target, you want to shrink quickly. In order to do this, division by 10 helps. For now we'll apply this only on multiples of 10, although we do have the division-by-10 table in Ahmose's papyrus. Try to do the following problem using only doubling and halving and see how difficult it becomes.

EXAMPLE: Divide 86 by 7400.

?	86
1	7400
$\overline{10}$	740
$\overline{100}$	74 ✓
$\overline{740}$	10 ✓
7400	1
$\overline{3700}$	2 ✓
$\overline{100}\ \overline{740}\ \overline{3700}$	

PRACTICE: Divide 31 by 2000.

ANSWER: $\overline{100}\ \overline{200}\ \overline{2000}$

All that's left to do is mix and match. In the next example we first divide by 10 because we want to get small quickly, and then we take the $\overline{\overline{3}}$ because it gets us closer.

EXAMPLE: Divide 660 by 45.

?	45
1	660
$\overline{10}$	66
$\overline{15}$	44 ✓
$\overline{660}$	1 ✓
$\overline{15}\ \overline{660}$	

See if you can do this next problem in four lines using both the $\overline{\overline{3}}$ and the 10 tricks.

PRACTICE: Divide 66 by 840.

ANSWER: $\overline{15}\ \overline{84}$

WHAT LIES AHEAD
Completions and Divisions

Time was of the essence for Alexander the Great. He had defeated the army of the great Persian Empire well over a year prior and the Persians were beginning to rebuild. The sieges of Tyre and Gaza had taken longer than expected. Fortunately, on his arrival in Egypt, he was welcomed as a liberator. Despite the need to return to Mesopotamia in order to meet the Persian forces once again, he ordered his armies to wait while he took a pilgrimage to the Temple of Amun in the oasis of Siwa. He told his men that there was something he needed to know about himself.

Perhaps he felt motivated by destiny. His mother's and father's sides claimed to be descended from Perseus and Hercules, respectively. Both of these fabled Greek heroes had sought divine guidance from the priests of Amun during their quests. Alexander's mother had instilled in him the notion that his true father was in fact Zeus, whom the Greeks identified with Amun. So it's no surprise that Alexander imitated the actions of his mythological ancestors.

As Alexander arrived at the temple, a priest greeted him as the son of Amun and took him into the holiest sanctum of the temple. In order for the priest to consult the god, a number of Egyptian men came in carrying a boat that was suspended on poles and within which was the god symbol of Amon. The men moved about, supposedly being directed by the god as the seer watched. He then interpreted their motions as omens, forming an oracle given only to Alexander. After leaving, Alexander simply declared that he had received the answer his soul desired.

In Egypt, divination doesn't seem to have had as strong an influence as it had in Mesopotamia or Greece. However, there were notable exceptions, such as the seers in the above story. For example, the priests of Imhotep were skilled in the interpretation of dreams. If Egyptians could truly foretell the future, I'm not sure they would have liked what they would have seen. They would be dominated by Greek rulers for about three hundred years. After that, less tolerant Roman leaders would hold the reins of power for over four hundred more. The Roman period would end shortly after the rise of Christianity, which finished the distinct Egyptian culture for good.

Yet I'm not sure they would have been surprised. The Egyptian culture had been in a state of decline since the rule of Ramses. They had suffered two periods of Persian domination, ended only by the arrival of Alexander. The Egyptians had grown so weak that their army consisted of Greek mercenaries whose pay drained the country's resources. The seat of power in Egypt continually moved south in an effort to stay a safe distance from the aggressive Mediterranean cultures. Eventually the pharaohs were Nubian kings who ruled lands that were originally south of Egypt's borders.

The lack of foresight in political matters stands in direct contrast to the foresight required in mathematical computation. As we've seen, Egyptian mathematics was not algorithmic. The practitioners had to use finesse rather than rote computation. Here I will show you that to perform Egyptian math effectively, you not only need to know where you are, but also where you are going. Being aware of what lies ahead is as important as any step in the calculation.

When Ahmose checked his value for $2 \times \overline{17}$, he paused in the middle and gave a "remainder." Here's how the division started.

?	2
1	17
$\overline{\overline{3}}$	$11\,\overline{3}$
$\overline{3}$	$5\,\overline{2}\,\overline{6}$
$\overline{6}$	$2\,\overline{2}\,\overline{3}$
$\overline{12}$	$1\,\overline{4}\,\overline{6}\checkmark$

Remainder $\overline{3}\,\overline{4}$

In order to understand what's being done, we need to notice what's missing. Ahmose wanted the right column

to add to 2. He got as far as $1\,\overline{4}\,\overline{6}$ and needed to know how much further he had to go. So he performed a completion to see what was left. Of course, completing $1\,\overline{4}\,\overline{6}$ to 2 is the same as completing $\overline{4}\,\overline{6}$ to 1.

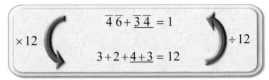

$$\overline{4}\,\overline{6} + \overline{3}\,\overline{4} = 1$$
$\times 12 \qquad \qquad \div 12$
$$3 + 2 + \underline{4+3} = 12$$

The remainder $\overline{3}\,\overline{4}$ is the amount needed to get to two. Ahmose quickly constructed 3 and 4 on the left-hand side of the tablet in steps 7 and 9. He then switched them in steps 8 and 10 to get the required $\overline{3}\,\overline{4}$ on the right. The full division looks something like this.

	?	2	
1)	1	17	
2)	$\overline{\overline{3}}$	$11\,\overline{3}$	$(1)\times\overline{\overline{3}}$
3)	$\overline{3}$	$5\,\overline{2}\,\overline{3}$	$(2)\div 2$
4)	$\overline{6}$	$2\,\overline{2}\,\overline{3}$	$(3)\div 2$
5)	$\overline{12}$	$1\,\overline{4}\,\overline{6}\checkmark$	$(4)\div 2$
	Remainder $\overline{3}\,\overline{4}$		
6)	2	34	$(1)\times 2$
7)	3	51	$(1)+(6)$
8)	$\overline{51}$	$\overline{3}\checkmark$	$(7)\leftrightarrow$
9)	4	68	$(6)\times 2$
10)	$\overline{68}$	$\overline{4}\checkmark$	$(9)\leftrightarrow$
	$\overline{12}\,\overline{51}\,\overline{68}$		

You can tell by looking at Egyptian multiplication that they always had a sense of what they needed to finish the problem. Their goal was to get one of the columns to add up to a given value. The numbers they created in that column varied depending upon what they required. They seemed to know how much more they needed in that column and then figured out how to generate that amount. Sometimes it's easy. If, for example, you need 10 and have 8, you want 2 more. But when fractions are involved, especially Egyptian fractions, it's not always clear what's needed. So if you need $10\,\overline{2}\,\overline{9}$ and have $8\,\overline{3}\,\overline{11}$, you're more or less forced to do a completion.

Let's say we are dividing $8\,\overline{\overline{3}}$ by 34. We'd probably start as follows:

?	$8\,\overline{\overline{3}}$
1	34
$\overline{2}$	17
$\overline{4}$	$8\,\overline{2}$

At this point we should realize that we're close. We need $8\,\overline{\overline{3}}$ and we've got $8\,\overline{2}$. So to finish the division, we need to determine how much more we require; therefore, we complete $\overline{2}$ to $\overline{\overline{3}}$.

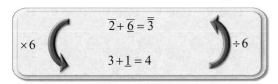

$$\overline{2}+\underline{\overline{6}}=\overline{\overline{3}}$$
$$\times 6 \qquad 3+\underline{1}=4 \qquad \div 6$$

We now know that we need $\overline{6}$ in the right column, so we finish the division by switching the first row and forming a $\overline{6}$ from the 1. Remember that the Egyptians didn't know how to divide by 6, so they would divide by 2 and then by 3. The division by 3 consists of first taking half and then taking $\overline{\overline{3}}$.

?	$8\,\overline{\overline{3}}$
1	34
$\overline{2}$	17
$\overline{4}$	$8\,\overline{2}$ ✓
Remainder $\overline{6}$	
$\overline{34}$	1
$\overline{68}$	$\overline{2}$
$\overline{136}$	$\overline{4}$
$\overline{204}$	$\overline{6}$ ✓
$\overline{4}\,\overline{204}$	

EXAMPLE: Divide 10 by 13 using completions.

We start off the same way as before.

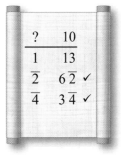

?	10
1	13
$\overline{2}$	$6\,\overline{2}$ ✓
$\overline{4}$	$3\,\overline{4}$ ✓

Now we're at $9\,\overline{2}\,\overline{4}$, which is very close to 10. So we need to complete $9\,\overline{2}\,\overline{4}$ to 10, which is the same as $\overline{2}\,\overline{4}$ to 1.

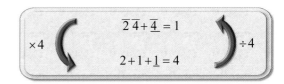

$$\overline{2}\,\overline{4}+\underline{\overline{4}}=1$$
$$\times 4 \qquad 2+1+\underline{1}=4 \qquad \div 4$$

This means we need $\overline{4}$ more, which is very easy to construct.

SOLUTION:

?	10
1	13
$\overline{2}$	$6\,\overline{2}$ ✓
$\overline{4}$	$3\,\overline{4}$ ✓
Remainder $\overline{4}$	
$\overline{13}$	1
$\overline{26}$	$\overline{2}$
$\overline{52}$	$\overline{4}$ ✓
$\overline{2}\,\overline{4}\,\overline{52}$	

In the following practice problem, repeatedly divide by 2 until you get the line $\overline{4}$, $2\,\overline{2}\,\overline{4}$, then perform a completion.

PRACTICE: Divide 14 by 11 using completions.

ANSWER: $1\,\overline{4}\,\overline{44}$

I was sloppy in the last example when I divided 10 by 13. I took half of 13 to get 6 $\overline{2}$, which is well under 10. I could have taken $\overline{\overline{3}}$ of the value to get 8 $\overline{\overline{3}}$, which is still under 10. We would now need 1 $\overline{3}$ to finish this division, both values are easy to make. Let's try another problem being more careful.

EXAMPLE: Divide 7 $\overline{2}$ $\overline{10}$ by 11 using completions.

The number 7 $\overline{2}$ $\overline{10}$ is close to 11, so taking half is probably too much. So we'll start off by taking $\overline{\overline{3}}$ of 11.

?	7 $\overline{2}$ $\overline{10}$
1	11
$\overline{\overline{3}}$	7 $\overline{3}$ ✓

We can tell we're close, so we need to perform a completion.

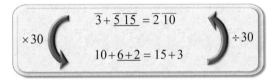

$$\overline{3} + \overline{5\ 15} = \overline{2\ 10}$$
$$\times 30 \qquad \div 30$$
$$10 + \underline{6+2} = 15 + 3$$

I'll now create the missing $\overline{5}$ and $\overline{15}$ by switching the 11, dividing it by 10, then both doubling and taking $\overline{\overline{3}}$ of this value.

SOLUTION:

?	7 $\overline{2}$ $\overline{10}$
1	11
$\overline{\overline{3}}$	7 $\overline{3}$ ✓
Remainder $\overline{5}$ $\overline{15}$	
$\overline{11}$	1
$\overline{110}$	$\overline{10}$
$\overline{55}$	$\overline{5}$ ✓
$\overline{165}$	$\overline{15}$ ✓
$\overline{\overline{3}}$ $\overline{55}$ $\overline{165}$	

In the next practice problem, perform the completion after you obtain the line $\overline{3}$, 7 $\overline{\overline{3}}$.

PRACTICE: Divide 8 $\overline{6}$ by 23 using completions.

ANSWER: $\overline{3}$ $\overline{46}$

Let's do a less trivial example.

EXAMPLE: Divide 35 $\overline{5}$ $\overline{10}$ by 80 using completions.

In this problem we start off by taking half of 80 to get 40.

	?	35 $\overline{5}$ $\overline{10}$	
1)	1	80	
2)	$\overline{2}$	40	(1) ÷ 2

Our answer 40 is close to 35, our target. If we take half of 40 to get 20, we way undershoot 35. So it seems to be a good idea to take $\overline{\overline{3}}$ instead of half. This gives us 26 $\overline{\overline{3}}$.

	?	35 $\overline{5}$ $\overline{10}$	
1)	1	80	
2)	$\overline{2}$	40	(1) ÷ 2
3)	$\overline{\overline{3}}$	26 $\overline{\overline{3}}$ ✓	(2) × $\overline{\overline{3}}$

Now 26 $\overline{\overline{3}}$ is a little more than 8 away from 35 $\overline{5}$ $\overline{10}$. This is fortunate since 8 is very easy to get from 80.

	?	35 $\overline{5}$ $\overline{10}$	
1)	1	80	
2)	$\overline{2}$	40	(1) ÷ 2
3)	$\overline{\overline{3}}$	26 $\overline{\overline{3}}$ ✓	(2) × $\overline{\overline{3}}$
4)	$\overline{10}$	8 ✓	(1) ÷ 10

The two checked values total 34 $\overline{\overline{3}}$. At this point we should realize that we are very close to the answer. However, it is very difficult to know exactly how far 34 $\overline{\overline{3}}$ is from 35 $\overline{5}$ $\overline{10}$. Hence it's time for a completion.

$$\overline{3} + \overline{2\ 10\ 30} = 1\ \overline{5}\ \overline{10}$$

$\times 30 \qquad\qquad \div 30$

$$20 + 15 + 3 + 1 = 30 + 6 + 3$$

So, all we're missing is $\overline{2}\ \overline{10}\ \overline{30}$. Luckily these are all easy numbers to obtain from 1 since the only factors in them are 2, 3, and 5. So we finish by creating the missing fractions.

SOLUTION:

	?	$35\ \overline{5}\ \overline{10}$	
1)	1	80	
2)	$\overline{2}$	40	(1)$\div$2
3)	$\overline{3}$	$26\ \overline{3}$ ✓	(2)$\times\overline{3}$
4)	$\overline{10}$	8 ✓	(1)$\div$10
Remainder		$\overline{2}\ \overline{10}\ \overline{30}$	
5)	$\overline{80}$	1	
6)	$\overline{160}$	$\overline{2}$ ✓	(5)$\div$2
7)	$\overline{800}$	$\overline{10}$ ✓	(5)$\div$10
8)	$\overline{1600}$	$\overline{20}$	(7)$\div$2
9)	$\overline{2400}$	$\overline{30}$ ✓	(8)$\times\overline{3}$
	$\overline{3}\ \overline{10}\ \overline{160}\ \overline{800}\ \overline{2400}$		

PAYDAY

Working Divisions in Parts

The scribe Amennakhte didn't do his job; he couldn't. The tomb workers of the Valley of the Kings had waited eighteen days after payday for their monthly wages, but Amennakhte had nothing to give them. The workers, hungry and disgusted, went to the back rooms of Thutmose III's temple and just sat down, beginning the first strike known in recorded history. The shouts and threats of Amennakhte had no effect—the workers just stayed where they were. The next day the scribe Pentaweret brought fifty-five loaves of bread, but apparently it was too little, too late. The workers complained of hunger and thirst. They had run out of fish and vegetables and needed new clothes.

In response, they were paid some but not all of their back wages. The chief of police sided with the workers and had promised to talk with the mayor of Thebes. Five days into the strike, the workers were each given half a sack of barley, used to brew beer, and the police managed to scrounge up a few rations, but the partial pay was not enough to sway the strikers. On the seventh day of the strike, they were finally paid that month's wages.

The next month came with no pay, and the strike started once again. One worker declared that if he were taken away, he'd begin to make plans to rob the tombs they had built. A letter arrived from the vizier himself. He explained that he had recently been appointed and that his predecessor was supposed to have paid them. When the new vizier went to the granaries supposedly containing the grain owed the workers, he had found them empty. It seems his corrupt predecessor had taken the grain on his way out. The best the new vizier could do was give the workers half of their due.

On the advice of the foreman, the crew took the half rations and immediately left work anyway. The scribe Amennakhte apparently had had enough. He stopped the crew at a guard post and began to threaten them. Now that they had been paid, he declared that there wasn't a court in the land that would not convict them. The workers changed their minds and went back to work. They were later paid directly by the mayor of Thebes. The ancient account of this story ends with the workers muttering in anger, claiming that they will exact revenge on those who have wronged them. Apparently they had some knowledge of the wrongdoings of their superiors and would soon make them public, including charges of tomb robbing, cattle theft, and adultery.

All of these events took place two years before the end of the reign of Ramses III, the last great pharaoh of Egypt. Earlier in his reign he had fought off the Peoples of the Sea, seafaring warrior nomads who had brought great disruption to the entire Mediterranean. He also had fended off a Libyan invasion. But Egyptian control outside its borders was gone. Seafaring technology had rendered their isolation moot. Gradually the great Egyptians would become subordinate to their more powerful neighbors. If this story is any indication, the corruption and disorder that weakened a once-mighty empire was already ongoing.

Let's examine through a simple example how wages were computed in ancient Egypt. Imagine two friends and I start a business and make $16\,\bar{4}$ bags of gold. Since I'm the only one in the group who knows math, I'm put in charge of dividing it up. Not trusting me, they demand a full share. When I agree, they smile approvingly. I then inform them that due to my accounting duties, I deserve a full $1\,\bar{4}$ share. In order to divide up the money, I first have to calculate the total number of shares. This is easy since the sum of 1, 1, and $1\,\bar{4}$ is clearly $3\,\bar{4}$. Now I need to find the value of a 1 share. This is done by dividing up the $16\,\bar{4}$ bags of gold between the $3\,\bar{4}$ shares.

?	$16\,\bar{4}$
1	$3\,\bar{4}$ ✓
2	$6\,\bar{2}$
4	13 ✓
5	

This means a 1 share, my friends' "fair" portion, is worth 5 bags of gold. My $1\,\bar{4}$ share can be computed by multiplying $1\,\bar{4}$ by the value of a 1 share.

$1\,\bar{4}$	?
1	5 ✓
$\bar{2}$	$2\,\bar{2}$
$\bar{4}$	$1\,\bar{4}$ ✓
	$6\,\bar{4}$

So my friends each get 5 bags and I get $6\,\bar{4}$ bags. The next time, we get $5\,\bar{2}\,\overline{12}$ bags of gold to split up. I begin my calculations just as before. There are still $3\,\bar{4}$ shares, so we divide $5\,\bar{2}\,\overline{12}$ by this to find the number of bags per share.

?	$5\,\bar{2}\,\overline{12}$
1	$3\,\bar{4}$ ✓
$\bar{3}$	$2\,\bar{6}$ ✓

At this point in the computation I know I'm close. I've got $5\,\bar{4}\,\bar{6}$ and need $5\,\bar{2}\,\overline{12}$. So I do a completion, but I don't finish it because I know a trick.

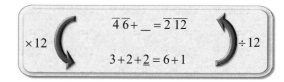

In the above completion I never return the 2 parts of 12 back into a $\bar{6}$. For one thing it wouldn't help me much. How could I make a $\bar{6}$ out of a $3\,\bar{4}$? By the operations known to Egyptians, there is no easy way. There are nice numbers in the first column, but I can't switch sums because of the entries in the right column. There is no nice way to switch $3\,\bar{4}$ to the other side. You can switch only single whole numbers or fractions, not mixed numbers. There are tricks that could make a whole number out of $3\,\bar{4}$, like multiplying by 4 first and then switching, but this doesn't work well in all such problems.

What you need to realize is that I already have all I need to know. I need 2 parts of 12. As long as I work in parts of 12, this is easy. The first line has a $3\,\bar{4}$ in the second column. In parts of 12, the 3 becomes 3×12, or 36, and the $\bar{4}$ becomes 3, for a total of 39. What's important about the 39 is that it is a whole number that can be switched. So we rewrite the first line expressing the second column in parts of 12 and resume the computation knowing we need 2 more.

If you still don't see why this works, think in terms of feet and inches. Let's reinterpret the previous problem as fence building. I want to build $5\,\bar{2}\,\overline{12}$ feet of fence. I know that $3\,\bar{4}$ feet costs a piece of gold. I use this to determine that $2\,\bar{6}$ feet costs $\bar{3}$ a piece of gold. So by spending $1\,\bar{3}$, I can build $5\,\bar{4}\,\bar{6}$ feet of fence. What I've just described is essentially the first two lines of the above division.

Now I have $5\,\bar{4}\,\bar{6}$ feet of fence and need $5\,\bar{2}\,\overline{12}$ feet. The obvious question is, how much more do I need? Specifically, I need to know how much larger $\bar{2}\,\overline{12}$ feet is than $\bar{4}\,\bar{6}$ feet. To simplify the problem by eliminating fractions, I decide to convert into inches. I know $\bar{2}\,\overline{12}$ feet is 6 and 1 inches and $\bar{4}\,\bar{6}$ feet is 3 and 2 inches. So I have 5 inches and need 7. Therefore I need 2 inches more. In the above math, I've intuitively done the completion. In the original problem I determined I needed 2 more parts of 12. In this problem I need 2 more inches out of the 12 in a foot. Notice that the numbers are identical.

The remaining question is how much will those 2 inches cost? I know that 1 bag of gold buys 3 $\overline{4}$ feet, but this is just 36 and 3 inches, for a total of 39 inches. So if 1 bag buys 39 inches, then $\overline{39}$ of a bag buys 1 inch. I double this to get $\overline{26}\ \overline{78}$ to get the cost of the remaining 2 inches. Finally I add this to the cost of the fence I've already bought for a total of 1 $\overline{\overline{3}}\ \overline{26}\ \overline{78}$ bags of gold. This finishes the problem.

Below is the multiplication. Note that we determined after the first two lines that we were short 2 parts of 12. The line that follows simply converts the 3 $\overline{4}$ into parts of 12, which is 39. The computation continues until the 2 parts are calculated.

?	5 $\overline{2}$ $\overline{12}$
1	3 $\overline{4}$ ✓
$\overline{3}$	2 $\overline{6}$ ✓
Rem. 2 of 12 parts	
1	39
$\overline{39}$	1
$\overline{26}\ \overline{78}$	2 ✓
1 $\overline{\overline{3}}\ \overline{26}\ \overline{78}$	

As we've seen, computations by parts can be thought of as unit conversions. The first part of the division is done in feet, and the completion tells us how much more we need in inches. So we finish the computation in inches. While we can't add some of the numbers in the second column because some are in feet and some are in inches, the first column always remains in "bags of gold," and hence all the entries are summable. The best part of this method is that completions, like any parts calculation, eliminate fractions and make the problems easier. It also uses the same completion that is needed to finish the problem anyway, so it requires almost no extra work. Here's one more example and a practice problem.

EXAMPLE: Divide 10 $\overline{3}$ by 1 $\overline{8}$.

SOLUTION: First we divide as usual until we get close to 10 $\overline{3}$.

?	10 $\overline{3}$
1	1 $\overline{8}$ ✓
2	2 $\overline{4}$
4	4 $\overline{2}$
8	9 ✓

The total is 10 $\overline{8}$, which is close to 10 $\overline{3}$, so we calculate how many parts we're short.

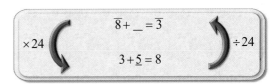

$$\times 24 \quad \overline{8} + \_\_ = \overline{3} \quad \div 24$$
$$3 + \underline{5} = 8$$

From the above computation we can see that we're short 5 parts of 24. We can now continue our above computation. We start from the first line, 1, 1 $\overline{8}$, but convert the right-hand side into parts of 24, giving 1, 27 since 1 $\overline{8} \times 24$ is 27. We now need to make the 5 missing parts in the right column.

?	10 $\overline{3}$
1	1 $\overline{8}$ ✓
2	2 $\overline{4}$
4	4 $\overline{2}$
8	9 ✓
Rem. 5 of 24 parts	
1	27
$\overline{27}$	1 ✓
$\overline{18}\ \overline{54}$	2
$\overline{9}\ \overline{27}$	4 ✓
9 $\overline{6}\ \overline{54}$	

$$1 + 8 + \overline{27} + \overline{9} + \overline{27}$$
$$= 9\ \overline{9}\ (\overline{27}\ \overline{27})$$
$$= 9\ (\overline{9}\ \overline{18})\ \overline{54}$$
$$= 9\ \overline{6}\ \overline{54}$$

PRACTICE: Finish the following multiplication by forming a completion and finishing in parts. Divide $8\ \overline{5}$ by $2\ \overline{3}$.

?	$8\ \overline{5}$
1	$2\ \overline{3}$ ✓
2	$4\ \overline{\overline{3}}$ ✓
$\overline{2}$	$1\ \overline{6}$ ✓
Rem.	

ANSWER: $3\ \overline{2}\ \overline{70}$

In theory, this method can be used to make short work of any division. For those of you who know algebra, this is essentially equivalent to multiplying through by the denominator of an equation to eliminate the fractions. In practice, this is not the case.

Consider the division of $6\ \overline{7}$ by $6\ \overline{21}$. In the very first line of the division we get close to the answer.

?	$6\ \overline{7}$
1	$6\ \overline{21}$ ✓

So far everything seems fine. All we need to do is complete $\overline{7}$ to $\overline{21}$.

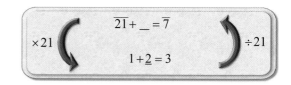

So we need 2 parts of 21. When we convert $6\ \overline{21}$ into parts of 21 we get 127. It looks easy until we try to double $\overline{127}$.

?	$6\ \overline{7}$
1	$6\ \overline{21}$ ✓
Rem. 2 of 21 parts	
1	127
$\overline{127}$	1
?????	2

The problem is we can't double $\overline{127}$ because it's not on our table. This method often produces huge numbers, creating fractions that are too big. How did the Egyptians deal with this problem? I have no idea. If we simply doubled $\overline{128}$ instead, we'd be off by less than one part in eight thousand. While this is not a big error, one of the attractions of the Egyptian method is that it provides exact answers. Using the sliced bread method, it's easy to see that $2 \times \overline{127}$ is $\overline{64}\ \overline{8128}$. The difficulty with this is that the fraction size is so large, it exacerbates the problem.

Unfortunately, just as we were examining the last method of Egyptian mathematics, we see a nearly fatal flaw. To be fair, a mixed number divided by a mixed number is problematic in any system. Even so, it's still sad that we seem to have reached the limits of Egyptian computation.

6
MISCELLANY

THAT'S A LOT OF STONE

The Volume of a Truncated Pyramid

Like father, like son. Khufu, a powerful pharaoh of the Fourth Dynasty, like his dad, Sneferu, had a thing for big pyramids. Sneferu made three separate pyramids, including the famed-but-disastrous bent pyramid. The lessons learned were not lost on the next generation. Khufu needed only one main pyramid built correctly the first time. While he used less stone than his father, it was all devoted to one structure, the Great Pyramid of Giza.

This mammoth structure remained the tallest man-made object for more than four thousand years, when it was replaced by the relatively flimsy Eiffel Tower. The pyramid was made of more than two million stone blocks, each weighing an estimated average of two and a half tons, totaling over ten times the weight of the Empire State Building. The nearly one hundred million cubic feet of solid stone had to be cut, moved, and placed during the twenty to thirty years of Khufu's reign.

The majority of the pyramid consists of the core structure, which is made of limestone that was quarried just south of the pyramid complex. The beautiful white limestone of the outer casing, now mostly gone, was obtained a few miles south and across the Nile. Limestone was used because it was readily available and relatively easy to cut compared to the harder granite and basalt. Especially strong stone, needed for things like columns and ceilings, was often quarried from places far up the Nile.

One of the first things you need to know when building a stone pyramid is how much stone you'll require. There's a simple equation to figure this out. You determine the volume of a box with the same base and height as the pyramid and then divide by 3.

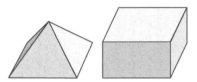

A pyramid with the same height as a box has $\overline{3}$ the volume.

In order to visualize this relationship, let's think like the ancient Greeks. They simply didn't use numbers for their theoretical math. Where we would say that the volume of a pyramid is ⅓ of the volume of the box with the same base and height, they would say they're in ratio 3:1. This ratio should not be thought of as a number but rather as a comparison. According to the Greek definitions, this simply means that three pyramids make one box. We will verify this with a model, for nothing helps understanding more than a physical demonstration.

For what follows, you can either study the diagrams or make your own models out of clay. If you choose to make your own, I suggest a commercial variety of clay like Play-Doh because you can use different colors for different parts. The models you make will never be perfect, but it helps to visualize how the parts fit together. The idea of the following demonstration is fairly simple. If two three-dimensional shapes can be made from the same parts, their volumes must be equal.

The goal is to take three pyramids and rearrange their parts into one box. The pieces we will use will be quarter pyramids. The cuts will be made vertically, perpendicular to the pyramids' bases.

If we quarter three pyramids, we get 12 pieces. We're cutting a square base into fours. This makes four half-size squares. You can use the square in the diagram below to get all of your squares to be the same size. Each quarter

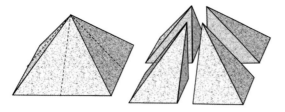

A pyramid cut into quarters.

pyramid has two triangles, pictured in brown above, from the interior of the original pyramid. These are vertical right triangles. For this demonstration to work, they need to be the same size as our square base sliced along the diagonal. The remaining two triangles, once part of the slanting surface of the pyramid, are created by filling in with the clay between the square and the two vertical right triangles.

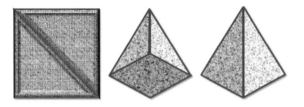

The square with two views of the quarter pyramid.

Now take four of the quarter pyramids and put them together to form a whole pyramid. Repeat making three pyramids.

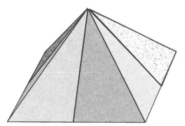

A pyramid made of four quarters.

We now need to show that these three pyramids can make one box of the same height and base. Note that each base of an original pyramid is twice as long as the square used to make the quarters. The square's side length is also the height of each pyramid. Disassemble the three

pyramids into twelve quarters and divide them up into four groups of three. Three of the four quarters can be placed together with the three square bases meeting at a point. Make sure that the tips opposite the bases all meet and that the other three sides of the cube is made completely of the right triangular sides.

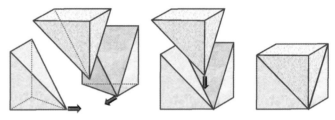

Three quarter pyramids make a cube.

You should now have four cubes. Arrange these into a two by two box. Convince yourself that this box has the same base and height as the old pyramids it was made from. We can now conclude that three pyramids make one box, hence the pyramid's volume is a third of the box with the same base and height.

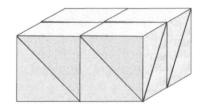

The four cubes as a box with the same base and height as the original pyramids.

The above argument works only if the height of the pyramid is the same length as the quarter bases. However the pyramids can be stretched vertically to match any ratio. Stretching a solid multiplies its volume by the proportion it is stretched. For example, assume we have a pyramid where the height is double the length of the square that forms the quarter's base. We know that when the height is the same, the pyramid is a third of the box. When both are stretched, both volumes double, so the new pyramid is still a third of the new box.

Did the Egyptians know of this relation? It seems probable that if they built pyramids, they'd know this.

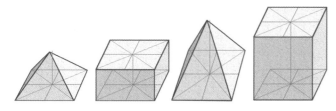

Both solids double, so the pyramid is still a third of the box.

A three-to-one relation is simple to spot. Did the Egyptians use something like our "clay" arguments to convince themselves of this relation? We always need to be careful. With our centuries of mathematical hindsight, modern mathematicians are capable of many unique and creative arguments. While we can't be sure what Egyptians knew about the volume of a pyramid, what's surprising is that they knew something even more complex.

The Moscow Mathematical Papyrus contains a surprising calculation for the volume of a truncated pyramid.

EXAMPLE: Consider a truncated pyramid with a 4-by-4 lower base, a 2-by-2 upper base, and a height of 6. What's the volume?

SOLUTION: Take the 2 and square it to get 4. Take the 4 and square it to get 16. Multiply the 2 and 4 to get 8.

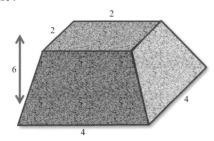

The squares and products of the base lengths.

Add the 4, 16, and 8 to get 28. A third of the height is 2. Multiply 2 by 28 to get 56, the volume.

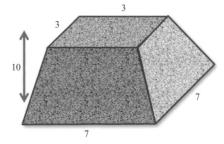

A third the height and the answer times the sum.

This is not a trivial computation and surprisingly it gives the exact solution. The general procedure is to square the base measurements, multiply them by each other, and then add up the three numbers. Take a third of the height and then multiply the two answers to get the volume. In the above example, the calculations are trivial but that's probably because the author wants us to focus on the method and not the computations. Here's an example with less trivial numbers.

EXAMPLE: Consider a truncated pyramid with a 7-by-7 lower base, a 3-by-3 upper base, and a height of 10. What's the volume?

SOLUTION: Take the 3 and square it to get 9. Take the 7 and square it to get 49. Multiply the 3 and 7 to get 21.

Add the 9, 49, and 21 to get 79. A third of the height is 3 $\overline{3}$. Multiply 3 $\overline{3}$ by 79 to get 263 $\overline{3}$, the volume.

PRACTICE: Consider a truncated pyramid with a 9-by-9 lower base, a 4-by-4 upper base, and a height of 8. What's the volume?

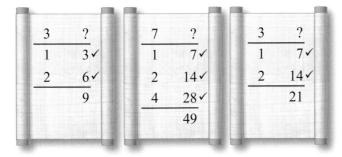

The squares and products of the base lengths.

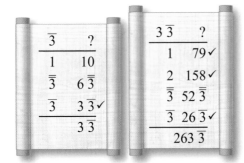

A third the height and the answer times the sum.

ANSWER: 354 $\overline{\overline{3}}$

So we're confronted with the question of how did the Egyptians know this method? Their solution corresponds to the following modern equation where a and b are the length of the bases and h is the height.

$$\text{Vol} = \tfrac{1}{3}\, h(a^2 + ab + b^2)$$

I've suggested in this book that the Egyptians may have stumbled across their estimates for π and other relations empirically. In other words, they looked at a few examples and guessed the relation. This is easy for relations like doubles or thirds, but I just don't see how the above relationship could be arrived at by guessing. My gut feeling as a mathematician is that they knew that this was the answer. They almost certainly must have derived it, but how?

Let's demonstrate the validity of this method using our clay models. It's difficult to say with any certainty that this is what the Egyptians did, but I want to show you that it can be done. This is not as wild an assumption as you might think. The ancient Greeks favored visual, geometric arguments over numerical ones. The Greeks also credited the Egyptians with the origins of their mathematics.

That claim is often dismissed immediately by historians. They argue that from a theoretical point of view, the Egyptians were not that advanced. That is an unfair opinion. We base all we that know about Egyptian mathematics on two ancient papyruses and a handful of scraps. One of the two papyri is incomplete. The idea that this is enough to judge the full mathematical capabilities of a civilization that lasted more than two and a half thousand years is a bit ridiculous. All we do know are some of their calculation methods, which admittedly don't demonstrate greatness; however, we need to realize that the Greeks had great respect for what the Egyptians did know.

Let's look at the calculation we performed to find the first volume. We took the side lengths of the bases, 2 and 4, multiplied them by themselves and together, and then added them up. This gives

$$(2 \times 2) + (4 \times 4) + (2 \times 4)$$

We multiplied this by $\overline{3}$ of the base of 6 to get the volume. We can express this as follows:

$$\text{Volume} = (\overline{3} \times 6)\,[(2 \times 2) + (4 \times 4) + (2 \times 4)]$$

As good Greeks, we should not use $\overline{3}$ as a number but rather as a proportion. This is arithmetically equivalent to writing the following:

$$3 \text{ volumes} = 6[(2 \times 2) + (4 \times 4) + (2 \times 4)]$$

Finally let's distribute the 6 to each of the three terms in the sum.

$$3 \text{ volumes} = (2 \times 2 \times 6) + (4 \times 4 \times 6) + (2 \times 4 \times 6)$$

Let's interpret these numbers geometrically. The $2 \times 2 \times 6$ can be viewed as the volume of a box with a base the same size as the top of the truncated pyramid and with a height of 6. Similarly the $4 \times 4 \times 6$ has a base the size of the bottom of the pyramids and also with a height of 6. Finally the $2 \times 4 \times 6$ has a "mixed" base and the same height.

Because the total is equal to 3 volumes, it has the same measure as three truncated pyramids. So our goal will be to make the three rectangles out of the parts of the three truncated pyramids.

We're going to need a few more of the clay pieces. First, keep three of the quarter-pyramid pieces we used in the first demonstration. We will need six "slants." Make them with two perpendicular sides with the dimensions of the following rectangle. The thin side is the same with as the square base of the quarters. The other length is arbitrary.

Three truncated pyramids equal three boxes.

Finally, we need three boxes. Use the following square for the base size and the above rectangle for the four sides. The long side must be equal to the long side of the slant and the height is equal to the short side.

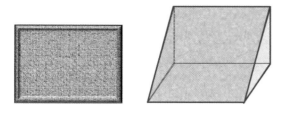

The slants have a base and a back equal in size to the rectangle.

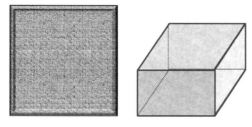

The boxes have the square as a base and sides equal to the bases of the slants.

Now we need to make the three quarter truncated pyramids. Each one will consist of one box, two slants, and one quarter pyramid. Fit them together as follows:

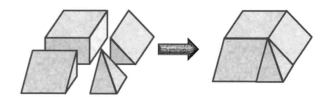

A quarter truncated pyramid made from four pieces.

We now need to disassemble the three truncated pyramids and reassemble them into three blocks with different bases. The first block has the same base as the top of the pyramid. This is easy because it's just the block used in the construction of our truncated pyramid.

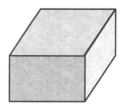

The first block.

The second block will also have a square base, which will be the size of the bottom of the truncated pyramid. To make this block, take all three quarter pyramids and reassemble them into a cube as we did above in the first demonstration. Then take a pair of slants and attach them along their sloping sides, making a box. Make two of these.

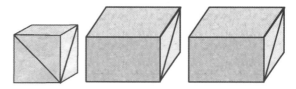

Three quarter pyramids and four slants make three blocks.

Then take one of the first blocks and the three new blocks just assembled and put them together as follows to get one large block:

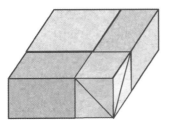

Three quarter pyramids, four slants, and a block make one large block.

Finally we need one last block with one side the width of the upper base and the other of the lower base. This can be done with the remaining block and two slants in much the same way as the last block was created.

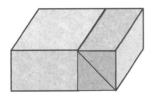

A block and two slants make the final box.

Let's recap to better understand the overall relationship. In our first mathematical example, we had a truncated pyramid with an upper base of 2×2, a lower base of 4×4, and a height of 6. The problem had us multiply 2 by 2, 4 by 4, and 2 by 4. This procedure recreates the two bases and creates a hybrid rectangle.

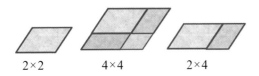

2×2 4×4 2×4

The products calculate the areas of the bases.

The procedure then had us multiply by $\overline{3}$ and the height 6. When we multiply by 6, we are essentially giving every base a height.

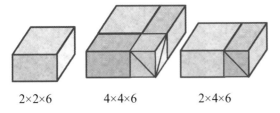

$2 \times 2 \times 6$ $4 \times 4 \times 6$ $2 \times 4 \times 6$

Multiplying by the height turns the bases into boxes.

We can then rearrange these "shapes" back into three truncated pyramids. So we get three pyramids that equal the sum of those three blocks. Hence one pyramid is one-third of the sum of the blocks. Algebraically we can represent this by multiplying by $\overline{3}$. Hence the value of $\overline{3} \times 6[(2 \times 2) + (4 \times 4) + (2 \times 4)]$ is the area of one truncated pyramid.

A third of three truncated pyramids is one pyramid.

So can we assume that the ancient Egyptians used such a method? There's absolutely no evidence to believe so and it doesn't conform to any known Egyptian method. However, there is no obvious method that could explain this. Clearly the Egyptians knew something. What they knew is, unfortunately, unknown.

GOING TO ST. IVES

Geometric Series

> As I was going to St. Ives,
> I met a man with seven wives,
> Every wife had seven sacks,
> Every sack had seven cats,
> Every cat had seven kits,
> Kits, cats, sacks and wives,
> How many were going to St. Ives?

This poem is generally interpreted as a trick question. It was never stated that the man with too many wives was actually going to St. Ives, so the only one going is the person asking the question. However, if we take the problem less literally, we end up needing to compute a sum. The sum starts with 1, representing the man. He has seven wives that need to be added. Since each wife has seven sacks, the number of sacks is 7 times the number of wives, or $7 \times 7 = 49$. Since each sack has seven cats, then number of cats is 7 times the number of sacks, or $7 \times 7 \times 7 = 343$, which can be abbreviated as 7^3. Each cat had seven kittens, so there are 7×7 cats $= 7^4 = 2401$ kittens. The total then becomes $1 + 7 + 7^2 + 7^3 + 7^4$, which is 2801.

The Egyptians had a similar problem. It read simply as a table rather than a poem. In what follows, spelt is a wheat-like substance and a hekat is a measure roughly equal to a quart.

Houses	7
Cats	49
Mice	343
Spelt	2,401
Hekat	16,807
Total	19,607

These are the same numbers we find in the St. Ives problem. The man had 7 wives, 49 sacks, 343 cats, and 2401 kittens. The Egyptians include one higher power of 7, namely 16,807, in their problem. It's possible to imagine a mnemonic story for this problem concerning a town with seven houses. Each house had seven cats. Each cat ate seven mice. Each mouse ate seven bags of spelt, and each bag contained seven hekat. So houses, cats, mice, bags, and hekat, how much in all. The answer is the sum of $7 + 7^2 + 7^3 + 7^4 + 7^5$, which they correctly give as 19,607.

What's interesting about the Egyptian problem is how they go about obtaining a solution. They take the sum of the houses, cats, mice, and spelt, which is 2800, and use this to find the total of houses, cats, mice, spelt, and hekat. Before I explain why their method worked, let's look at their computation. They first add 1 to the above to get 2801 and then they multiply by 7, getting the following:

7	?
1	2,801 ✓
2	5,602 ✓
4	11,204 ✓
	19,607

You can get the entire series in this way. The sum starts with 7. We then take this, add 1, and multiply by 7 to get $(7 + 1) \times 7 = 56$. This is the sum of $7 + 49$, the first two terms.

We can repeat the process with 56 by adding 1 and multiplying by 7 to get $(56 + 1) \times 7 = 57 \times 7 = 399$, which is the sum of $7 + 49 + 343$.

EXAMPLE: Given $7 + 49 + 343 = 399$, use the Egyptian trick to find $7 + 49 + 343 + 2401$.

SOLUTION: $(399 + 1) \times 7 = 2800$

PRACTICE: Given $7 + 49 + 343 + 2,401 + 16,807 = 19,607$, use the Egyptian trick to find $7 + \ldots + 16,807 + 117,649$.

ANSWER: 137,256

In order to understand why this works, let's look at a simpler problem. Instead of letting the numbers grow each time by a factor of 7, let's let them double. This will keep the numbers and my diagrams smaller. So now we get

Houses	2
Cats	4
Mice	8
Spelt	16
Total	30

We can find these numbers using the Egyptian trick. We still add 1 but then multiply by 2 instead of 7. Starting with 2 we get $(2 + 1) \times 2 = 6$, which is the sum of $2 + 4$. Since our sum starts with 2, let's look at two mice. Each mouse has two bags of spelt. Note that there are six things, just like we calculated above.

2 mice + 4 bags = 6

As before, we can get the next sum by adding 1 to 6 and then doubling. So we get $(6 + 1) \times 2 = 14$. We can think of the addition of 1 as the cat that ate the two mice. Adding the cat to the diagram is equivalent to adding the 1.

Adding the cat adds the 1 to the 6.

The doubling gives us two cats, each of which ate two mice. Algebraically we get 2×7, but pictorially we double by simply making two copies of the previous diagram. Notice that each mouse still has two spelt and the cats each have two mice.

Multiplying by 2 gives us two cats.

Note that we started with two mice and now have two cats. Looking at the rows it's apparent that we have $2 + 4 + 8$, the next sum. We can now repeat the procedure. First we add 1, which represents the home of the two cats. This gives $14 + 1 = 15$.

Adding 1 represents the house.

Now we double, creating the two houses. Arithmetically

this is $15 \times 2 = 30$. Looking at the rows, we see that this means that $2 + 4 + 8 + 16$ is 30.

Doubling makes two houses.

Hopefully you can see that the pattern continues and the trick will always work. The same procedure works for the original problem, but it has too many objects to easily represent in a diagram.

AHAH!

Inverses

Delusion is a wonderful thing. Life doesn't always give you what you want but as long as you believe it does, happiness is yours. Historians and historians of mathematics have great debates about the past. Sometimes I disagree with the prevalent opinion. In these cases I'm always right, and that is because I choose to believe this. As long as I don't leave my personal bubble, everything is wonderful. Occasionally I'm confronted with horrid things called "facts," and I'm forced to adjust my belief system, but I quickly become right once again as I now believe that I believe the truth.

Oddly enough this is precisely the method Egyptians use to solve some mathematical exercises called *ahah* problems. *Ahah* is an Egyptian word, and I like the mental image of an ancient solving a problem and exclaiming aha! I'm sure it doesn't mean this, and I haven't bothered to look up the true definition because it would ruin the magic of the word. You may think this is intellectually lazy, but trust me, I'm right in this just as I am about everything else I believe in.

The first ahah problem in Ahmose's papyrus asks what number and its seventh is 19. The answer is the ugly number $16\,\overline{2}\,\overline{8}$. I can't easily determine its seventh, and it's far from obvious that when I add the seventh I get 19. Life has given us a problem with an ugly answer, so we pretend the answer is nice. The obvious delusional thing to do is declare that the answer is 7. Ahmose and I choose 7 because its seventh is trivial. A seventh of 7 is 1, and when we add them, we get 8.

$1\,\overline{7}$	?
1	7 ✓
$\overline{7}$	1 ✓
	8

7 and its $\overline{7}$ is 8.

This is where those unfortunate facts get in the way. We were supposed to get 19, not 8. We now need to adjust our answer to fit reality. In order to do this, we need to scale up our answer to fit the actual question.

In order to understand what we're about to do, consider the statement "6 and its half is 9." This statement is scalable. If we "double" it, we get the new statement "12 and its half is 18." We could just as easily triple it and get "18 and its half is 27" or halve it to get "3 and its half is 4 and a half." All such scaled versions of our theorem are true.

In our original problem we got an answer of 8 when we wanted 19. The question we need to ask is, what number will scale 8 up to 19? The answer is remarkably simple, it's 19/8, since if we multiply by 8, the 8s cancel, leaving 19. To an Egyptian 19/8 is not a number but a problem to be solved.

?	19
1	8
2	16 ✓
$\overline{8}$	1 ✓
$\overline{4}$	2 ✓
$2\,\overline{4}\,\overline{8}$	

19 ÷ 8 gives the scaling factor.

So we get a scaling factor of $2\,\overline{4}\,\overline{8}$. Since our delusional answer was 7, we need to scale it up by $2\,\overline{4}\,\overline{8}$.

7	?
1	$2\,\overline{4}\,\overline{8}$ ✓
2	$4\,\overline{2}\,\overline{4}$ ✓
4	$9\,\overline{2}$ ✓
	$16\,\overline{2}\,\overline{8}$

Scale up the false answer by $2\,\overline{4}\,\overline{8}$.

And we now have the answer of $16\,\overline{2}\,\overline{8}$.

EXAMPLE: A number and its $\overline{5}$ is added and it is 14. What is the number?

SOLUTION: Assume the number is 5.

Calculate 5 and its $\overline{5}$ to get 6.

$1\,\overline{5}$	?
1	5 ✓
$\overline{5}$	1 ✓
	6

5 and its $\overline{5}$ is 6.

Scale 6 to 14 by calculating 14 ÷ 6, which is $2\,\overline{3}$.

?	14
1	6
2	12 ✓
$\overline{3}$	4
$\overline{3}$	2 ✓
$2\,\overline{3}$	

14 ÷ 6 gives the scaling factor.

Scale up the assumed answer of 5 by a factor of $2\,\overline{3}$ to get the answer $11\,\overline{\overline{3}}$.

5	?	
1	$2\,\overline{3}$	✓
2	$4\,\overline{\overline{3}}$	
4	$9\,\overline{3}$	✓
	$11\,\overline{\overline{3}}$	

Scale up the false answer by $2\,\overline{3}$.

PRACTICE: A number and its $\overline{2}$ is added and it is 16. What is the number?

ANSWER: $10\,\overline{\overline{3}}$

The above method is often referred to as *false position* and is still taught in some parts of the world. I personally prefer the term "delusional math," but oddly enough no one else agrees with me. It works when all terms are given proportionally. In the above practice problem, we get "its $\overline{2}$." This is 50%, a proportion. Even the phrase "a number" is proportional because it represents 100% of the number. False position does not work with constants. For example, if the problem read "a number plus 7," it could not be solved with false position. The term 7 is a number and when scaled is no longer 7. Similarly other mathematical terms, like "squares," won't work with false position.

It will, however, work with subtraction of proportional terms. Let's work with a number from which we remove an eleventh. As before, we will pick 11 as our guess since its eleventh is easy to calculate.

EXAMPLE: An eleventh is removed from a number and 3 remains. What is the number?

SOLUTION: Assume the number is 11, it's eleventh is 1 and when removed leaves 10.

Scale 10 to 3 by calculating $3 \div 10$, which is $\overline{5}\,\overline{10}$.

?	3	
1	10	
$\overline{10}$	1	✓
$\overline{5}$	2	✓
$\overline{5}\,\overline{10}$		

Scale the assumed answer of 11 by a factor of $\overline{5}\,\overline{10}$ to get the answer $3\,\overline{5}\,\overline{10}$.

11	?	
1	$\overline{5}\,\overline{10}$	✓
2	$\overline{3}\,\overline{5}\,\overline{15}$	✓
4	$1\,\overline{5}$	
8	$2\,\overline{3}\,\overline{15}$	✓
	$3\,\overline{5}\,\overline{10}$	

PRACTICE: A ninth is removed from a number and 20 remains. What is the number?

ANSWER: $22\,\overline{2}$

USELESS OR THEORETICAL

Arithmetic Series

How can you tell if a civilization has reached the pinnacle of mathematical inquiry? Ask modern theoretical mathematicians, and they will reply in unison, "If it's useless!" Of course they would probably phrase it in a way to make it sound less ridiculous. Perhaps they would say "if they studied mathematics for mathematics sake," or "if their souls were unfettered by the bounds of the mundane." Here we have two diametrically opposed views of the same thing, and as we will see, both sides have valid arguments.

This opinion is fairly ancient. The great philosopher Plato scorned intellectuals who used their mathematical knowledge for personal gain. This is a very noble sentiment until you realize that Plato was already rich and hence a little hypocritical in condemning the poor working for money. However, this opinion is not without merit. If you ignore ideas simply because they have no obvious use, you may be missing something potentially important. The first people to study electricity did so out of mere curiosity. It was "used," by and large, to give people small shocks and make their hair stand on end. If these initial useless experiments had never happened, the modern technological age might never have arrived. There's even some justification for doing things that will never be practical. People still climb Mount Everest because "it's there." In mathematics the mere intellectual challenge is often enough to make it worthwhile.

While I agree with the above argument, it sometimes gets carried to an illogical extreme in modern academia. There, "practical" is just a dirty word. Doing something useful is scorned and reduces the value of the mathematics in their eyes. The word "application" often means nothing more than mathematics that is used to create more mathematics. Both sides in this argument run the risk of missing valuable insights by ignoring the ideas generated by the other.

The theoretical point of view overlooks the fact that most of the mathematics in history was inspired by practical concerns. Both calculus and trigonometry were invented to create astronomical models. Napier invented logarithms to shorten the time it took to perform astronomical calculations. Even theoretical mathematics owes its existence to "practical" concerns, at least practical in their creators' eyes. The ancient Pythagoreans learned mathematics in order to understand God. They noticed that much of the world organizes itself mathematically, particularly in the harmonies of music. The Pythagoreans believed that if mathematical relations were favored by God in his great creation, then surely one needs to study math to gain spiritual appreciation of this.

Thousands of years later, Bertrand Russell and Alfred Whitehead tried to put all mathematics in a logical framework. In his youth, Russell logically "proved" the nonexistence of God. While his proof would not convince many people, it left him shattered and he became determined to fill the void left by the absence of God with logic. By attempting to give even the most basic mathematical notions firm logical grounding, he sought to restore purpose and order to the universe. Apparently "practical" is in the eye of the beholder, and inspiration is valuable no matter what its source.

Ultimately I would argue for balance. Advocates for both sides want to dismiss the views of the other. Whether you attack "mundane" or "useless" mathematics, you're limiting the creation of new ideas. This is never good. In the end it often hurts both sides because history is filled with practical math that inspired theoretical mathematics and abstract mathematics that eventually obtained a surprising application.

The Egyptians have now entered the ranks of the useless or inspired, depending upon your point of view. Oddly enough, the Egyptians felt the need to phrase one particularly "useless" problem as a "practical" exercise. However, the problem was so contrived that it's nigh impossible to occur in the real world, and even if it did happen once, it hardly seems worth learning a special method to solve it. The solution, however, is as delightful and simple as it is inspired, making it a worthy intellectual exercise.

The problem deals with 10 men being paid 10 hekat of barley. The unusual part of this exercise is that the salaries of the men are arranged from smallest to largest and each man is paid $\overline{8}$ hekat of grain more than the man before. The problem is contrived because Egyptian wage differences were proportional, not additive. So a share of $2\,\overline{2}$ is 25% more than a share of 2, not $\overline{2}$ more. Even if there had been an occasion when the wages were separated additively, the odds that the differences were all exactly the same is ridiculously small.

Before I solve the above problem, let's look at an easier one. Consider the wages of 4, 6, and 8 hekat. These three numbers represent an arithmetic progression because it starts with some number, 4, and then increases by the same amount each time, in this case 2. The Egyptians realized that the middle number is special. Consider the average of the three numbers. Algebraically this means the sum divided by the number of numbers. The sum is $4+6+8$ which is 18. Since there are three numbers in the sum, we divide 18 by 3 to get 6. I personally don't like arithmetic procedures because they often mask the simple ideas behind the objects. I prefer to think of the average as a "fair distribution." With $4+6+8$, there are a total of 18 things split between three numbers. The above distribution isn't fair because the first person gets 4 and the last gets 8, but it would be fair if they all got 6. Then the sum $6+6+6$ is still the full 18 objects, but now the distribution is equitable. Hence, we can think of the average as a fair distribution.

In the above example, the average of 4, 6, and 8 was 6, precisely the middle number. Let's see if this is a coincidence. Consider the arithmetic progression 5, 8, 11, 14, and 17. The sum is 55 and when divided by the number of numbers, 5, we get 11. Once again this is the middle number. This will always happen in arithmetic progressions provided there is a middle number.

In order to understand why, we need to look at the problem from the perspective of the middle person who has 11. He looks at the two people next to him, who have 8 and 14. He sees that the person with 14 has 3 more than he does and the one with 8 has 3 less. The difference is the same in both cases because it's part of an arithmetic progression that goes up each time by 3. He decides that the one with 14 has 3 too much and convinces her to give 3 of hers to the person with 8. The girl with 14 now has 3 less, or 11. The man with 8 now has 3 more, also giving him 11. So the three middle people all have 11. Finally the

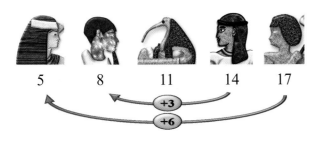

With two payments, everyone has 11.

middle person notices that the first and last persons have 6 less and 6 more than he. After some convincing, the last person gives 6 to the first person, and now everyone has a fair distribution of 11. Hence the average is 11.

Once you know the middle person gets the average, these problems are trivial to solve.

EXAMPLE: Seven people receive a total of 56 loaves of bread distributed so that each person gets 2 more than the before. How much does each person get?

SOLUTION: The middle person gets the average of $56 \div 7$.

?	56
1	7
2	14
4	28
8	56 ✓
8	

The middle person gets 8. The three above each get 2 more: 10, 12, and 14. The three below each get 2 less: 6, 4, and 2.

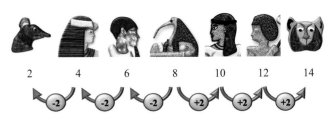

The other six are steps of 2 away from the middle person.

PRACTICE: Five people receive a total of 35 loaves of bread distributed so that each person gets 3 more than the one before. How much does each person get?

ANSWER: 1, 4, 7, 10, and 13

The Egyptians apparently thought this was too easy and dealt with the case where there is an even number of people. Consider having four people in a row. The above method doesn't work because there is no one in the middle. What you need to do in this case is imagine a pseudo-person between the second and third persons.

EXAMPLE: Four people receive a total of 28 loaves of bread distributed so that each person gets 3 more than the one before. How much does each person get?

Now in this case the "middle person" gets 7 loaves. Of course there is no middle person but, as we can see from the diagram below, the middle spot is $1\,\overline{2}$ people away from the end. Each jump is 3 long, so to find the distance from the 7 to the start we need to know how far $1\,\overline{2}$ jumps of 3 are. So let's compute $1\,\overline{2} \times 3$.

One and a half jumps from the 7 to the start.

$1\,\overline{2}$	?
1	3 ✓
$\overline{2}$	$1\,\overline{2}$ ✓
	$4\,\overline{2}$

So the total numerical distance is $4\,\overline{2}$ from 7. Hence the start is at $7 - 4\,\overline{2}$, or $2\,\overline{2}$. Now we go up 3 for each of the persons after the first. The solution, without the multiplication and division, would appear as follows:

SOLUTION: The average is $28 \div 4 = 7$. The middle is $1\,\overline{2}$ jumps of 3 from the start. Since $1\,\overline{2} \times 3$ is $4\,\overline{2}$, the first person gets $7 - 4\,\overline{2}$, which is $2\,\overline{2}$. Adding 3 to get the rest, the numbers of loaves are $2\,\overline{2}$, $5\,\overline{2}$, $8\,\overline{2}$, and $11\,\overline{2}$.

The Egyptians actually found the last person's wage and subtracted their way down the list. Since fractional subtraction can be cumbersome, I chose to do it additively. Let's do the actual problem in Ahmose's papyrus.

His problem deals with 10 people getting a total of 10 hekat of grain. Notice how easy the average is to compute. We can see the instructor trying to get the students to focus on the more important ideas by not letting confusing computations distract their attention. With 10 people in a row, there are 9 jumps from the first person to the last. So from the first person to the center it's half of 9, or $4\,\overline{2}$. Here's the problem. To give you a good model for more difficult questions, I've included the work for divisions that even the Egyptians would have omitted.

Ten people start $4\,\overline{2}$ jumps from the center.

EXAMPLE: Ten people get 10 hekat of grain where each person gets $\overline{8}$ more than the one before. How much does each person get?

SOLUTION: The average is $10 \div 10$, which is 1.

?	10
1	10 ✓
1	

The middle is $4\,\overline{2}$ jumps of $\overline{8}$ from the start.

$4\,\overline{2}$	?
1	$\overline{8}$
2	$\overline{4}$
4	$\overline{2}$ ✓
$\overline{2}$	$\overline{16}$ ✓
	$\overline{2}\,\overline{16}$

Since $4\,\overline{2} \times \overline{8}$ is $\overline{2}\,\overline{16}$, the first person gets $1 - \overline{2}\,\overline{16}$, which is $\overline{4}\,\overline{8}\,\overline{16}$.

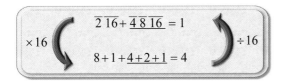

Adding $\overline{8}$ to get the rest, the wages are $\overline{4}\ \overline{8}\ \overline{16}, \overline{2}\ \overline{16}, \overline{2}$ $\overline{8}\ \overline{16}, \overline{2}\ \overline{4}\ \overline{16}, \overline{2}\ \overline{4}\ \overline{8}\ \overline{16}, 1\ \overline{16}, 1\ \overline{8}\ \overline{16}, 1\ \overline{4}\ \overline{16}, 1\ \overline{4}\ \overline{8}\ \overline{16}$, and $1\ \overline{2}\ \overline{16}$.

PRACTICE: Six people get 33 loaves of bread. If each one gets $1\ \overline{2}$ more than the one before, how much does each person get?

ANSWER: $1\ \overline{2}\ \overline{4}, 3\ \overline{4}, 4\ \overline{2}\ \overline{4}, 6\ \overline{4}, 7\ \overline{2}\ \overline{4}$, and $9\ \overline{4}$

TOO EASY IS JUST RIGHT

Arithmetic Series and Proportions

Students make mistakes and get stuck. They're supposed to. If they already knew everything, they wouldn't need to be students. Every teacher faces this dilemma on a daily basis. The only question is how to respond to their difficulties. One tried-and-true strategy is to respond with a simpler problem that uses the same ideas as the one they're having trouble with. When they grasp the notion in the easier problem, they can transfer the ideas to the harder one.

I teach a college-level course on introduction to proofs. One of the concepts students need to understand is that of negation. It's basically saying the opposite of something. If you say something's red and I say it's not red, then I've negated your statement. Suppose I ask a student to negate the following:

$$\text{If } x \in S \cap T \text{ then } \Psi(x^2) = 10.$$

The student may stare blankly, confused by the symbols and notation, let alone by the methods of negation that I've been teaching. However, I know that they know the answer. They just don't realize it yet. What I need to do is isolate the simple idea embodied in this problem from the complicated abstract language.

So I ask them this: "Imagine I go to you and say that if you lend me $20 today, I'll pay you back tomorrow. But I lied, and remember negations are about lies, the opposite of truth. You're mad because I lied. How do you know I lied?" They think about it and respond, "Because you didn't pay me back." I ask them if they're sure. Then I ask, "Would you be mad at me if you never lent me any money?" "Of course not," they reply. "So what has to happen?" I ask. "I have to lend you the $20 and you don't pay me back."

Whether they realize it or not, they've just negated a statement. I then write the following on the board:

Statement: If LEND, then PAY BACK.
Negation: LEND and don't PAY BACK.

I ask what happened, in a logic sense, to "PAY BACK." They'll reply, "Since you didn't pay me back, it got negated," I ask, "Was LEND negated?" "No, I did lend you the money," they'll answer. "What happened to the 'if … then…'?" I ask. "It got replaced with '… and… .'" Now I ask them to look at the original problem.

$$\text{If } x \in S \cap T \text{ then } \Psi(x^2) = 10.$$

I remind them that the first part was left alone, the second negated, so the "if…then" becomes "and." By now they are ready to write the answer below, knowing the negation of = is ≠.

$$x \in S \cap T \text{ and } \Psi(x^2) \neq 10.$$

My point in all of this is that new complicated problems can often be solved using simple methods that are already known. Seeing a simple problem right before the similar complicated one can trigger ideas and lead to a solution. The simple problem needs to be transparent and so easy to understand that the mental focus is on the method, not on the other aspects of the question.

Ahmose has another arithmetic progression problem on his papyrus that's so poorly explained that there is debate as to how the problem is solved. This time we're splitting up 100 loaves of bread among five men whose shares are in arithmetic progression. But instead of giving the difference in shares, he says the shares of the first two are a seventh of the top three.

The solution seems to come from nowhere. It's simply announced that the difference between the shares should be 5 $\overline{2}$, and then the values of the shares are given as 23, 17 $\overline{2}$, 12, 6 $\overline{2}$, and 1 for a total of 60. This solution is then scaled upward as in the ahah problems to make the total 100. Everything that is done is correct, but nowhere in the text is it explained where these numbers come from. When we add up the top three shares—23, 17 $\overline{2}$, and 12— to get 52 $\overline{2}$ and we add the bottom two—6 $\overline{2}$ and 1—to get 7 $\overline{2}$, it's possible to see, although not immediately obvious that 7 $\overline{2}$ × 7 is 52 $\overline{2}$, just as the problem promised. There are many opinions as to how these numbers were obtained, and I have my own guess.

The key to understanding this exercise is to look at the problem just before it. This problem is easy. In fact it's much too easy for its location in the book, but that's the point. The lessons from the easy problem can be used to solve the far more complicated one just after it.

The easy problem is to distribute 100 loaves of bread between 10 men such that the first 4 get half and the last 6 get half. Clearly half of 100 is 50, so all we need to do is divide 50 by 4 to determine what the first 4 men get and then divide it by 6 to figure out what the last 6 get. It's a simple whole-number division.

In order to understand the role of the simple question, let's first see how it compares with the problem after it. Both exercises deal with the distribution of 100 loaves of bread and both break the people into two groups. The easy problem has a group of 4 out of 10, and the difficult problem has a group of 2 out of 5. This is exactly the same proportion, a concept central to Egyptian mathematics that is used, for example, in computation by parts. The author is begging the student to draw parallels between the two.

The easy problem says the small group gets half the total and the difficult problem says that the large group's share is 7 times the small group's share. These may seem different on the surface, but in fact they are basically the same conditions. In order to see this you need to realize that if you have 7 times more than I do, then I have an eighth of the total. Just think in parts. If I have one part, then you have seven. There is a total of eight parts, of which I have one. Hence, I have an eighth. The easy problem gives the small group $\overline{2}$ of the total while the difficult problem gives them $\overline{8}$ of the total. Calculations using these two fractions are almost exactly the same. One's division by 2 and the other is repeated division by 2.

The main difference between the two problems is that the easy problem gives equal amounts to those in a certain group, whereas the hard one gives out the bread in an arithmetic progression. These conditions may seem radically different, but they differ largely in semantics. If I have 12 items to be divided up between three people, I could fairly give each person 12 ÷ 3, or 4. If I was dealing with an arithmetic progression, I would still divide 12 by 3, but I would interpret the 4 as the average amount of each, not the exact amount. Once we're talking "average," we're really discussing the share of the middle person.

In the difficult problem, the number of loaves was changed from 100 to 60. I suspect the number 60 was selected simply because it's easy to work with. You can take $\overline{2}$, $\overline{\overline{3}}$, and $\overline{10}$ of it and still get whole numbers. The Egyptians clearly favored 60. Out of the last 15 entries on Ahmose's times-2 table, a third start with $\overline{60}$. It's easy to break into parts and has many factors. Although 100 isn't bad for the number in this exercise, most other numbers would be quite cumbersome to use, and the author is trying to teach us general methods that should work with any quantity. Hence, we do the hard parts of the problem with easy numbers.

Let's work through the problem using 60 loaves instead of 100. In the problem there are five people in arithmetic progression. The average person gets 60 ÷ 5 loaves, which is 12.

?	60	
1	5	
10	50	✓
2	10	✓
12		

60 ÷ 5 is the middle person's share.

Since we have five people, the middle person is the third one in line. This means that the third person gets the average, which is 12 loaves of bread. Remember that the last three get 7 times as much as the first two; therefore, the first two get $\overline{8}$ of 60, which is 7 $\overline{2}$.

$\overline{8}$	?
1	60
$\overline{2}$	30
$\overline{4}$	15
$\overline{8}$	7 $\overline{2}$ ✓
	7 $\overline{2}$

60÷8 is the share of the first two persons.

Since the first two get a total of 7 $\overline{2}$, the "middle person" between them gets half, which is 3 $\overline{2}$ $\overline{4}$. We need to know how much more the middle person of the whole group gets than the middle person of the first two. Completing 3 $\overline{2}$ $\overline{4}$ to 12, we get a difference of 8 $\overline{4}$. Looking at the following diagram, it's easy to see that the person getting 12 and the one getting 3 $\overline{2}$ $\overline{4}$ are one and a half jumps apart.

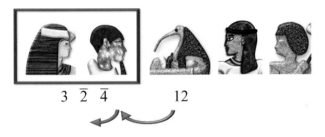

3 $\overline{2}$ $\overline{4}$ 12

There are 1 $\overline{2}$ jumps from the middle person of the five to the middle person of the first two.

So 1 $\overline{2}$ jumps goes a total distance of 8 $\overline{4}$ away, and hence each jump goes 8 $\overline{4}$ ÷ 1 $\overline{2}$. The answer to this division is 5 $\overline{2}$, exactly what Ahmose's text told us the answer would be.

So this is perhaps where the 5 $\overline{2}$ came from. Now that we know the middle and the jump, we can simply add 5 $\overline{2}$ to and subtract 5 $\overline{2}$ from 12 to get the other four values,

?	8 $\overline{4}$
1	1 $\overline{2}$ ✓
2	3
4	6 ✓
$\overline{2}$	$\overline{2}$ $\overline{4}$ ✓
	5 $\overline{2}$

giving 17 $\overline{2}$ and 23 going up, and 6 $\overline{2}$ and 1 going down. We now know the five shares if there are 60 loaves. It's now time to scale up the loaves to 100. We do this by multiplying each answer by 100÷60, which we will now compute.

?	100
1	60 ✓
$\overline{\overline{3}}$	40 ✓
	1 $\overline{3}$

We multiply each of the five shares by 1 $\overline{\overline{3}}$ as follows. (I will omit the obvious 1 $\overline{\overline{3}}$ × 1 computation.)

1 $\overline{\overline{3}}$	?
1	6 $\overline{2}$ ✓
$\overline{\overline{3}}$	4 $\overline{3}$ ✓
	10 $\overline{\overline{3}}$ $\overline{6}$

1 $\overline{\overline{3}}$	?
1	12 ✓
$\overline{\overline{3}}$	8 ✓
	20

1 $\overline{\overline{3}}$	?
1	17 $\overline{2}$ ✓
$\overline{\overline{3}}$	11 $\overline{3}$ ✓
	29 $\overline{6}$

1 $\overline{\overline{3}}$	?
1	23 ✓
$\overline{\overline{3}}$	15 $\overline{3}$ ✓
	38 $\overline{3}$

Hence, each person gets 1 $\overline{\overline{3}}$, 10 $\overline{\overline{3}}$ $\overline{6}$, 20, 29 $\overline{6}$, and 38 $\overline{3}$. If you examine the above numbers carefully, you might notice that each number is 9 $\overline{6}$ higher than the one before it. The sum of the first two is $12\,\overline{2}$, and the sum of the last three is $87\,\overline{2}$. I'll leave it up to you to determine if the last three are truly seven times greater than the first two. Let's try another one without all the explanation.

EXAMPLE: Six people get paid 69 loaves of bread given out in an arithmetic progression. The last three get three times as much as the first three. How much does each person get?

SOLUTION: Assume they get a total of 60 loaves of bread. The average share is $60 \div 6$, or 10. Hence the "middle person" gets 10.

?	60
1	6
10	60 ✓
10	

The first three get a 1 share to the last three's 3 share and hence get $\overline{4}$ of the total. $\overline{4}$ of 60 is 15.

$\overline{4}$	?
1	60
$\overline{2}$	30
$\overline{4}$	15 ✓
	15

Since there are three in the first group, the middle person gets $15 \div 3$, or 5, loaves of bread.

?	15
1	3 ✓
2	6
4	12 ✓
5	

5 ⟵ ⟶ 10

The "middle person" gets 10, $1\,\overline{2}$ steps away, the middle of the first three gets 5.

The middle person of the first three makes 5 less than the middle person of all six and is $1\,\overline{2}$ jumps away, and hence each jump is $5 \div 1\,\overline{2}$, or $3\,\overline{3}$.

?	5
1	$1\,\overline{2}$ ✓
2	3 ✓
$\overline{3}$	$\overline{2}$ ✓
$3\,\overline{3}$	

We know the second person gets 5, to which we can add and subtract $3\,\overline{3}$ to get the others, giving: $1\,\overline{\overline{3}}, 5, 8\,\overline{3}, 11\,\overline{\overline{3}}, 15,$ and $18\,\overline{3}$. We need to scale up our answer to 69 loaves of bread by multiplying by $69 \div 60$, which is $1\,\overline{10}\,\overline{20}$.

?	69
1	60 ✓
$\overline{10}$	6 ✓
$\overline{20}$	3 ✓
$1\,\overline{10}\,\overline{20}$	

Now we multiply $1\,\overline{10}\,\overline{20}$ by each answer to get the adjusted shares.

$1\ \overline{10}\ \overline{20}$

1	$1\ \overline{\overline{3}}$ ✓
$\overline{10}$	$\overline{10}\ \overline{15}$ ✓
$\overline{20}$	$\overline{20}\ \overline{30}$ ✓
	$1\ \overline{3}\ \overline{4}$

$1\ \overline{10}\ \overline{20}$

1	5 ✓
$\overline{10}$	2 ✓
$\overline{20}$	4 ✓
	$5\ \overline{2}\ \overline{4}$

$1\ \overline{10}\ \overline{20}$

1	$8\ \overline{3}$ ✓
$\overline{10}$	$\overline{\overline{3}}\ \overline{10}\ \overline{15}$ ✓
$\overline{20}$	$\overline{3}\ \overline{20}\ \overline{30}$ ✓
	$9\ \overline{2}\ \overline{12}$

$1\ \overline{10}\ \overline{20}$

1	$11\ \overline{\overline{3}}$ ✓
$\overline{10}$	$1\ \overline{10}\ \overline{15}$ ✓
$\overline{20}$	$2\ \overline{20}\ \overline{30}$ ✓
	$13\ \overline{3}\ \overline{12}$

$1\ \overline{10}\ \overline{20}$

1	15 ✓
$\overline{10}$	$1\ \overline{2}$ ✓
$\overline{20}$	$\overline{2}\ \overline{4}$ ✓
	$1\ 7\ \overline{4}$

$1\ \overline{10}\ \overline{20}$

1	$18\ \overline{3}$ ✓
$\overline{10}$	$1\ \overline{3}\ \overline{10}\ \overline{15}$ ✓
$\overline{20}$	$2\ \overline{3}\ \overline{20}\ \overline{30}$ ✓
	$21\ \overline{12}$

Hence, they get $1\ \overline{3}\ \overline{4},\ 5\ \overline{2}\ \overline{4},\ 9\ \overline{2}\ \overline{12},\ 13\ \overline{3}\ \overline{12},\ 17\ \overline{4},$ and $21\ \overline{12}$.

PRACTICE: Ninety loaves of bread get distributed between five people in arithmetic progression. If the upper three get four times as much as the lower two, how much does each person get?

ANSWER: 6, 12, 18, 24, and 30

PRACTICE: Sixty-six loaves of bread get distributed between five people in arithmetic progression. If the upper two get two times as much as the lower three, how much does each person get?

ANSWER: $1\ \overline{3}\ \overline{10}\ \overline{30},\ 7\ \overline{3},\ 13\ \overline{5},\ 19\ \overline{15},$ and $24\ \overline{\overline{3}}\ \overline{5}\ \overline{15}$

These problems are tricky; they would be in any mathematical system. However, the easy problem before them, if carefully understood, holds the key to the solution.

THE PYLONS OF THE NIGHT

Cycles in the Doubling Table

Starting each dawn, the sun god Ra sails across the daytime sky. In the morning he is young, but by sunset he is old. At dusk, the solar barque he sails passes between two mountains entering a long, dark valley called the Duat. This is the land of the nighttime sun. It's the Egyptian underworld, quite possibly the origin of the Christian conception of hell. It's populated by gods, demons, and the souls of the dead, both damned and those searching for salvation. The Duat is divided into twelve distinct regions, each demarcated by fortified pylons. The pylons represent the twelve hours of night that Ra must pass through in order to be born again in the eastern horizon.

Ra in his night boat protected by two attendees and a serpent.

There are two main sources that describe Ra's journey through the night. They are based on two competing mythologies, one in which Ra is central and the other promoting Osiris. As a result, they differ somewhat in their description of the twelve hours. The sources are *Things in the Duat* and *The Book of Pylons*, and the following is a mix of the two.

As Ra approaches the land of the night, he is as dead as a living god can be. The book *Things in the Duat* refers to him simply as flesh. He is joined by the goddess Lady of the Boat. She is the embodiment of the first hour of the night and will act as his guide through this region. Each hour he will have a new guide to help him on his way. In the first hour, the boat approaches a giant hall fifteen miles long. Ra is allowed to enter by ape doormen who open the giant gate, and he is greeted by a heavenly choir. The hall is filled with the souls of the dead awaiting entry onto the boat, hoping that they can spend eternity with their beloved god.

At the end of the first hour, Ra's boat approaches a giant pylon. It's guarded by gods and a giant serpent. Inside is a hallway lined with spear tips. At each corner is a *ureaus*, a cobra-like creature who spits fire on unwelcome intruders, and a representation of the power of the gods. Similar pylons separate each hour of the night. As Ra moves into the second hour, he is greeted by more souls of the dead. These spirits are still making their way to the great hall of the first hour in hopes of joining Ra. As they bask in the rays of the nighttime sun, they temporarily come back to life. Ra addresses the souls and gives them food, water, and divine air to breath.

The third hour is called "the country of those that slay." This land is filled with the souls of the damned. Many are unwillingly dragged from their tombs here by the gods of the underworld. In this hour, the gods enact divine justice, spitting fire on and hacking up the souls that have been deemed the enemies of Ra.

The fourth hour is the outer region of Seker's domain. He is the hidden divinity, never seen by mortal or god. There are no longer gods on the banks of the river. There are only monstrous serpents, many with multiple heads, human body parts, or wings. They scour the land for evil souls to consume. The realm is so dark that the light of the sun god is insufficient illumination. The boat spawns a fire-breathing serpent head to light the way.

The fifth hour is the heart of Seker's domain, where his body lies buried under a mound guarded by sphinxes and serpents. The boat needs to be dragged over the sand of the mound. At the top is a scarab beetle, one of the holiest of Egyptian symbols. In modern times, this insect is known, ignobly, as a dung beetle. It rolls a ball of feces containing its eggs and pushes it into a hole where its young hatch. The Egyptians saw this process as collecting inert matter and transforming it into new forms of life, a power usually reserved for gods. The beetle's appearance is symbolic of Ra's receiving the germ of life, giving Ra the potential of rebirth.

The scarab.

In *The Book of Pylons*, the sixth hour is spent in the Hall of Osiris, the mummified lord of the Duat. In this version, Ra doesn't even appear in this realm. According to the priests of Osiris, this is the location of the hall of judgment. Here Osiris holds court, determining the fate of the souls of the dead.

In the remaining gates, Ra and his company must defeat the darkness in order to be reborn. The night is personified by the great serpent Apep and his host of beasts, born of the primordial chaos. In the seventh hour, the boat runs aground while being confronted by the beast Neha-hra. Isis, the wife of Osiris and mother of Horus, casts powerful magic to get the boat in motion. A band of gods pins down Neha-hra with their daggers as Ra passes safely.

In the eighth hour, Ra is in the home of Kheti, a giant serpent that spits fire. His role is to seek out and destroy the damned called "the enemies of Osiris." He has the power to destroy them, body and soul.

In the ninth hour, Apep himself makes an appearance. A horde of men, women, and monkey magicians rush to Ra's aid, casting protective spells while other gods restrain Apep with ropes. Men with pikes appear, stabbing the great beast repeatedly. By the tenth hour, Apep is fastened to the ground with chains as the star gods begin to appear announcing the near arrival of the dawn. By the eleventh hour, Appep is finally sliced into pieces, finishing off the darkness just before the first light.

The coils of Apep.

1	$\overline{19}$
2	$\overline{12}\ \overline{76}\ \overline{114}$
4	$\overline{6}\ \overline{38}\ \overline{57}$
8	$\overline{3}\ \overline{19}\ \overline{38}\ \overline{114}$
16	$\overline{\overline{3}}\ \overline{12}\ \overline{76}\ \overline{114}\ \overline{19}\ \overline{57}$
32	$1\ \overline{3}\ \overline{6}\ \overline{12}\ \overline{19}\ \overline{38}\ \overline{76}\ \overline{114}$
64	$3\ \overline{6}\ \overline{12}\ \overline{19}\ \overline{38}\ \overline{57}\ \overline{76}\ \overline{114}$
128	$6\ \overline{3}\ \overline{19}\ \overline{57}$

$\overline{19}$ doubled repeatedly.

In *Things in the Duat*, the eleventh hour is spent in the land of fire. It shows one more scene of the torment of the damned. It contains five fiery pits, each overseen by a dagger-wielding, fire-breathing woman. In the first pit, the condemned smash their own brains with axes. In the second, the bodies of the damned are demolished. The other pits are where the soul, shadow, and heads of the enemies of Ra are destroyed.

Finally in the twelfth hour, the boat enters the tail of a great serpent named Life of the Gods and travels through its body. The boat of Ra emerges out of the mouth of the beast into the eastern heavens. Ra once again is young and is ready to repeat his journey through the daytime sky.

Many people mistakenly believe that the Egyptians were consumed with the thought of death. But the time and effort they spent on their tombs was not with the aim of housing their dead bodies but rather of preparing them for a new life. The notion of rebirth is clearly portrayed by the story of the night. It's reflected in their view of monarchy. The living pharaoh was a representation of Horus. When he died, he became Osiris, living in a new realm. Horus would then be reborn in the new pharaoh and the cycle continued. Consistency and order, called *Ma'at*, is maintained.

There's a sort of rebirth in the endless cycles that occur in Egyptian mathematics. Consider the fraction $\overline{19}$ and what happens when it is consistently doubled. The following is computed using only simplifications from the doubling table and the even-fraction rule.

Although the above seems complicated, the same fractions—$\overline{\overline{3}}, \overline{3}, \overline{6}, \overline{12}, \overline{19}, \overline{38}, \overline{57}, \overline{76},$ and $\overline{114}$—keep appearing, dying, and being reborn in an endless cycle. No matter how far we go, there will never be any fractions but these nine. Although more and more fractions appear, since there are only nine, eventually they'll repeat. So perhaps we'll get $\overline{38}$ and $\overline{38}$. Of course, it is easy to simplify as $\overline{19}$. If we were to get an odd fraction, like $\overline{57}$ and $\overline{57}$, we simply do a table look up and replace it with $\overline{38}\ \overline{114}$. Note that if we take $2 \times \overline{3}$ and $2 \times \overline{3}$ for granted, there are only two table look-ups because the only odd fractions are $\overline{19}$ and $\overline{57}$.

If I were to change Ahmose's table arbitrarily, the group wouldn't be so tight. Say, for example, I changed the table entry of $2 \times \overline{19}$ to $\overline{10}\ \overline{190}$, then the doubling would look like this:

1	$\overline{19}$
2	$\overline{10}\ \overline{190}$
4	$\overline{5}\ \overline{95}$
8	$\overline{3}\ \overline{15}\ \overline{60}\ \overline{380}\ \overline{570}$
16	$\overline{3}\ \overline{10}\ \overline{15}\ \overline{190}\ \overline{285}$

$\overline{19}$ doubled with new value.

I have to stop here since there's no entry for $\overline{285}$ on Ahmose's table. One change in the table, and we go from two table look-ups to five odd fractions—$\overline{5}$, $\overline{15}$, $\overline{19}$, $\overline{15}$, and $\overline{285}$—one of which isn't even on the table. We'd get even more odd fractions if I changed the entries for one of the other four odd fractions. The tables, I believe, are designed to cycle so that each multiplication uses as few fractions, odd ones in particular, as possible.

To demonstrate this I'm going to create what I call an odd-fraction family diagram. Consider the fraction $\overline{15}$. When doubled, it becomes $\overline{10}\,\overline{30}$. Both of these numbers are easy to double, becoming $\overline{5}$ and $\overline{15}$ respectively. These are both odd fractions, so I make the following graphic depiction.

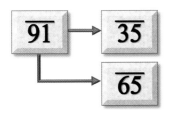

The arrow from $\overline{15}$ to $\overline{5}$ represents the $\overline{10}$ that reduces to the odd $\overline{5}$ through doubling. The arrow from $\overline{15}$ to itself indicates that the $\overline{30}$ reduces back to the original $\overline{15}$. Let's try another.

EXAMPLE: Represent the $\overline{91}$ node of an odd-fraction family tree.

SOLUTION: $2 \times \overline{91} = \overline{70}\,\overline{130}, 2 \times \overline{70} = \overline{35}, 2 \times \overline{130} = \overline{65}$.

So we get

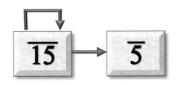

Try this one, and remember that you keep doubling the pieces until they become an odd fraction.

PRACTICE: Represent the $\overline{49}$ node of an odd-fraction family tree.

ANSWER:

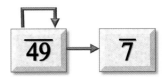

I consider the fractions $\overline{2}, \overline{3}$, and $\overline{\overline{3}}$ trivial to double, so when one of these is reached I will mark the box with an asterisk (*) to indicate that no complicated math is required from that point forward.

EXAMPLE: Represent the $\overline{23}$ node of an odd-fraction family tree.

SOLUTION: $2 \times \overline{23} = \overline{12}\,\overline{276}$.
$2 \times \overline{12} = \overline{6}, 2 \times \overline{6} = \overline{3}, 2 \times \overline{276} = \overline{138}, 2 \times \overline{138} = \overline{69}$.

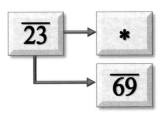

Now these are not complete odd-fraction family trees. We looked at only one number. We want all the arrows that come out of every number in our diagram. For example, when we created the node for $\overline{15}$, one arrow went to $\overline{5}$ and the other back to $\overline{15}$. However, there are no arrows leaving $\overline{5}$, forcing us to find $\overline{5}$'s arrows too. If an arrow out of $\overline{5}$ led to another odd, we would need to keep going until no new odds were generated. Since $2 \times \overline{5}$ is $\overline{3}\,\overline{15}$, one arrow goes back to $\overline{15}$ and the other goes to *. There are no new odd fractions, hence we are done.

The complete $\overline{15}$ family tree.

Notice how small the diagram is. These represent the only odd fractions we will obtain by the continual doubling of $\overline{15}$. So when multiplying by $\overline{15}$, unless we change the fractions with some simplification trick, the only odd fractions we'll get are $\overline{5}, \overline{3}$, and $\overline{15}$ itself.

EXAMPLE: Find the complete odd-fraction family tree for $\overline{11}$.

SOLUTION: $2 \times \overline{11} = \overline{6}\ \overline{66}, 2 \times \overline{6} = \overline{3}, 2 \times \overline{66} = \overline{33}$.
$2 \times \overline{33} = \overline{22}\ \overline{66}, 2 \times \overline{22} = \overline{11}, 2 \times \overline{66} = \overline{33}$.

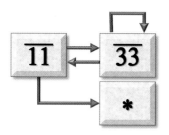

Sometimes we can use previous answers to complete a problem.

EXAMPLE: Find the complete odd-fraction family tree for $\overline{77}$.

While solving this example, we're going to run into $\overline{11}$, whose tree we already know, so we can "cut and paste" the previous answer into the current problem.

SOLUTION: $2 \times \overline{77} = \overline{44}\ \overline{308}, 2 \times \overline{44} = \overline{22}, 2 \times \overline{22} = \overline{11}$.
$2 \times \overline{308} = \overline{154}, 2 \times \overline{154} = \overline{77}$.

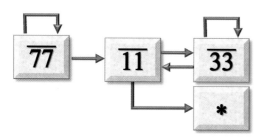

PRACTICE: Find the complete odd-fraction family tree for $\overline{21}$.

ANSWER:

PRACTICE: Use the previous answer to find the complete odd-fraction family tree for $\overline{63}$.

ANSWER:

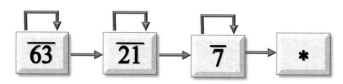

The point I want to make once again is that these trees are remarkably small. At the beginning of this section, I made one change and created a number whose tree would have been larger than any other we've made. Most trees are small. There are a few bad apples, particularly those that make odd fractions off the doubling table. But most of them cycle between the same odd fractions, keeping the total number of different fractions from growing without bound.

I personally suspect that this was intentional on the part of the creators of the table. The pattern seems clear, but we need to be careful. Patterns are a part of math and occur naturally. It's possible that the pattern we see is an unintended consequence of the method they used to create their doubling table. However, the connection between elements of the doubling table is not the only synchronicity we see in Egyptian math.

As we have seen, the Egyptian Mathematical Leather Roll has many seemingly unrelated identities. Consider the identity $\overline{7}\ \overline{14}\ \overline{28} = \overline{4}$. These numbers come out of the doubling of $\overline{7}$. Consider the family tree of $\overline{7}$ enhanced to have all fractions that arise by doubling it. I've put the numbers on the arrows. So the arrow that leaves $\overline{7}$ and goes back to itself does so by doubling $\overline{28}$ twice. Hence the numbers $\overline{28}$ and $\overline{14}$ appear on the arrow. The new diagram shows all fractions generated by $\overline{7}$ except the three trivial numbers $\overline{3}, \overline{\overline{3}}$, and $\overline{2}$.

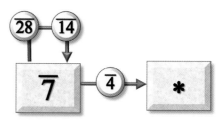

The fractions generated by the tree of $\overline{7}$ are in $\overline{7}\ \overline{14}\ \overline{28} = \overline{4}$.

Notice that these numbers are precisely the ones in the identity. Because these fractions are likely to remain in many problems where $\overline{7}$ is continually doubled, this identity can help simplify these.

Now let's consider the identity $\overline{18}\ \overline{27}\ \overline{54} = \overline{9}$. Looking at the fractions on the left, we see that doubling reduces them to $\overline{9}$ and $\overline{27}$. Let's look at the family tree for $\overline{27}$.

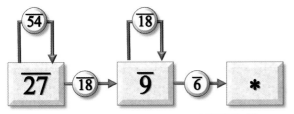

The family of $\overline{27}$ has the fractions of $\overline{18}\ \overline{27}\ \overline{54} = \overline{9}$.

EXAMPLE: Draw the augmented complete family tree of $\overline{49}$ and show that it contains the fractions on the left-hand side of $\overline{28}\ \ \overline{49}\ \ \overline{196} = \overline{13}$.

SOLUTION: $2 \times \overline{49} = \overline{28}\ \overline{196}, 2 \times \overline{28} = \overline{14}, 2 \times \overline{14} = \overline{7},$
$2 \times \overline{196} = \overline{98}, 2 \times \overline{98}\ = \overline{49}, 2 \times \overline{7} = \overline{4}\ \overline{28}, 2 \times \overline{4} = \overline{2}.$

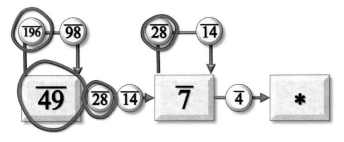

All the elements of $\overline{28}\ \ \overline{49}\ \ \overline{196}$ are circled and each appears at least once.

Here's a practice problem with one of the more involved family trees.

PRACTICE: Draw the augmented complete family tree of $\overline{75}$ and show that it contains the fractions on the left-hand side of $\overline{25}\ \overline{50}\ \overline{150} = \overline{15}$.

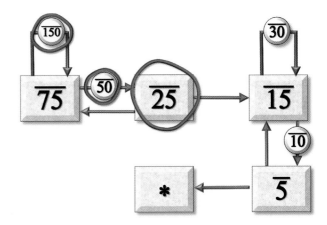

Almost all of the entries in the Egyptian Mathematical Leather Roll can be thought of in this light, although there are a few that can't, such as $\overline{15}\ \overline{25}\ \overline{75}\ \overline{200} = \overline{8}$. However, the first three fractions can be generated by the family tree of $\overline{75}$. If we allow taking half in addition to doubling, the $\overline{200}$ can be derived by twice taking half of $\overline{50}$, which is in the tree.

It appears that the Egyptian system is carefully thought out. The process is fine-tuned for exceptional efficiency.

7

BASE-BASED MATHEMATICS

Stop reading this chapter! For God's sake, aren't you listening! I know you're still reading and I'm beginning to get annoyed. I have half a mind to turn this book around and go home. You'd be left with a hundred or so blank pages and who would be sorry then?

I'm faced with a dilemma. Most authors try to write to please their audience. Up to this point I've tried fairly hard to work a little color and humor into a book that might otherwise consist of little but a few dry mathematical procedures. I've tried to make the mathematics relatively easy and painless to learn. However, things have to change. You see, one of the main points of the book is that Egyptian mathematics stands up in comparison with modern methods. So far I've spent my pages touting the Egyptian system, but to complete the argument I now have to downplay our modern base system. To do this, I need to show you some dreadful mathematics. I'll still try to trick you into reading what's to follow. I'll throw in an amusing anecdote or an interesting bit of trivia, but in the end, my goal is to make you miserable. Ah, I see you're still reading. I respect your choice even though I don't fully understand it.

There's another problem. You're too conversant with modern methods. You've forgotten how hard they are to learn, and you probably take all the nuances for granted. Just as it's difficult for a native-born English speaker to appreciate how difficult their first language is to learn, I would have trouble convincing you how difficult your mathematical methods are.

So I'm going to disguise what you already know so that it seems new to you. I'm going to make the smallest change. I'm going to teach you the mathematical procedures of the ancient Babylonians, whose system is probably the origin of our system today. The primary difference is in what they used as a base. In base-ten mathematics a 1 followed by a 0 means 10. The Babylonians essentially used base-sixty mathematics, where they would read "10" as 60. I'll change the symbols we use to write numbers, and I'll modify the way multiplication tables are organized. All of these changes will conform by and large to the way the Babylonians did their computation, but they are basically cosmetic.

If you are like my students in my history of math classes, this small change will have an enormous effect. Instead of being able to solve some multiplications using Egyptian methods in thirty seconds, in Babylonian they can take half an hour. Frustrated students check and recheck their work, unable to find their one careless mistake. They groan and beg me not to put a Babylonian multiplication on the final exam. Unfortunately, their pain makes my point.

WHEAT, BONES, AND TIME

Tally Sticks

Almost ten thousand years ago, humans living in the Middle East were confronted with a new problem. This relatively new species had burst from Africa between fifty and a hundred thousand years prior and quickly dominated the globe. These humans were smarter and more adaptable than their humanoid relatives and quickly drove the latter to extinction.

Before this period, human development occurred rather slowly. Six million years ago, our genetic line separated us from other monkeys. We learned to walk four million years ago. Two and a half million years ago, we started using stone tools as our brain size began to expand. One and a half million years ago, we mastered fire as our larynx began to drop, indicating that we had some form of primitive speech. We find that the earliest settlements were made around three-quarters of a million years ago, the same time that stone axes began to be standardized. A half million years ago, the nerve to our tongue thickened. Together with a corresponding growth in brains size, these suggest that human language was becoming more complex. All of the above are big milestones for humanity, but they're all measured in millions of years.

The new humans changed at an unheard-of pace. Everywhere they went, they adapted to their environment. They adapted to forest, desert, island, mountain, and jungle terrains. Those near the poles adapted to cold, and those near the equator adapted to heat. Their tools, clothes, and habits changed constantly to suit whatever situation they encountered. In contrast, previous humanoid species had changed slowly. If you were to look at two ax blades separated by a half million years, you would hardly notice the difference unless you were an expert. The evolution of the early species of humanoids seemed to be driven by genetic mutations, a process that takes considerable time. The evolution of modern humans was driven by changes in ideas that could be altered by a flash of inspiration.

For most of the time since these humans spread from Africa, the world had been in an ice age. We tend to associate ice ages with winter and assume it must have been difficult to survive. Actually the opposite is true. The ice age caused the forests of the world to recede and be replaced with plains. Grasses grow much faster than trees and, being close to the ground, are easy to eat. As a result, huge herds of animals roamed the plains of the ice age. This is in direct contrast to the tiny scattered animals that inhabited the forests predating the ice age. Take down one herd animal in the open plains and you could feed a tribe. Chase a squirrel in the forest for hours and you would get a light lunch for yourself. Meat was plentiful in the ice age. In many ways, at least when food is scarce, meat is far more nutritious than plants, so human hunters thrived.

The ice age peaked a mere twenty thousand years ago and began to recede at different rates in different locations during the next ten thousand years. The thriving human population found their main food source disappearing with the plains. The forests and other terrains that replaced the plains simply could not sustain the relatively large populations that had grown during the abundant years. Humans had to adapt or die, but these humans specialized in adaptation.

In the hills near modern Israel, there grew a grass that produced hard little seeds. Some adventurous, or perhaps desperate, humans decided to eat the seeds. They were surprisingly nutritious, containing a concentrated food source intended for the seed. Today we refer to this grass as wheat.

Up to this point, humans were hunter-gatherers. They travelled from food source to food source as the seasons progressed. This routine became more and more difficult as humans filled the globe. What's the point of walking to a new area if it has already been picked over by another band of humans? The wheat didn't immediately stop the humans from wandering, but it did slow them down. The wheat grew on hills tall enough to be noticeably cooler as one went up. Hence the seasons were offset, and the wheat would produce seeds at different times at different altitudes. This enabled the hunter-gatherers to stay in one spot for a fair stretch of time.

Then something wonderful happened. The wheat began to change to the benefit of humans. As strange as that sounds, similar processes have happened many times in the course of biological evolution. Have you ever wondered why fruit comes in so many bright colors and tastes so good? It's this way because the plant "wants" the fruit to be eaten. Originally the food of the fruit evolved to feed the seeds. Certain animals adapted to eat the fruit. Although this would be bad for many seeds, some survived the trip through the animal's digestive tract. The seeds would find themselves far away from their parent and in a nutritious soil that only plants and flies could love.

Being dispersed far and wide is actually a great benefit for seeds. Typically, fruit falls off their parent tree and lands nearby. So a parent and all its descendants would be located in a fairly small area. A wandering herd, a dry spell, or a fire could spell doom to an entire family tree of some plant. Dispersal of seeds over a great distance would not only protect a genetic line from disaster, but it could also find new habitats for growth.

Hence fruits that were eaten had a survival advantage over those that simply lay where they fell. Evolution favored fruits that were more likely to be consumed. Animals had to want to eat them; hence, the fruits had to be nutritious and tasty. Animals also had to be able to find them, so the fruits turned bright colors when the seeds were fully developed. It's not that the fruit consciously changed, but the plants that possessed these traits had a greater chance of reproducing. In a similar way, flowers produced nutritious pollen and bright colors that were attractive to bees.

Wheat underwent similar modifications. When the hunter-gatherers carried the wheat back to the settlements, they inadvertently dropped seeds, enabling these plants to spread quickly. As a result, the wheat changed. The seeds became larger because the bigger ones were more enticing to humans and more likely to be picked. The connections to the plant evolved to become brittle, so most seeds would fall off when the forager gave them a good shake. As the wheat altered, it became more productive for the hunter-gatherers to collect it, in turn causing humans to harvest more. As a result, the wheat changed faster, prompting humans to collect even more of it.

Presumably some humans noticed that wherever they put seeds, wheat would eventually grow. Some bright human probably decided to throw a bunch of seeds near his home. Why should they walk miles when they could get their food to grow nearby? The more these humans relied on wheat, the less they had to move. This situation made it worthwhile to build sturdier shelters and make tools that were too heavy to carry easily in the typical hunter-gatherer life. In turn they could specialize in wheat production, and eventually farming became a way of life.

Wheat has one special property that enabled it to change the world. Vegetables wilt, fruits get moldy, and meat gets rank, but I would bet that you have a bag of flour in your kitchen that has been sitting there unrefrigerated for weeks, months, or possibly years. Basically, it doesn't go bad. Most of the hunter-gatherers still around today don't stock food; they usually collect what they eat each day. If they tried to do otherwise, they might get sick from eating spoiled food. But wheat gathers could collect more than they needed and store it for later use, as long as they could keep it dry and protected from vermin. Hence, wheat could be used to feed the hunter-gatherers as they waited for the next food source to arrive.

At this point, humans faced a new problem. Here's how my over-overactive imagination depicts their dilemma:

> The wheat storage pit is three feet in diameter and the collected wheat is two feet deep. If a person eats two cups of wheat each day and there are sixty-eight people in the tribe, will the wheat last until the migration of the gazelles, which is estimated to occur in eighteen days?

I can see many of you cringing at the sight of a word problem. At least it doesn't involve two trains heading toward each other at different speeds.

Up to this time, humans never had anything that required intricate accounting. They had to carry everything they owned, and you don't need a calculator to keep track of a knife, a spear, and two strings of beads. This is not to say that humans never used numbers to keep track of anything. We know they'd been doing so for tens of thousands of years. Archeologists have located a number of bones covered with hundreds of notches. It's presumed that these are tally sticks and that the cuts represent a count of something.

What does a hunter-gatherer have in such large quantities? There's only one thing we can think of. It's something that even the poorest among us usually have in abundance—and that's time. Everyone needs to keep track of time, even hunter-gatherers. They might have needed to know how long until a particular plant ripens or a herd animal migrates through a given pass. They might have needed to know how many days there were before the next full moon, when they've agreed to meet other tribes for a festival. Or they might have needed to know when to start building warmer shelters for the coming winter. It perhaps should come as no surprise that numbers related to time, like the number of days between a new and full moon, occur more than statistically would be expected based on ancient artifacts we have found.

Tally sticks work well for the measuring of time. If my clan walks three days to the shore to collect shell fish, I can make three notches in my tally stick to record this. If we spend two days there, I can add two more notches to update my records. By counting the five notches, I perhaps realize that there are nine more days to the full moon, when we are expected to meet in the valley.

Three notches and two more make five notches.

These notches work great for time, but not for wheat. Time only goes forward. More days equal more notches. The quantity of wheat, on the other hand, goes up and down. Some days we gather wheat, and our supply goes up. Other days we eat wheat, and it goes down. Adding notches for more wheat is easy but removing them is not. All we can do easily on a tally stick is add. The other operations required by the new stores of wheat need something else.

PEBBLES OF MUD

Basic Calculi Operations

The domestication of wheat increased the computational needs of the early humans. We would expect them to have developed new mathematical methods, and the archeological record backs this up. In particular, throughout the Middle East we find small clay tokens originating from this period shortly after the rise of farming. We now refer to these objects as *calculi*, the Latin word for "pebbles." They were an elegantly simple solution to the farmer's accounting problems. The farmers had invented mobile mathematical symbols.

Symbols were not new to the human race. The new species that emerged from Africa consistently demonstrated symbolic behavior. They made the first cave paintings. We're all familiar with the ancient paintings of bison and other creatures. It may seem obvious to us today, but this was a big step forward in the development of the human mind. Where most creatures would see colored pigments scratched on a wall, we see a bison. We accept that color

patterns can have symbolic value, and we can see and contemplate what literally is not there.

Most people are less familiar with the cave paintings that consist of dots, lines, and other abstract shapes. These actually constitute the majority of the cave art but are less known simply because they're less attractive and hence are less likely to appear in books or in a documentary.

Most prehistoric cave art is abstract.

These demonstrate that humans could use symbols in a purely abstract way.

It's from the same period that we find the first concrete evidence for jewelry, music, and religion. All three are highly symbolic in content. Jewelry goes hand in hand with status. Wearing rocks and shells on strings has no intrinsic value other than the notions it conveys. For example, various articles might represent the office of tribal chief or shaman, or they might represent kinship with a particular clan. The oldest known musical instrument, a flute, originates from this period. Music is nothing but semiregular vibrations in the air. It's the symbolic value we attach to them that makes melodies sad or exciting.

Religious ceremonies and practices are also symbolic. There is no immediate survival benefit to activities like refraining from certain foods on certain days; however, such practices are representative gestures of devotion and sacrifice. The rituals of religion cause the participants to focus on the meaning of life and their place in the universe.

I've often wondered if this new symbolic thought is what enabled humans to dominate the entire globe so quickly. In many ways symbolism goes hand in hand with the ability to think in the abstract. To contemplate problems you haven't yet encountered or to compare solutions you haven't yet tried must have been a powerful survival tool. It's perhaps what allowed them to adapt so quickly to every situation they encountered.

Think about the bone notches in tally sticks. If a hunter carves a notch in a bone for each deer killed, then the cuts take on a symbolic value. He doesn't see five notches but rather five slain animals. He can compare this value with those from previous years and decide if it's enough food to last the winter. This abstract representation allows him to visualize and plan for a situation that doesn't yet exist.

On the surface, calculi don't seem much better than notches carved in a bone or dots painted on a wall, but the latter objects once placed cannot be moved. The power of calculi comes from their mobility. Today, at an early age we memorize that four and three is seven. This knowledge is completely unnecessary in the world of calculi. If I had four sheep, I'd know this by having four calculi. If I got three more, I'd record this by picking up one calculus for each new sheep. I'd now have the original pile of four and a new pile of three.

Four sheep and three more sheep represented in calculi.

If I wanted to know "How many is this many and this many?" pointing in turn to each pile, I could answer the question by pushing the piles together and saying, "It's this many." As with all calculi operations, computation is performed by moving the tokens around.

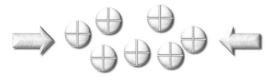

The two piles pushed together are the sum.

Subtraction is just as easy. Suppose we wanted to take four from nine. In order to represent this we'd have two piles, one with nine calculi and the other with four.

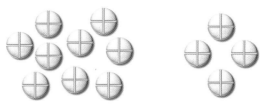

A pile of nine calculi and four calculi to compute 9 − 4.

We can then pick up each of the four calculi, one at a time. As we do this we also pick up one of the nine for each one of the four we remove. After we're done, there are five calculi remaining, representing the answer to the subtraction 9−4.

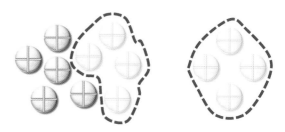

After removing four calculi, five remain so 9 − 4 is 5.

Multiplication can be performed by making repeated groups of a given number. For example, 3×5 can be computed by making three rows of five. The 15 calculi in the diagram represent the solution to the product.

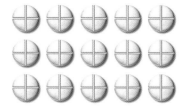

Three rows of five calculi compute 3×5.

Division can be performed by literally dividing the calculi between a number of piles. For example, if we wanted to divide 17 by 3 we could start with a pile of 17 calculi and three empty bowls. I could then pick up 3 calculi out of the pile and place 1 in each bowl. I would repeat this process of picking up and distributing 3 calculi as long as I could. Eventually there would be fewer than 3 left, forcing me to stop. The number in each bowl is the quotient with the number left over being the remainder. When we divide 17 tokens into three bowls, we would get three bowls of 5, and 2 left over. So the solution to $17 \div 3$ would be 5 remainder 2.

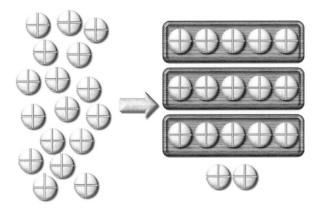

Seventeen calculi divided into three groups of 5 with 2 left over means $17 \div 3 = 5$ remainder 2.

In each case, all the computations are performed by moving the calculi around. It's the mobility of the calculi that makes them so useful. The other great advantage of calculi is that computations with them are completely intuitive. When you want to combine two flocks of sheep,

you combine two piles of calculi. When you want to divide up grain between three families, you divide calculi into three piles. The operations of calculi mimic the activities in the real world that they symbolically represent.

As they had many times before, the new humans adapted. When presented with the computational need brought about by the storage of grain, they quickly came up with a symbolic solution. Perhaps this solution was inevitable because symbolic thinking appears to be what humans do best.

COINS AND LARGE QUANTITIES

Basic Base Mathematics

Success breeds success. As humans farmed, they stored more grain, giving them time off from the relentless pursuit of a day's meal. They could use this time to make stronger homes or better weapons, but they could also use the time to be better farmers. They could make new farming tools, like the plow, for turning up fresh soil, or scythes, for quickly harvesting wheat. They could dig ditches for irrigation or build better storage facilities to protect their grain from vermin. These activities made farming more productive, increasing the surplus of grain that gave them more time for these and other activities.

The size of the grain surplus grew and grew, and in turn so did the complexity of accounting. The calculi they used still functioned in theory but became impractical due to the quantities involved. Some ancient mathematician devised an elegant solution. Why not make more than one

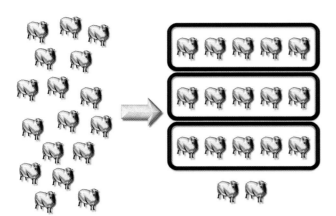

Dividing sheep into pens is exactly the same as dividing with calculi.

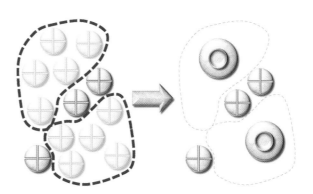

Two groups of five can each be replaced by two "5" calculi.

type of calculi where one type of calculi represents a group of another? For example, we could make slightly different sized or shaped calculi to represent five basic types of calculi. So if I had 13 calculi, I could replace 10 of them with 2 new 5-valued calculi. Hence we could refer to 13 as two 5s and 3.

As quantities continue to grow, we might run into the same problem as before, but this time getting too many "5" calculi. Once again we can create a new calculus worth five of the "5" calculi. When we get five or more "5" calculi, we can begin to replace them with the new, higher-valued calculi, "25s."

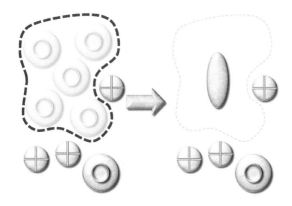

Five "5s" gets replaced with one "25" calculus.

Since five calculi get traded up for one higher-valued calculus, we call this system *base five*. Working with this type of calculi is equivalent to working in base-five mathematics.

While most people would probably feign ignorance of base-five mathematics, you can't fool me. You know it backward and forward. If I show you the following base-five calculi, you might hesitate in telling me its value.

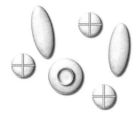

But now tell me how much the following is worth.

If you're familiar with American currency, I'd be willing to bet that within a few seconds you knew it was 58 cents. Note that the calculi pictured above the coins also have a value of 58. Both diagrams have two value-25 tokens, one value-5 token and three value-1 tokens. I even put them in the same relative position. The truth is, you've been using calculi all your lives. You've simply been calling them coins. The only difference between early calculi and modern coins is that coins have a monetary value.

Let's say you're asked to write 47 as a base-five number. The solution is trivial because you already know base-five calculi. Just express 47 cents in pennies, nickels, and quarters. We know that we can make 47 cents using 1 quarter, 4 nickels, and 2 pennies. This means that 47 is written 142 in base five. Note that I simply took the number of coins from largest to smallest.

1   4   2

Forty-seven cents can be written as 142 in base five.

We can think of our base ten in exactly the same way. Dollars, dimes, and pennies are base-ten calculi. This is because it takes 10 pennies to make a dime and 10 dimes to make a dollar. The number 47 can be expressed as 4 dimes and 7 pennies. Hence it's written as 47 in base ten. As we can see, numbers can be written in more than one way, depending upon the base you choose. To avoid confusion, I'll sometimes put a subscript indicating the base. So if I write 32_5, you'll know I mean 32 in base five. Let's do some simple conversions.

EXAMPLE: Express 68_{10} in base five and base-five calculi.

We first need to write the number in terms of quarters, nickels, and pennies. Then we write the values from the largest to the smallest denomination.

SOLUTION:

2   3   3$_5$

PRACTICE: Express 107_{10} in base five and base-five calculi.

ANSWER: 412_5

If we want to convert from base five to base ten, we simply need to think in coins.

EXAMPLE: Convert 142_5 into base ten.

SOLUTION: 1 quarter, 4 nickels, and 2 pennies is worth

$$(1 \times 25) + (4 \times 5) + 2 = 47_{10}$$

PRACTICE: Convert 321_5 into base ten.

ANSWER: 86_{10}

Adding with calculi is just like counting loose change. Suppose we want to add 132_5 to 141_5. The first thing we need to do is to turn them into calculi. I'm going to do this on a counting board that organizes the calculi by value.

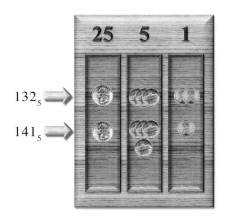

The two numbers expressed on the counting board.

In order to add the pennies, we simply push the piles together. I'm going to slide them down to the bottom.

The pennies get "added."

Note that we never actually need to know that $1+2$ is 3 when working with calculi. The 3 just appears when the coins are merged. When we add the nickels, we run into a slight problem. If we just pushed the piles together, we'd end up with seven nickels. While this is technically correct, it's not simplified. No good scribe would leave five or more nickels on the counting board since they could be more efficiently described with quarters. We'll perform the "addition" by picking up five nickels and replacing them with one quarter in the next column. The remaining two nickels get slid down to the bottom of the board.

The seven nickels turn into one quarter and two nickels.

All we're physically doing is making change; however, a modern mathematician would say that we are "carrying the 1." Think about it. If you add 15 and 28, you first add the 5 and 8. This gives 13, which we interpret as "3 carry the 1." If we were to combine piles of 5 and 8 pennies, we could replace 10 with a dime. The 3 remaining pennies are the 3, and the dime is the "carry the 1."

All that's left to do is combine the three quarters. This gives us the final answer of 3 quarters, 2 nickels, and 3 pennies, which is 323_5.

323_5 →

The solution remains at the bottom.

Addition with base calculi is simply exchanging coins when you have too many. The base is the number of coins that make the next higher one. So five pennies become one nickel and five nickels become one quarter.

PRACTICE: Using penny, nickel, and quarter calculi, add $214_5 + 123_5$.

ANSWER: 342_5

In these problems we've been writing "answers" like 342_5. You need to remember that in a calculi system, the calculi are the answer. Calculi predate writing by thousands of years. If you were asked how much you have, you might hold out your hand showing the calculi and say "this many" in the same way a child holds up their fingers when asked how old they are.

Above, we've used the coin analogy to describe calculi. The relationship between the two is far closer than you might realize. Calculi were not just used to calculate but also were often used to represent values. Imagine that I offer to pay you three sheep if you deliver some jewelry to a neighboring city. I promise to pay you when I get back, so I put three sheep calculi in a bowl. These tokens now represent the promise of the sheep. Like coins, they have value.

Three sheep calculi in a bowl.

There's a serious problem with this method. Assume that when you get back from delivering the jewelry, I claim that I only promised two sheep. You look in the bowl and there are two sheep calculi. You know I cheated you. You know I removed one calculus, but how do you prove it? If there is a next time, you may insist that you keep the bowl, but now how do I stop you from adding calculi to the bowl? The mobility of calculi that makes it so good for calculating makes it terrible for record keeping.

Around 3500 BCE, someone in either Sumer or its neighbor Elam came up with an elegant solution. Take the sheep calculi and trap them in a clay ball. We each sign the ball using our cylinder-seals, a thumb-size object that imprints a picture when rolled onto wet clay. Once the clay is dry, the only way I could change the number of calculi is to break open the ball and then create a new ball around the altered tokens. However, your seal would not appear on the new ball, and I could be charged with tampering with a contract. When you arrive back from your delivery, we would proceed to break the ball in front of witnesses, and the contract would be confirmed. This ball is called a *bulla* and was used in Sumer and Elam for a few hundred years.

The only problem with bullae is that the clay ball literally hides the number of calculi it contains. If you forget how much is in the ball, you can't break it open to see how many calculi are inside. By doing this you could be accused of tampering with the contract. In order to solve this problem, another simple innovation was created. Before being placed inside the clay ball, the calculi were pressed into the clay leaving their imprints. So your contract now consists of three sheep tokens in a ball, with the

A bulla marked with three sheep calculi and the "signatures" of Mr. Heart and Ms. Sun. Baked inside are three sheep calculi.

imprint of three tokens on the outside and the seals of the parties involved.

In less than a hundred years, people realized that the calculi inside were redundant. The impressions of the calculi preserved the relevant information and the seals authenticated them. So instead of putting calculi in balls, they just put the impressions onto flat clay tablets. Over time, symbols were added to give details of the contract. In our example, we might draw a picture of two feet, meaning that you were to travel. We might then use a symbol for the jewelry you were to deliver. Finally we might put a symbol of payment near the sheep impressions. As this system became more complex, it developed into writing, originally a mere accounting tool for mathematicians.

MAYANS, GODS, AND NUMBERS

Different Bases

Time is very, very heavy. So heavy, in fact, that it needs to be carried by the gods. At least this was the belief of the ancient Mayans. It is difficult to know exactly what the Mayans knew or believed. Their culture rose and declined long before the arrival of the Europeans in the Western Hemisphere. In a misguided effort to stamp out heresy, Catholic missionaries burned most of the few surviving books they found. This left modern archeologists just a few paper documents and many obscure carvings with which to reconstruct a great civilization.

The Mayan numbers that remain are not those of the civil service we find in Egypt but rather the careful time keeping of the ancient high-priest astronomers. Their enumeration of time is a strange mix of number systems

and cycles to which the priests attached what are now poorly understood meanings.

The most holy calendar was the *sacred almanac*. Each day in the almanac consisted of two numbers. The first ran from 1 to 13 and represented the gods of the upper world. The second ran from 1 to 20 and referred to a god, holy object, or sacred animal. So everyday would be represented by a pair of sacred things that determined the rituals of the day. Each day combination had its own connotation. Different days were good for different things and acted much like a modern horoscope. The day on which a Mayan was born supposedly told a lot about that person.

The thirteen gods were sometimes represented by an actual picture of the god but more often by a number. The Mayans used dots for 1s and lines for 5s, so three dots and a line would represent the number 8. The twenty sacred objects were depicted pictorially with a hieroglyph.

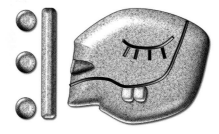

The sacred almanac date 8, Cimi.

We can't think about this as sort of a month–day pair. In such a system we would expect March 5 to be followed by March 6. In the Mayan sacred almanac, both symbols progressed each day. So if we used a similar system, March 6 would be followed by April 7. As strange as this may sound, we do something similar. Note that Tuesday the sixth is followed by Wednesday the seventh. The days of the week cycle every seven days independently from the days of the month, which restart about every thirtieth day. In much the same way, the thirteen gods of the upper world rotated each day, taking turns watching over the world independently of the holy objects, which did the same in a cycle of twenty days. While this may appear as a strange system, knowing the day immediately tells what divine forces are in control of the universe.

Another of their calendar systems, the Mayan long count, is more like a traditional system of enumeration. Time was counted in days starting on some day in the distant past. Just as with calculi, large groups of days were gathered into units. Twenty days formed a *uinal*, the Mayan equivalent of a month. Both the days and the objects of the sacred almanac cycled every 20 days, and the two would always be in sync. This meant that by knowing the day, they also knew which holy object influenced the day. Every 18 uinals formed a *tun*. Since a tun is eighteen groups of twenty, it is $18 \times 20 = 360$ days.

Note that this makes a tun almost a year, and this correspondence was probably intentional. Their solar year calendar consisted of 18 months of 20 days and a short "month" of 5 days, making 365 days. So the bottom two digits of the long count mimicked the number of the month and day of the solar calendar. Every year the long count would slip 5 days ahead, ignoring the five cursed days of the month that "has no name."

The rest of the long count acted like a traditional base-twenty system. Twenty tuns made a *katun*, 20 katuns made a *baktun*, and 20 baktuns made a *pictun*. Thus, a pictun is about 20^3, or 8000, years long. So the long count acted like a huge number. The number pictured below is written in much the same way that a long count would appear on a monument, except it would have had pictures of the gods as representations of kin, uinal, tun, and so on either next to or carrying the numbers. The digits go from smallest to largest as one reads upward. The two bottom digits are 6 kin and 2 uinals. Kin is the day, so it basically reads "the sixth day of the second month." The 6 also tells us the sacred object of the day. As a number, this represents $6 + (2 \times 20)$, or 46.

The tun, katun, baktun, and pictun can be thought of as the larger digits of the whole number. A tun is 360 days, so 14 tuns is 14×360, or 5,040 days. A baktun is 20 tuns, making it 360×20, or 7,200 days. So the five katuns is $5 \times 7,200$, or 36,000 days. While a baktun is $20 \times 7,200$ days, the shell, the world's first 0, tells us that there are none of them. Finally the one pictun is $20 \times 20 \times 7,200$ days, which is 2,880,000 days. The total of these numbers is 2,921,086 days.

The above representation has added functionality. The top four digits can be thought of as a base-twenty

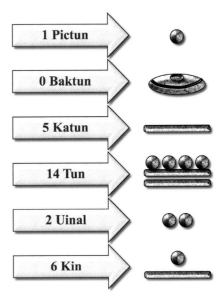

The number 2,921,086 in the long count.

representation for the number of years. Hence it can be interpreted as

$$(1 \times 20^3) + (0 \times 20^2) + (5 \times 20) + 14 = 8,114 \text{ tuns}$$

We have a modern tendency to expect systems to be systematic. We want our bases to be consistent. However, we see the Mayan numbers as a strange mix of bases 13, 20, and 18, while their numbers themselves are an alternating base of 5 and 4. But we can also see that there are very good reasons for this. The base thirteen kept track of the upper god watching over them; the base twenty followed the sacred objects, and the base eighteen gave them the ability to estimate the number of years in a long count.

There is a trade-off. Multiple bases complicate the rules of computation. Hence, it becomes a question of priority. The Mayans clearly gave priority to keeping track of the divine calendar over simplicity in operations. In the narrow sense, the lesson here is that no number-system base is intrinsically the "correct" base. The best value can vary even within one culture. In a broader sense, the lesson is that number systems are like tools. You can't say which tool is best unless you know what you need to do.

Bases are arbitrary. So how do we work in an arbitrary base? It's actually fairly easy if you don't think about the numbers but rather of the calculi. The base of a digit is simply the number you exchange in order to get a larger-valued calculus.

Let's work in base four. All calculi systems start with a token with a value of 1. Since four of these calculi make the next higher calculus, this new token is worth four 1s, which is of course, 4. The next token is worth four of these new tokens, which is four 4s, or 16. Continuing on in the same way, four 16s is worth 64. There's a simple pattern. We simply multiply a token by the base, in this case 4, to get the next token. We can apply this rule to any base. For example, if in our last section we needed a token whose value was larger than a quarter, we would simply multiply the base, 5, by the value of the quarter, 25, to get 125. Hence we would need a coin whose value is 125 cents.

Multiply by the base to get larger calculi values.

EXAMPLE: Convert the following tokens into base four and base ten.

Note that this number is essentially already in base four. We simply note the number of calculi from the largest to the smallest. The base-ten number can be determined by adding up the calculi or by multiplying the number of each calculi type by its value.

SOLUTIONS: In base four: 1312_4

In base ten:

$$(1 \times 64) + (3 \times 16) + (1 \times 4) + 2 = 64 + 48 + 4 + 2 = 118_{10}$$

PRACTICE: Convert the following tokens into base four and base ten.

ANSWERS: 2131_4 and 157_{10}

Adding doesn't really change with the base except in the number of calculi you need to make 1 higher. In base four, it of course is 4.

EXAMPLE: Using base-four calculi, add $322_4 + 123_4$.

The first step is to convert these numbers into calculi and place them on the counting board. We go from right to left, starting with the smallest denomination.

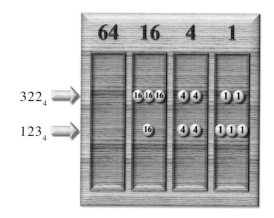

Then we group by fours, creating a "carry" that is one value higher for each four. The solution for this problem is 1111_4. Note that we get four 16s, which means we need a calculus of higher value than what we started with. This is no problem since we merely need to multiply our 16 calculi by 4 to get a value of 64 calculi. As before, every one of the original calculi gets carried or slid to the bottom to form the answer.

SOLUTION:

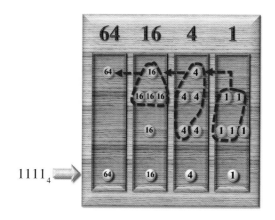

PRACTICE: Using base-seven calculi, add $351_7 + 132_7$.

ANSWER: 513_7

The Mayans seem to have been the first culture to discover the idea of "zero." Other cultures, like the Babylonians, came close, and all cultures have the concept of "nothing," but the Mayans seemed to have used it like a number.

People are often surprised that the concept of zero came so late in the intellectual development of humanity, but they really shouldn't be. Zero is paradoxical. To believe in zero, you need to accept that nothing is in fact something. Imagine I were to take you outside and ask you to describe the trees. I first ask, "What color are they?" and you reply, "Green." I then ask, "How big are they?" and you answer, "Huge." Finally I ask, "How many?" and you say, "Five." What would your response be if I asked you what color the unicorns were? You'd probably think I was nuts and say, "But there are no unicorns." Yet if I were to ask how many unicorns are there, you might say, "Zero." In some ways, zero is the only adjective that applies to things that are not there.

Ask a mother how old her child is, and she might reply, "She's exactly one year old." However, if you had asked her the same question one month earlier, she probably would have replied, "Eleven months." If the baby's age is less than one year, it is in fact, zero years old; however, we never say that. Zero is not a natural concept. It just doesn't make sense to describe things that are not in existence. So when something "isn't," we use qualitative descriptions like "nothing" instead of the quantitative "zero." We need to be trained by teachers to think of zero as a number.

There's much speculation on why cultures like the Mayans invented the number 0. Here's one of many possibilities. Recall that each month-god carries the days. There are 20 days in a month, so after 20 days, a new god takes over. It's natural to see this god as a representation of the number 20. If he carries a 5, then he and the number form 25. So what does he carry on his first day? If he carries a 1, then he and the number form 21. How do we get the number 20 out of this system? The month-god has to be

carrying a 0. Hence, there's a religious need to see nothing as something.

The Hindus independently discovered zero on the other side of the world. They had the unusual habit of writing their mathematics including numbers as poetry. If we used the same system, we might say "eyes, seasons" to represent the number 24, since we have two eyes and four seasons of the year. As strange as this may sound there was an extremely practical reason for doing it this way.

It's very easy when copying numbers to make a mistake. Look at 92649732 and 9264932. They almost look the same. If they weren't so close together you may not notice that the second number is missing a 7. It's harder to miss an entire word than it is to miss one digit. Furthermore, poetry has rules. You leave a word out, and you can get bad poetry that doesn't have the right number of beats per line. Hence poetic numbers were extremely accurate when old records were copied from generation to generation.

When we write a check, we might write a value of "eight thousand, four hundred, and five" or "three thousand and six." Notice that the second one has no hundreds. We don't write any hundreds because there are none. But if a Hindu tried to write three thousand and six as "tricycle, hexagon," it would appear to be the number 36, not 3006. They would need to write something like "tricycle, void, emptiness, hexagon" to properly express this number. Being forced to vocalize nothing as something, perhaps, led to the idea of zero as a number.

There is no zero in calculi. For example, the number 18_{10} when expressed in base-four calculi is

However, if we tried to express it in base-four numbers, it looks like 102_4. The zero represents the missing 4-valued calculi between the 16s and the 1s. We need to remember that calculi existed some five thousand years before writing was invented. To a calculus mathematician, there is no "zero 4s," just a 16 and two 1s. Consider the following:

EXAMPLE: Using base-seven calculi, add $205_7 + 164_7$.

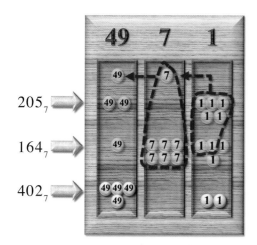

The answer, 402_7, contains a zero, which represents the lack of value-7 calculi. Early in the age of writing, the Mesopotamians were at a loss on how to record such a number. They started by leaving a gap, just as is seen on the counting board. So 402 was basically written 4 2.

PRACTICE: Using base-three calculi, add $120_3 + 211_3$.

ANSWER: 1101_3

ROMANS ≠ MATHEMATICIANS

Alternating Bases

Years ago, two of my history of mathematics students asked if they could cowrite their term paper on Roman mathematics. I told them that I didn't know anything about the subject and suggested that they pick a more common topic. However, they insisted as they wanted to tie it into their Latin studies. They were both excellent students, and I was confident that they would come up with something interesting. About two weeks later, they came to me for help. I could see the panic in their eyes as they explained that they could barely find anything to fill the required twenty pages. I told them to do what they could, and I would take the difficulty of the subject into account when grading their paper. They handed in a paper that went into excessive detail on the Roman number systems and calendars and nothing else.

Why couldn't they find anything? The Romans seem to hold the unique position in history to be the only major civilization to have had no mathematical accomplishments. Most textbooks mention this fact and move on. Some have harsh criticisms of the Romans. Others feel the need to say something nice. Their statements are to math history as "she has a nice personality" is to blind dating.

That's not to say the Romans had no impact on mathematics. They killed the greatest mathematical mind of the ancient world, Archimedes. They also killed one of the first historically known great female mathematicians, Hypatia. They burned down the library of Alexandria, the greatest intellectual center in the Mediterranean. Actually, they burnt it down a couple of times. Shortly before the fall of Rome, fueled by the rage of religious intolerance, they drove most of the Greek intellectuals out of the western Roman Empire. The fleeing Greeks took their books and wisdom with them. This helps explain why Europe suffered more than a thousand years of ignorance after the collapse of Rome. Ironically, it was the return of Greek texts that reignited Europe's intellectual progress.

You're probably wondering why I bother mentioning the Romans at all. I need to discuss alternating bases, and the Romans used the only such system moderns are remotely aware of. The Romans, like most cultures, used a straight line to represent 1. This is believed to be derived from the ancient use of tally sticks. Hence they represented 2 by two vertical lines and 3 by three lines. These tally cuts are represented by the letter *I*.

I, II, III

But too many vertical lines blur together in the mind's eye. Today, many people keep tallies with similar slashes. However, every fifth slash is made through the previous four. This has the effect of turning five individual slashes into one group, which the mind perceives as five. Dice work in the same way. When we roll a six, we don't see six dots on the die but rather two rows of three dots, which we interpret as six.

We see a slashed group and two groups of three, not five and six.

It's possible that the Romans originally made tallies in groups of five in the same way we do today. Eventually, they may have realized that the four vertical slashes were unnecessary and tedious to draw. So, it's suspected that they may have simply stopped writing the first three, creating the V shape we know today as the Roman number for five.

When the first three lines are omitted, the five becomes a V shape.

Similarly, two groups of five may have been "slashed together" to signify a group of ten. Two of these lines may have formed the X of our Roman numerals.

The two grouping slashes of ten lines form an *X*.

However the Romans derived these symbols, they present a dilemma. What base is the Roman system? Since five Is make a V, it appears to be base five. However, since two Vs make an X, it now appears to be base two. The next symbol in Roman numerals is the L, which represents 50. Since five Xs make an L, it's now base five again. Two Ls make a C, the Roman numeral for 100. It continues to dyslexically bounce between two and five until the Romans run out of symbols. Ten Xs make two Ls, which in turn make one C. If we ignore the intermediate Ls, this is just like a base-ten system in which ten of one type of calculus make one of the next higher. In fact, any pair of symbols two away from each other can be exchanged at a rate of ten to one.

As strange as this double-base system might seem, it's actually more common in the ancient world than single-base systems. The Chinese used almost the same system as the Romans. The Mayans, when they didn't draw their numbers as gods, alternated between base five and four. The Sumerians, as we shall see, alternated between base ten and base six. The bases change from culture to culture, but for some reason the alternation seems to be almost universal.

Addition in an alternating base is the same as in a single base except that the exchange rates change from column to column. As we've said above, five Is, Xs, and Cs make a V, L, and D. At the same time, two Vs, Ls, and Ds make an X, C, and M. The difficulty is in keeping the two groups straight.

$$V = IIIII$$
$$X = VV$$
$$L = XXXXX$$
$$C = LL$$

The exchange rates for Roman numerals.

There is one difference between calculi and standard Roman numerals. Calculi are additive, so we can express numbers like 6 as VI, the sum of a 5 and a 1. However, Roman numerals sometimes use a subtractive notion. For example, 4 is usually written IV. The I, appearing before, instead of after, the V, is *minus one*. So IV is read as 5 – 1. There are no "negative" calculi, so we will express 4 as the additive IIII.

Let's add the numbers 19 and 28 in Roman numerals. We first need to express them in Roman calculi. The first number, 19, is XVIIII. Note that we will not use the usual notation of XIX. Similarly, 28 is XXVIII. First, we place these numbers on the counting board.

The I column is base five. Since we have at least five calculi in this column, we can cash five of them in for a V. However the V column is a base-two column, so two Vs are exchanged for an X. The X is another base-five column, and since there are not five of them, we are done.

Little changes for longer examples. The columns alternate between five and two for the sake of carrying. Notice that in the following example, the even rows, counting from right to left, exchange two for one. Similarly, the odd rows exchange five for one.

EXAMPLE: Add CCLXXXVI and LXXXVIII as Roman calculi.

ANSWER:

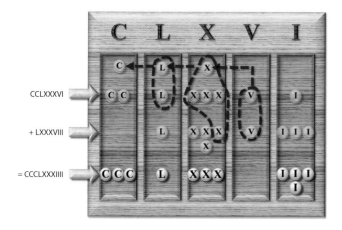

Try the following example. You could use pennies, nickels, and dimes for Is, Vs, and Xs, respectively. You simply need two other token types for Ls and Cs. You could use buttons, game pieces, poker chips, or whatever. You might want to make a counting board out of paper with a C, L,

X, V, and I on the top. It's not necessary but helps to organize the calculi.

PRACTICE: Add CXXVIII to LXXXXVIIII.

ANSWER: CCXXVII

THE LEGACY OF SUMER

Sumerian Calculi

Long forgotten, the ghosts of Sumer forever linger in our cultural memory. Stories, like Noah and the flood, were passed down from generation to generation for at least five thousand years. Western tradition places the mythological birthplace of humanity, the Garden of Eden, near the Tigris and Euphrates Rivers, which by, no coincidence, run through Sumer. This should perhaps come as no surprise since Abraham, the founder of the religious traditions of Judaism, Christianity, and Islam, originated from the city of Ur, once the mighty seat of a Sumerian empire. But even in Abraham's day, the Sumerians were long gone.

Sumer had been completely forgotten until the advent of modern archeology. As scholars dug up the remains of ancient Mesopotamian cultures like the Assyrians and the Babylonians, they found references to an older civilization. The ancient Mesopotamian scribes worked in two languages, their own and that of Sumer, even long after it ceased to be spoken. They kept the old traditions alive in much the same way as medieval scholars maintained and studied Latin texts more than a millennium after the fall of Rome. Sumer is the oldest known civilization. They invented many of the underpinnings of society that we take for granted, like writing, the plow, and the wheel. So much of what we are began in Sumer.

The Sumerians, like others in the ancient world, used calculi. Like the Romans, they used an alternating base system, but instead of five and two, they used ten and six. This makes their system essentially base sixty. Although there is rampant speculation, no one knows why the Sumerians selected such a large base. They used a small clay cone to represent the number 1. Ten of these made up a small clay ball. So the number 25 in Sumerian calculi would appear as follows:

Twenty-five in Sumerian calculi.

The lower base is ten, so ten 1s make a 10 calculus. Since the next base is six, six 10s make a 60 calculus, which looks like a bigger version of the 1 calculus. The number 134 expressed in Sumerian calculi would consist of two 60s, one 10, and four 1s.

One hundred and thirty-four in Sumerian calculi.

The base of the calculi now alternates back to ten, so the next calculus is composed of ten 60s and hence has a value of 600. The calculus for 600 looks just like the one for 60 except that the former has a small round indentation in it that is believed to represent the round 10 calculus. This was probably used to indicate that the 600 calculus, like the 10 calculus, were both base six. So we could express the number 671 as follows:

Six hundred and seventy-one in Sumerian calculi.

This system could represent numbers up to, but not including, 3600; however; this number wasn't large enough. As Sumer attained the status of a civilization, its towns grew into city-states of unprecedented size. Their mathematical needs likewise grew, and they needed larger and larger numbers.

Although all ancient civilizations were generally huge cities, modern anthropologists tend to define them instead by their complexity. The evolution to civilization is complex and difficult to explain. Various theories often involve enormous flow charts, and no two academics seem to agree on which theory is correct. Let me try to explain my primitive understanding in terms of a story.

Imagine we live in a prehistoric farming village where every family is essentially the same. The strong hunt, those of average strength or burdened with child-care duties farm, and the weak and infirm perform less taxing duties, like sewing and pottery making. You and I are reasonably trustworthy and good with numbers, so we've been put in charge of the grain storage pit. The number of families has grown, and we need more farmland to sustain them; however, all of the amply watered plots have been taken. You get an inspired idea: why not bring water to the dry, unused plots of land.

We gather the strongest members of our tribe and ask them to dig a channel from the river to the fallow regions of parched soil. They want to help, but they worry that if they stop farming and hunting long enough to do this, their families will go hungry. I ask each of them to bring a large pot to the storage pit where I fill each with some of the excess grain. I tell them this should be enough to feed their families while they dig, and they proceed to construct the irrigation channels.

When they're done, something wonderful happens. The quality of our farmland has improved dramatically and our grain crops are huge. The small surplus that was used to pay the diggers has resulted in an enormous surplus. We literally have too much grain and don't need as many farmers. This time I have an idea. The last time we traded with the neighboring tribe, I noticed that they exchanged more than usual for the pots you made. This is no surprise since everyone knows you make the best pottery. So I ask you to make all the pots next time. You're about to object when you remember how we gave the diggers some of the excess grain. You are willing to make the pots provided you're given enough grain to make up for the time you lost growing wheat for your family.

Normally you only make a few pots a year, but I've asked you to make thirty, and this job changes the way you go about your task. You know where the best clay is located, but it is fairly far away. For three pots the trip isn't worth it, but it is for the thirty you are now going to make. For the same reasons, you decide to make the pots with three different paint colors instead of the usual one, taking the time to collect a variety of pigments. As you make the pots, you experiment and try different methods. Some are failures, but others come out surprisingly well. The practice you get making all the pots increases your already superior skill.

On the next trading day, the neighboring tribe stares in amazement at your beautiful pottery, and they trade us their best pieces of obsidian for your miraculous handiwork. From the obsidian, our hunters make the best spears they've ever had, and the following hunt is the most successful ever. Once again, we have more food than we know what to do with. Having learned from our past experiences, we look for other chores to be done by whoever is best at doing them. The best spear maker, vegetable grower, and leather tanner are asked to only do their specialty. From the ever growing surplus, we provide them with everything else they need. As we assign and manage all these operations, we find it more and more difficult to farm and hunt ourselves. So we compensate for our lost time from the surplus, becoming the world's first full-time leaders.

Our society thus fragments. Initially we were all hunters and farmers. We're now construction workers, craftsmen, leaders, and so on. Each fragment specializes, increasing their productivity. The excess production generated can be used in turn to enable another fragment to specialize and increases productivity once more. The resulting abundance of resources enables the tribe to grow and grow. Civilization, as defined by the fragmenting of society into specialized units, is like a chain reaction given the proper conditions. Towns of a few thousand blossom into cities with populations of a hundred thousand or more.

Managing these cities required a monumental effort. The Sumerian sky watchers, who used simple math to track the time, evolved into managers of the storage pits and eventually formed the priestly bureaucracies that micromanaged the giant cities of the Sumerians. As we've seen, writing was invented to keep track of the transactions of the Sumerian economy. The numbers they needed to account for were unprecedentedly large. So the 600-valued calculus was insufficiently small.

Recall that the 600 calculus was a large cone imprinted with a circular dot. The dot indicated that it was a base-six calculus, so six of them made the next higher calculus. Hence the new calculus was worth 6×600, or 3600. The Sumerians represented it with a large ball.

One 3600 is made of six 600s.

The large ball has no dot, so it is a base-ten calculus, and so ten of them make the next calculus, which has a value of 10×3600, or 36,000. This calculus looks just like the 3600 calculus but with a dot. Being base six, the next calculus, if it existed, would have a value of 6×3600, or 216,000, more than a fifth of a million. This is how the Sumerians dealt with the enormous quantities that arose from the growth of their civilization.

Addition with Sumerian calculi is just like any other calculus addition. You simply need to take care to remember which tokens are base ten and which are base six. To make this easier, I've put numbers between the columns of the counting boards indicating the exchange rate between the two.

EXAMPLE: Add ⌾𝄃𝄃𝄃𝄃𝄃𝄃∞∞∞ + 𝄃𝄃∞

SOLUTION: 𝄌𝄃𝄃𝄃∞∞

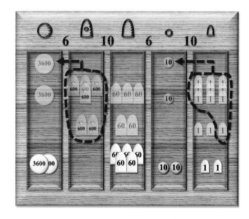

PRACTICE: Add 𝄌𝄌𝄃𝄃𝄃∞∞ + 𝄌𝄌𝄃𝄃∞∞

ANSWER: ⌾𝄃𝄃𝄃∞∞∞

THE STYLUS IS MIGHTIER THAN THE SPEAR

Babylonian Numbers

The rise of mathematics in Sumer grew hand in hand with the increasing needs of their ever-growing cities. Remember that the natural resources of Mesopotamia were mud, mud, and mud. A large work force needed an increasing number of tools. Copper from Oman and tin from Afghanistan had to be imported to make bronze for these implements. Wood was exceedingly rare and had to be brought in from places like Lebanon. The growing upper class required jewels, obtained from places like Iran, to display their status in cities grown so large that few would know personally who was important and who was not.

The organization of Mesopotamia into city-states along the Tigris and Euphrates was not particularly conducive to massive trade. The priest-rulers in need of goods could either trade with their nearest neighbors or send out long-range expeditions. The first option was exceedingly expensive. For raw materials to travel any distance, they needed to go from city to city passing through a series of middlemen, each of whom had to make some profit. The accumulated cost was prohibitive. The second option was dangerous. A trade caravan had to leave the small sphere of influence of their home city. Other cities had little incentive to protect these merchants from robbers or barbarian raids. They might even take the goods for themselves.

The geography of Mesopotamia made the trade routes still more important. Sumer was situated between rich eastern and western cultures. To the east there was India, Susa, and the cultures of the Persian Gulf. To the west there was Egypt, the Minoans, the cities of the Levant, and other Mediterranean cultures. There was enormous potential for east-west trade between these groups. To the north of Mesopotamia there were impassable mountains. To the south there were unforgiving deserts. The cities of Mesopotamia lay on one of two east-west rivers, which were essentially the only trade routes. There was great potential for vast riches to be accumulated through trade, yet the question remained, how could it be done effectively. Unfortunately, the answer was far too simple.

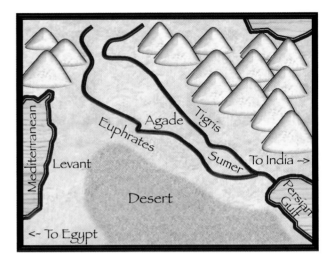

Mesopotamia

Warfare has always plagued humanity. When neighboring groups had disputes on their boundaries, there would be a clash, a few people would die, and the losers would back off. However, it is at this time, the second half of the third millennium BCE, that wars became more systematic and common. When the army of one city-state subdued another, they would no longer take just a little territory or a bit of plunder. The victorious leader would now demand submission and leave representatives and soldiers in order to acquire trade rights to their victim's stretch of river. Taking just one other city doubled the length of the route they controlled and essentially doubled the trade profits. If you could double your profits by holding two cities, why not triple it by ruling three? How about four? Cities with no common boundary or real disputes now fought regularly for control of a stretch of the river.

The traditional leader of a Sumerian city was an *en*, the high priest. *En* literally means "the spouse of a god," through which the priest was joined in ritual marriage. In times of strife, the elders of a city would select a *lugal*, literally a "big man," to lead a band of warriors. As warfare became more prevalent, the lugal rose in status. Gradually we see the title of the king change from *En* to *Lugal*. The priestly class could manage a great city, filling it with arts, industry, and rituals, but what good are these to a city about to be conquered and plundered? The need for security in these desperate times overruled all other considerations. The intellectual class that had ruled for thousands of years had been edged out by the warrior class, and this marked out the beginning of the end of the Sumerians.

Sargon of Agade, the first great Mesopotamian emperor, was Semitic, not Sumerian. The Semites are not one people but a collection of several peoples whose languages have a common origin. Today, the cultures of the Middle East speak Semitic languages. Legend has it that Sargon was born of a high priestess, who by tradition had to remain celibate. To avoid scandal, she placed Sargon in a reed basket caulked with pitch and set her son down the Euphrates River. He was found by a water-drawer and raised as a son. Many believe this story is the antecedent to the story of Moses. In any case, this tale is almost certainly a fictional attempt to give Sargon royal blood, justifying the kingship held by a commoner.

Sargon started as a gardener in the city of Kish but managed to rise to the level of royal cupbearer, a high official in the king's court. When Kish was conquered by the Lugal of Uruk, Sargon fled and founded the city of Agade, where he raised his own army. After he defeated his nemesis from Uruk, Sargon went on to conquer city after city. His empire stretched from the Mediterranean to the Persian Gulf, including parts of Iran. His own records describe it as being almost a thousand miles long.

Imagine the job of the mathematician in this new empire. In a city-state, there might have occasionally been the need for numbers in the thousands and, rarely, totals in the ten thousands, but what numbers would there have been in the new empire? How many containers of grain were required to feed an army of 90,000 men for three months? Having different symbols for each place-digit in a number is confusing. For small numbers, there are only a few symbols, most you use only on occasion, but for large numbers, you need an extensive collection of different symbols. The Semites used base 10, not 60 like the Sumerians, so they could not use the Sumerian symbols as they found them. The less educated Akkadians simply didn't have the means to express, let alone deal with, the quantities required for their empire.

A new form of numbers was created as the scribes in the Sumerian tradition adapted to their Akkadian masters. The result is an unusual but simple mix of the numbers of the two cultures. For numbers from 1 to 99, the Akkadian numbers are written just as the Sumerian numbers. For example, 37 is three 10s and seven 1s. So we need three 10 symbols, ⟨, and seven 1 symbols, ⟨, and 37 would look like the following: ⟪⫲. For numbers from 60 to 99, we see traces of the Sumerian base sixty that use a large ⟨ to represent 60. So 85 is ⟨ ⟪⫲.

For numbers above 100, the rules change. The word *ME*, means "hundred," so when we see the number ⫲ ME ⟪⫲, we read it as "two-hundred and forty-six," or 246. Notice that this is almost exactly the way we write checks. The word for thousand is *LIM*. We can now interpret ⟪⫲ LIM ⫲ ME ⟨⟪⫲ as twenty-one thousand, four hundred, and seventy-four, so it's 21,474. They also recognized that ME could be used for hundreds of thousands. So ⫲ ME ⟪ LIM ⫲ ME ⟪⫲ is four hundred and twenty thousand, two hundred, and thirty-eight, giving 420,238. Using this notation we can easily write numbers up to but not including one million. By introducing LIM.LIM, which is literally a thousand thousands, or a million, we can now go up to a trillion, satisfying the numerical needs of the greatest empires.

The priest-mathematicians lost their place as the rulers of Mesopotamia, but their role remained just as vital. They went from ruling cities to managing vast empires; however, the good old days were over. The Mesopotamian empires were horribly unstable. In general, the benefits accrued only to the ruling city. As a result, rebellion was commonplace. To reduce the effect of revolution, an emperor would often tear down the protective walls of his vassal cities so they would not be able to protect themselves from his armies, but doing so made the empire particularly vulnerable from the outside. Barbarian tribes and free cities could easily plunder the virtually defenseless cities. The emperor would then need to lead an army to put down the invasion. If he won, the empire would survive for a while. If he lost or was seriously weakened, his vassal cities would turn on him, and the empire would break up into independent city-states, each vying to be the next seat of the new empire.

Sargon's empire lasted about three generations, until it was taken down by the Gutians, barbaric warriors whom the scribes described as ape-faced savages. Oddly enough, the cyclical transition back and forth between empire and chaos was remarkably stable. Dynasties would rise and fall. Barbarians would overtake the region but then assimilate. Each emperor, whether he was from an ancient Sumerian city or a barbaric tribe from the fringes, would need his empire managed. The only ones who could do this were the scribal class, the remnants of the priest-rulers, and they kept their knowledge and traditions from falling apart.

The period of the Gutians was chaotic as kings rose and fell. On one tablet the question is asked, "Who was king?" and the reply was the sarcastic, "Who was not?" The city of Uruk was finally capable of throwing off the Gutian menace. Uruk's reward was to be almost immediately conquered by the Sumerian city of Ur. This began what is called the Third Dynasty of Ur and is the last time a Sumerian city rose to dominance in Mesopotamia. The Gutians were replaced by a Sumerian empire with the capital city of Ur. However, the Akkadian people and culture had infiltrated Sumer gradually over the centuries, and by this time, Sumerian was probably a dead language.

Although no one used Sumerian as their everyday language, it had not been forgotten. The wisdom of the ancient Sumerians was venerated by the Akkadians. Akkadian priests would dress up in Sumerian garb for important rituals. They even used the Sumerian written language for their texts on subjects like religion and magic. Akkadian scribes all had to learn to read and write in Sumerian. This was no easy task. Today when we learn to speak a foreign language, we need to memorize thousands of words. The Akkadian scribes had to do this too, but with an added difficulty. When we write a foreign language, once we know their alphabet, we can at least approximate the spelling of any word, but at the time of the Third Dynasty, the alphabet had not yet been invented. The language used separate symbols to represent each syllable. So while we have to memorize fewer than thirty letters, the Akkadian scribe had to memorize hundreds of syllables, each written with its own sign. Hence the Akkadian student spent years learning Sumerian just to give certain subjects their proper respect according to ancient traditions.

One of the subjects venerated in this tradition was mathematics. When an Akkadian scribe wrote about how to solve a problem, he often wrote in Akkadian, but he always did the mathematical computation in Sumerian. This was not a trivial change, because the Sumerians used base 60, not base 10 like the Akkadians. Imagine growing up and learning all of your math in a number system that's different from all the numbers used outside of your school. You would literally need to convert each answer to use it for everyday life.

The Akkadians not only adopted the system, they improved it ever so slightly in a crucial way. Consider the Sumerian translation of the number 163. When we break it into base sixty, we get $(2 \times 60) + 43$. The Sumerian mathematician had separate symbols for 60, 10, and 1, giving 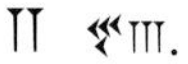. The Akkadians replaced it with 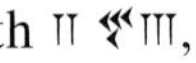, using the same symbol for 60 as for 1. This may not seem like a big step, but it's extremely important. When we write the number 252, we use exactly the same symbol for two and for two hundred. We know the difference between the two simply by the location of the 2s in the number. This system enables the use of the same ten symbols, 0 to 9, to express numbers of any size. Because of the standardization, the same rules apply. So, for example, if you know how to add two-digit numbers, you also know how to add thirty-digit numbers by repeating the same processes.

It's not clear why they made this change. Perhaps the similarity in symbols made the alteration more obvious. Perhaps it's because they, oddly enough, used the same spoken word for 60 as for 1. Perhaps it was because lazy scribes simply found it inconvenient to make two different-sized wedges using the same stylus. It probably wasn't a conscious decision to adopt a place-value system because they didn't change the other symbols to be uniform.

The Akkadians were then replaced by the Amorites. By this time, the scribes realized that they didn't need so many symbols. Imagine that a scribe performs a calculation on his counting board and gets the solution 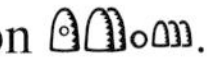.

Since the day is over, he writes down the number, intending to finish his computations the next day; however, when he returns in the morning, he realizes that he left all

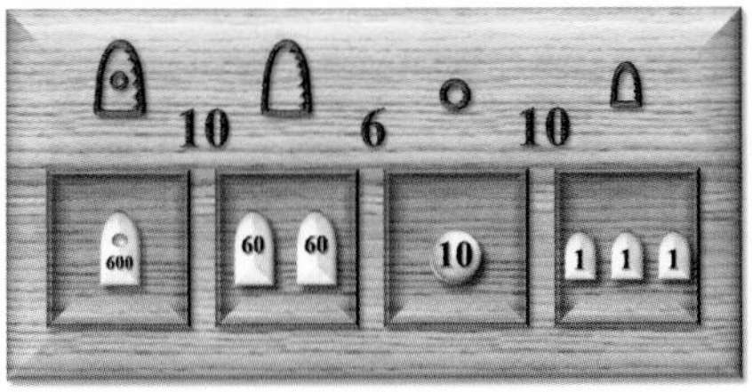

but his one-valued calculi at home. Not being sure of what to do, he places them down on the counting board exactly where the old calculi used to be.

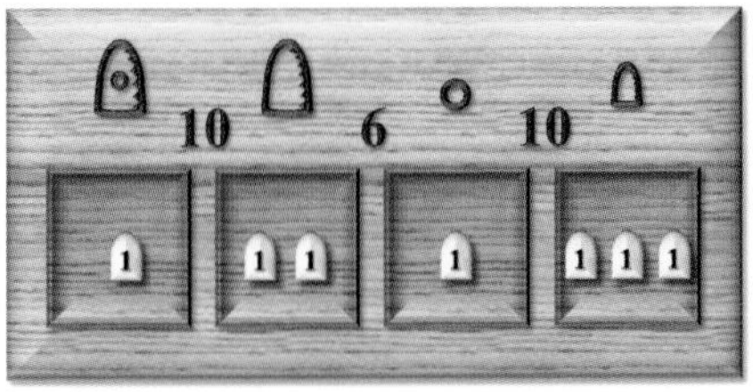

Much to his surprise, he realizes that there really is no problem at all. He can determine the value of each calculus by simply noting the column it is in. He adds, subtracts, and multiplies just as he did before. When he picks up the calculi to replace with larger-valued calculi, he simply needs to note where he's picking them from to determine if he needs six or ten.

He later realizes that he can apply this same notion when writing numbers. He decides to represent all base-ten calculi with the symbol for 1, 𒁹, and all base-six calculi with the symbol for 10, 𒌋. So he can express the calculi number ⩀∩∩∘ᴖᴖᴖ as 𒌋𒈫 𒌋𒐈. The symbol 𒌋 represents both the ⩀ and the ∘, but he can tell which is which by their position within the number, just as he could tell the value of the calculi by their position on the board.

EXAMPLE: Convert the Sumerian calculi ◎◎◯◯◯⩀∘∘∘ᴖᴖᴖᴖᴖᴖ into Babylonian cuneiform.

SOLUTION: We just replace the calculi with the appropriate symbols, 𒌋 or 𒁹, depending upon whether they are base six or ten, respectively. Note that there are no ∩ calculi, so there will be no 𒁹 symbols between the 𒌋 and the 𒌍.

𒎙𒐈 𒌋 𒌍𒐋

PRACTICE: Convert the Sumerian calculi ◎◯◯◯◯◯⩀⩀∩∩∩ᴖᴖ into Babylonian cuneiform.

ANSWER: 𒌋𒐊 𒎙𒐈 𒈫

We can convert these numbers into base ten by thinking of them as base-sixty numbers. You simply group leading 𒌋s with their corresponding 𒁹s. For example, 𒎙𒐋 is interpreted as 26, a single base-sixty digit, not a 20 and a 6. For larger numbers, just separate these digits. So 𒈫 𒌋𒐌 can be thought of as two base-sixty numbers, 2 and 17. The 2 is in the sixties place and hence has the value of 2×60, or 120. The 17 is in the ones place and is just 17. So the final value is just $120 + 17$, or 137.

EXAMPLE: Convert 𒁹 𒌍𒈫 𒌋𒐍 into base ten.

SOLUTION: The digits are 1, 32, and 18, so it's equal to $(1 \times 60^2) + (32 \times 60) + 18 = 3600 + 1920 + 18 = 5538_{10}$.

PRACTICE: Convert 𒌋𒈫 𒐏𒐈 𒎙𒐎 into base ten.

ANSWER: $45{,}809_{10}$

When adding, the Amorites may have used a counting board. They only needed one type of token, whose value would be determined by its location on the board. I'm going to use two types of tokens, one for 𒁹 and the other for 𒌋, to keep the bases straight. We need to remember that in any column, ten 𒁹's make a 𒌋 and six 𒌋's make a 𒁹. Here's a sample addition.

EXAMPLE: Add 𒌍𒐊 𒎙𒐌 + 𒐏𒐎 𒐐𒐊 as calculi.

SOLUTION:

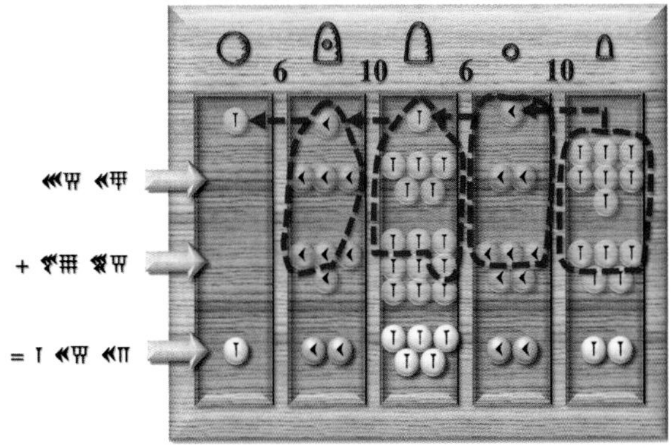

Unfortunately there is little physical evidence that they used counting boards; however, there is some indirect evidence. One Babylonian tablet lists the equipment used by various professionals. For the accountant, it lists *Ges-Dab-Dim*, which translates roughly into "wooden formulating board," and later lists reed bits. Since they only needed one or two types of calculi, reed bits would have made a simple substitute for the clay tokens used in earlier times. They would also help explain the disappearance of calculi at about this time.

Wood and reed bits wouldn't survive the nearly four-thousand-year journey to the present as well as baked clay tablets and tokens. We also need to remember that sizable pieces of wood were scarce in Babylon and that the board would have cost a small fortune, like the leather scrolls of the Egyptians. Old tablets would have probably been reused or recycled, reducing the number, and hence the probability, that we would find one. We always need to be wary of arguments based on slight evidence. It's possible that at this time they added without calculi, memorizing additions in the same way we learn 5 + 7 = 12. However, counting boards and their cousins, abaci, would make a reappearance in classical times and remain in common use until the invention of the electronic calculator. Hence, it's not outrageous to assume that the Babylonians used them too.

THE BULL OF HEAVEN

Babylonian Fractional Values

Anu, king of the heavens and all the gods, had had enough. For years mortals had been praying to him to stop Gilgamesh, warlord of Uruk, from ravishing their wives and daughters. But now Gilgamesh and his companion, the wild man Enkidu, had stolen great cedar trees from Anu's forests. In doing so they had killed Humbaba, the giant whom Anu had charged with the protection of his trees. Now his daughter Ishtar, goddess of lust and war, begged her father for vengeance against Gilgamesh, the only mortal to spurn her advances. In Anu's mind, there was only one thing to do. He unleashed the Bull of Heaven.

This fearsome monster didn't attack people but rather the very earth itself, opening great rifts, which swallowed up a hundred men at a time. In the fierce battle that ensued, Enkidu managed to grab the tail of the Bull of Heaven, giving Gilgamesh time to cleave the beast in two. After the loss of his prized bull, Anu no longer wanted to kill Gilgamesh. He wanted to make him suffer.

Gilgamesh, mythical warlord of Uruk.

The *Epic of Gilgamesh* is the oldest known literature. It was written by the Babylonians, who strung together a series of Sumerian tales, giving them a narrative cohesion. While my introduction portrays the epic as an action novel, the work actually examines how humans deal with death.

The gods decide to kill Enkidu in revenge, but they don't kill him quickly. Instead they give him visions so that he is aware of his fate, and he is shown the bleak underworld to which his soul will be condemned. As Enkidu slowly dies, he relates his grim dreams to Gilgamesh, who in turn is overcome with dread. After the death of his friend, Gilgamesh is determined to avoid the same fate. To do so, he seeks out Ut-napishtim, the only man to have achieved immortality.

Gilgamesh's quest literally takes him to the end of world, passing through strange alien lands along the way. When he finds Ut-napishtim, the old man relates, in a tale that is almost identical to the myth of Noah and his ark, how he achieved immortality. Ut-napishtim tells Gilgamesh of a time when the gods decided that humans, who had overpopulated the earth, made far too much noise and hindered the sleep of the gods. Enlil, the king of the gods and lord of the winds, decided to destroy humanity and sent a series of plagues. Ea, the god of freshwater and wisdom, realized how foolhardy this was. Humans were created to serve the gods, bringing them food and fulfilling all their needs. So he warned Ut-napishtim, who in turn made sacrifices to other gods, who protected the humans from Enlil's attacks.

Enlil ordered Ea to flood the world, but Ea refused. When Enlil decided to do it himself, all Ea would do was warn Ut-napishtim, advising him to build a giant boat, which he did. The craft covered an acre and was six decks high, and he filled it with silver, gold, seeds, wild and domesticated animals, craftsmen, family, and friends.

After the flood, most of humanity was dead and had ceased feeding the gods, who were seized with ravenous hunger and realized their mistake. When Ut-napishtim's boat landed, Ea advised him to immediately make food sacrifices to the gods. Enlil's hunger outweighed his anger at being tricked and he let Ut-napishtim and his family live. For his services, Ut-napishtim was given food of the gods, which made him and his wife immortal.

So Ut-napishtim tells Gilgamesh that immortality is not for him to bestow but instead is a gift from the gods. Gilgamesh is not satisfied with this answer and pressures Ut-napishtim to tell him where he can find this magical food. Ut-napishtim ultimately tells Gilgamesh what he wants—the food is a plant found at the bottom of the sea. The gods, however, refuse to grant Gilgamesh immortality, and they have a snake sneak up and eat the small plant Gilgamesh has acquired. The hero, being doomed to human mortality, returns to Uruk in defeat.

One of the things that fascinates me about this story is how it begins and ends. It actually doesn't start with Gilgamesh but rather with a narrator who, after briefly extolling some of the achievements of our hero, goes on and on about the walls of Uruk. The walls, built by Gilgamesh, surround three square miles. The walls of the great cities of Mesopotamia were usually huge. The tale, interestingly enough, ends with the same ode to the walls of Uruk.

Go upon the wall of Uruk ... and walk around.
Inspect the foundation platform
and scrutinize the brickwork!
Testify that the bricks are baked.
And that the Seven Councilors must have laid
its foundations.

The prologue and epilogue of the story seems to suggest that Gilgamesh is wrong in his belief that he failed to achieve immortality. The tale is narrated in front of the walls that he built hundreds of years prior. He has achieved immortality through his accomplishments in life, if not in mere flesh. The ironic part of this tale is that the walls of Uruk are all but gone, yet another character of the tale thrives to this day. It's the Bull of Heaven, whom you know as the constellation Taurus the Bull.

Each culture invents its own customs, gods, and myths, so why do we, the Sumerians, Babylonians, Greeks, Egyptians, and Romans all call the exact same group of stars the Bull? Many have suggested that ancient people looked at the night skies and used their imagination to make up stories about what they saw. This is nonsense. It's foolish to believe that the combined imagination of all of those cultures couldn't conjure up a single image beside a bull.

There is a particular band of stars that circle the earth. They are grouped into constellations, including Taurus and Gemini, and are collectively called the zodiac. In fact, the Greeks, Egyptians, Romans and the modern western world use most of the same constellations as those of the Mesopotamians.

It's amazing how little has changed in more than four thousand years. Gemini, the twins, were actually sibling giants who guarded the gates of hell. We've all seen pictures of Aquarius, usually depicted as a man carrying two jugs of water. In fact he is the god Ea, the same deity who saved humanity from the flood. He's also the god of fresh water and is depicted by the Mesopotamians with a river

What's left of the Bull of Heaven.

flowing through him. We still draw these streams today, but we've forgotten what they actually are. Likewise, only the front half of Taurus is drawn in modern as well as ancient times, although most don't realize that the bull has lost its backside to Gilgamesh's sword.

We need to remember that to the ancients, stars were not a fanciful field of pretty lights, but an ultra-accurate clock and calendar. To tell the date and time, you need to know exactly where any one star is and how long it has been since sunset. It's easy to find a star on a map. Each star has a longitude and latitude, which place its location using the grid lines on a map. The problem is that there are no lines in the night sky, making a star difficult to find and, even worse, its location nearly impossible to describe once it is found. To tell time, you need to say exactly where the star is, and "up there" doesn't suffice.

The lack of grid lines in the night sky is a serious problem, yet there actually are, in fact, two such lines. The horizon is one, and the path of the sun is the other. Due to the spinning of the earth, the sun appears to go around us once each day. If we could stop this motion, we would notice that the sun still goes around us, once each year, and we could track its motion through the relatively motionless stars.

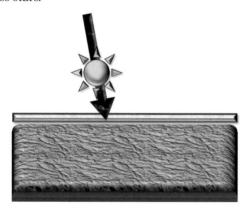

The path of the sun and the horizon form two distinct lines in the sky.

If we had a way to measure where the sun is within the stars, we could easily know the exact date. Hence, we need a ruler in the sky. A ruler is nothing but a number of well-spaced marks on some item, like a stick. The night sky is already filled with such marks—the stars. To measure where the sun is, we can use the stars lying in its path, the zodiac.

On a typical ruler, inches are marked with large thick lines and the eighths are smaller. This enables us to make measurements like 7 and ⅜ inches, rather than ⁵⁹⁄₈ of an inch. Having larger groups makes data easier to deal with in the same way Egyptian fractions make good approximations. Ancient humans first kept track of time using full moons. Since there are about twelve full moons a year, they broke the zodiac ruler into twelve constellations that measured the months. When the sun entered a new constellation it marked the beginning of a new month.

The stars of each constellation of the zodiac were used to mark individual days. There were not enough bright stars for each day, but there were enough to interpolate. If, for example, the sun was between the stars that marked the eighth and eleventh days of the month, and it was more than half way, it must be the tenth. It's almost impossible to remember a hundred stars, but the pictures of the zodiac acted as a mnemonic device. It's much easier to remember things like "the right foot of Taurus" than it is to remember each star's location individually.

The sun measured on the zodiac ruler.

The way to read the ruler is to wait for dusk. Shortly after the sun sets, you'll be able to see a few stars at the location where the sun went down. Theses stars are part of the zodiac and change from day to day. Hence it's possible to tell the day of the year by the stars you see. The star watchers, known in Babylon as the magi, were able to predict the coming of the seasons, the time of plantings, and so on.

It's difficult to know how ancient this knowledge is. The Bull of Heaven's name is written in Sumerian, the world's first written language. We find bull imagery in the world's first towns, five thousand years before Sumerian civilization. We even find paintings of bulls in the prehistoric cave paintings of France. Initially, it takes no science or math, but rather memory, to use the zodiac as a calendar, suggesting that the practice could be very ancient, even by Egyptian standards. If you can remember the pictures of the twelve constellations and the days associated with their various parts you can track time on any clear day. Armed with nothing but a tally stick, you can begin to predict the coming of yearly events. To others in your tribe, you appear to see the future simply by watching the signs of the immortal sky gods.

Omens were very important to Mesopotamian divination. The Mesopotamians maintained huge lists of omens and their meanings. Presumably, if Mesopotamian magi witnessed an unusual occurrence, they would assume that it was a message from the gods. If it occurred with some event, they would record both the omen and the event so they would know what the omen meant should it happen again. As time progressed, omens in the sky became more and more important to the magi. In particular, the motion of the planets was important. The solar system is shaped like a disc. To anyone within the disc, the sun and planets would all appear to move along the brim. Hence all the planets appear to move within the same band of stars as the sun. Hence the magi could use the same zodiac signs to map the motion of the planets.

In order to watch the night sky and record these messages from the gods, the magi posted watches. Each night was divided into three shifts manned by a scribe, who recorded what he saw, when, and where it happened. At some point, the Mesopotamians realized that the motion in the heavens was fairly regular. This meant it was possible not only to see the omens but also to predict them months in advance. So the Mesopotamians began to create mathematical models of the motion of the planets. To do this they needed accurate measurements. The zodiac allowed them to find stars, but now they needed a numerical measure that they could use to generate values to be used in the calculations.

To find a position of a star, you need to measure an angle. If you go outside and point to a star, your arm forms an angle with the ground. This angle can be recorded to locate the star in the sky. Easier still, if you have good records, you can determine how high a star is by simply knowing which stars are on the horizon.

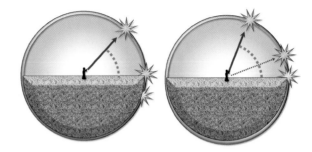

Knowing which star is on the horizon can be used to determine the angle to another star.

The Mesopotamians needed a system to measure these angles, but they already had one in place. The three watches of the night essentially cut the night sky into three pieces. The stars on the horizon at the start of the first watch would be a third of the way up at the end of the watch. Similarly, they would be two-thirds of the way around at the end of the second watch, and down at the horizon at the end of the third watch.

The three watches cuts the night sky into three pieces.

So the Mesopotamian night sky was composed of three watches. Of course, numbering the positions in the sky as 1, 2, or 3 was not nearly accurate enough to fix the position of a star or planet. They needed fractional values. Where would a star be if it started the night on the horizon one-fifth of a watch after the first watch ended? We might be inclined to describe this numerically as 1, and this is essentially what they did.

One and one-fifth of a watch from the horizon.

SOLUTION:

$$\text{𒈨.𒐼𒐊} \times 60 = 2\tfrac{48}{60} \times 60$$

$$= (2 \times 60) + (\tfrac{48}{60} \times 60)$$

$$= 120 + 48$$

$$= 168$$

PRACTICE: How many degrees is 𒐕.𒌋𒈨 watches?

ANSWER: 72 degrees

Do the numbers in the practice problem look familiar? Recall that 1⅕ watches is 𒐕.𒌋𒈨. But the Babylonians would write this as 𒐕 𒌋𒈨, which looks like 72, precisely the number of degrees in 1⅕ watches. We still use the watch system of measurement; we've just forgotten that there is supposed to be a decimal point that was never actually written. Since they're the same system, it's trivial to go back and forth. Hence, 𒈨.𒐼𒐊 watches is 𒈨 𒐼𒐊 degrees, which translates to (2 × 60) + 36, or 156 degrees.

It's not hard to see why this works. When we multiply by 10 in a base-ten system, we move the decimal point one place. For example 1.27 × 10 is 12.7. Similarly, when we multiply by 60 in a base-sixty system, we also move the decimal point. So 𒐕.𒌋𒈨 × 60 becomes 𒐕 𒌋𒈨. Hence multiplying by 60 to convert watches into degrees just slides the decimal point over.

The celestial sky spins as the night progresses, just as the sun circles the sky during the day. This means that the magi needed to keep careful track of time as well as of the angle. The desired accuracy could require more than one digit behind the decimal point, so they might need a number like 𒐕.𒌋𒌋𒐖 𒌋𒌋 hours. Each digit represents a value one-sixtieth of the one before it. That's why the 𒌋𒌋𒐖 is 24⁄60. So the 𒌋𒌋 has a value of a sixtieth of a sixtieth. This is a 3600th since 60 × 60 is 3600. Hence we interpret the 𒌋𒌋 as 30⁄3600.

EXAMPLE: Translate 𒌍𒐖 . 𒌋𒌋𒐖 𒐼𒐕.

SOLUTION: $36 + \tfrac{15}{60} + \tfrac{41}{3600}$

Something strange happens if we convert the above times into hours, minutes, and seconds. Our answer of $36 + \tfrac{15}{60} + \tfrac{41}{3600}$ becomes 36 hours, 15 minutes, and 41

However, the Mesopotamian system has more in common with our decimal system than with our fractional one. In decimals, we describe 1⅕ as 1.2. The .2 represents two tenths, or equivalently 2⁄10, which reduces to the ⅕ we see in 1⅕. Of course the Mesopotamians would not use tenths but rather sixtieths, since they had a base-sixty system. The fraction ⅕ is equal to 12⁄60. So, just as we write 1 2⁄10 as 1.2, they would write this number, 1 12⁄60, as something like 𒐕 . 𒌋𒈨.

The Mesopotamians didn't actually have a decimal point. They would write the number as 𒐕 𒌋𒈨. While this number looks just like 72, they probably knew that it was 1.2 from the context of the situation. In this book, I will be very un-Mesopotamian and include the point for clarity.

EXAMPLE: Translate 7½ into Babylonian.

SOLUTION: $7\tfrac{1}{2} = 7\tfrac{30}{60} = \text{𒐌.𒌍}$

PRACTICE: Translate 31⅔ into Babylonian.

ANSWER: 𒌍𒐕.𒐼

Now let's translate these Babylonian angles into modern degrees. Three watches make up the night sky, which in turn makes half a circle. Since a circle is 360 degrees, half is 180 degrees. So we need to divide 180 by 3 to get 60 degrees, the angle of one watch. So two watches has 2 × 60, or 120 degrees, and it's not hard to see that we simply need to multiply the number of watches by 60 to find the number of degrees.

EXAMPLE: How many degrees is 𒈨.𒐼𒐊 watches?

seconds. Notice that the numbers match exactly. That's because we measure time in Babylonian. We just don't realize it. The translation between Babylonian and modern time is trivial. We can translate 2 hours, 15 minutes, and 40 seconds by just writing their Babylonian equivalents, putting the decimal point after the 2, giving ⟑.⟨⟆ ⟆. This is similar to our writing 5.29 dollars as 5 dollars, 2 dimes, and 9 pennies. Minutes and seconds, just like dimes and pennies, are mathematically just numerical digits following the decimal point.

How is it that after thousands of years we're still measuring angle and time exactly as the Babylonians? Imagine that you come from a culture that knows little math and no real astronomy. If you wanted to know these subjects, you would have to learn them from a culture that does. So you get a Babylonian text and work through the examples, slowly gaining an understanding. By the time you really knew what you were doing, you'd be used to doing it the Babylonian way.

Since you come from a base-ten culture, you might consider translating the Babylonian system to base ten and do the work in your system. But remember that no one does "decimal" work in fractional values because these ideas have just arrived in your region. So no one else would appreciate the conversion. Then you realize that every math table and star chart you have uses the base-sixty decimals, and for them to be workable, you would need to convert everything. As I said, after years of study, you've become used to the system, and habit, oddly enough, is habit forming. So you take the path of least resistance and just let things be. So the Babylonians taught the Greeks whose books taught the Arabs and Hindus and whose work eventually traveled back to Europe in a journey of a few thousand years. This tale is even stranger when you realize that the Babylonian masses used base ten, whereas only intellectuals used the Sumerian base sixty for computations. Hence, the tools of astronomy, both the zodiac and the base-sixty system, were carried on from generation to generation, spanning the entire age of civilized man.

How do you describe a Babylonian base-sixty decimal digit to persons in a culture who don't even have the concept in their native base ten? Imagine them confronted with ⟑.⟨⟆ ⟨⟆ hours. They recognize the ⟨⟆ as 35 of something and ask their teacher what it means. The teacher tries to explain that it's less than 1, but the students once again ask how thirty-five things can be so small. The teacher replies, "They are small things," which in Latin is *minuta*. The students then ask, "What are the 28 things after the 35?" "They are even smaller," the teacher replies; the second small things. In Latin this is *secunda minuta*, and hence our minutes and seconds were born out of an improper understanding of Babylonian numbers.

THE TABLETS OF NIPPUR

Digit Multiplication

The Sumerian city of Nippur is ancient, even by Sumerian standards. According to myth, Nippur was the home of Enlil, the same god who flooded the earth. Many of the early Sumerian tales involving gods took place in Nippur, which, according to legend, seems to have existed since the beginning of time. Although Nippur seems to have never exerted great political or military power, it was held in great reverence by the Mesopotamians. Kings and gods, in the form of divine statues, regularly traveled to Nippur to get the blessings of the great Enlil.

In this city and south of the temple called the Mountain House lived a priestly scribe in a small brick home. Here he taught the next generation of scribes how to read, write, and do mathematics. The students learned from tablets made of clay, which they copied to help them remember the contents. Most of the tablets consisted of lists of words used to teach reading. In addition, there were some tables of unit measurements, a few Sumerian myths, and some math texts the students needed to learn.

Anyone who's gone to school knows that paper accumulates rapidly. Whether it's homework, practice, or notes, there's just too much of it to keep it all. Now imagine you live in Babylonian times, in which each sheet of work is a clay tablet about the size of a hand and far from paper thin. It's not hard to see why the scribe had a waterproofed bin that he presumably used to recycle the student's work.

The scribe's house was eighty years old, and the walls needed some serious repairs, especially new bricks to replace the more dilapidated ones. He had already used up all of the bricks available to him, but he had an inspiration: he began to repair his home with the recycled student homework. He used about 1500 pages, and it is the discovery of this veritable gold mine of archeology that tells us the majority of what we know of Mesopotamian mathematics.

The Babylonian schoolchildren needed tablets in order to multiply. Today, young students memorize their times tables, whereas back then, they made a series of multiplication tables. We use the digits 0 to 9 and hence have to know how to multiply these ten digits by each other. The Babylonians used the digits ⌶ to ⪫ and ⟨ to ⪩ and had to know the operations of these fourteen symbols. Below I have representations of a times-6 and times-20 tables.

The times-6 and times-20 tables.

But my tables are inauthentic, for a number of reasons. My tables run from 1 to 10 and then jump by 10s up to 50. True Babylonian tables ran from 1 to 20 before jumping by 10s. We will see that these numbers, 11 to 19, are not needed and that I simply wanted to keep my tablets a reasonable size. I've also used the symbol ⸱ liberally. Babylonians would only use this zero-like symbol between digits to represent a gap. I've put them before and after digits to represent missing terms. This would be similar to our writing "4×5 is 2 ⸱" and "2×3 is ⸱ 6."

The first table is the times-6 table. We can tell this by looking at the first row, ⌶   ⪫, which tells us that ⌶ times 6 is ⪫. Hence, you can always tell what table you have by looking at the second column of the top row, because it will have the value of 1 times the number.

EXAMPLE: What is ⪫ × ⪝?

To answer this question we go to the times-⪫ table and move down the first column until we see the ⪝. The number to the right of it is ⸱ ⪨ ⪝, which, if we ignore the ⸱, translates to 48. This should come as no surprise, since we all know $6 \times 8 = 48$.

ANSWER: ⪨ ⪝

PRACTICE: What is ⪫ × ⪫?

ANSWER: ⪩ ⪝

We can use the same table to determine ⪫ × ⪩. Looking in the row with ⪩, we see ⪝ ⸱. The ⪝ is in the 60s column. Translating, this tells us that $6 \times 40 = 4 \times 60$. or equivalently, 240. Note that the Babylonians would have simply given ⪝ as the answer. The reader would have to assume that 6×40 isn't 4, but rather 60, times larger, a fairly obvious conclusion. The reason for the ⸱ symbol is that the table is composed of two columns. I'm misusing the Babylonian symbol to indicate an empty column. Two columns are required because some products produce two-digit numbers. Let's see this using the times-⟨⟨ table.

EXAMPLE: What is ⟨⟨ × ⪝?

As before, we go down the first column of the ⟨⟨ table until we find ⪝. Looking across from it, we see ⼌ ⟨⟨. The leading 2 is in the 60s place and the ⟨⟨ is in the 1s place. Translating into base ten we get

$$⼌ \; ⟨⟨ = (2 \times 60) + 20 = 140$$

The answer should again be no surprise because 20×7 is 140.

ANSWER: ⼌ ⟨⟨

PRACTICE: What is ⟨⟨ × ⪩?

ANSWER: ⟨⪫ ⪨

The reason I inserted the ⸝ symbol was to deal with problems that occur in exercises like the following:

EXAMPLE: What is 𒐚 × 𒐜𒐘?

Going back to the times-𒐚 table, we see that the answer is 𒐈 ⸝. The Babylonians would have simply written 𒐈. When we translate this to base ten, it becomes $6 \times 30 = 3$, which appears to be nonsense. However, we need to realize that the 3 is in the 60s place, so it really represents 3×60, or 180, which is the correct answer to the problem. So, to make the table easier to read, I always inserted a digit in the 60s and 1s places; where there is none, I put the symbol ⸝.

Now that we know how to read the tables, let's use them to perform digit multiplications.

EXAMPLE: Multiply 𒐜𒐚 by 𒐕𒐚.

Note that the first number is 𒐜𒐚, and we already have available the 𒐜 and the 𒐚 tablets. If you're familiar with algebra, you might remember FOIL, the acronym for "first, outer, inner, last." "First" represents the first parts of the 𒐜𒐚 and the 𒐕𒐚 multiplied, which is 𒐜 × 𒐕 and from the times-𒐜 table can be seen to be 𒐈 𒐜.

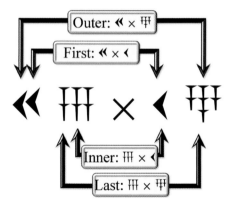

The products of FOIL.

The other products of the FOIL can each be found on one of the two tables above and are listed below.

| First: | 𒐜 × 𒐕 | = | 𒐈 | | 𒐜 |
| Outer: | 𒐜 × 𒐚 | = | 𒐈𒐕 | | 𒐜 |

| Inner: | 𒐚 × 𒐕 | = | 𒐕 | | ⸝ |
| Last: | 𒐚 × 𒐚 | = | | ⸝ | 𒐜𒐈 |

These numbers can now be put on a counting board as follows:

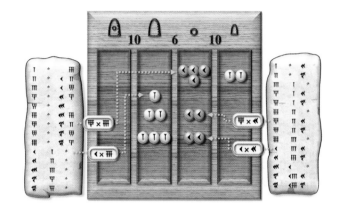

Finally, we simplify our answer giving the following:

SOLUTION:

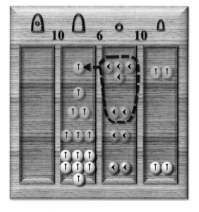

In order to multiply numbers beside 𒐜𒐚, we need other tables. Here are two more tables to help you do the next problem. The full set is in the computation tables at the front of the book. Feel free to copy them for your own use.

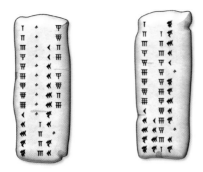

The times-50 and times-4 tables.

EXAMPLE: Multiply 𒐏𒐕 by 𒐖𒐗.

SOLUTION:

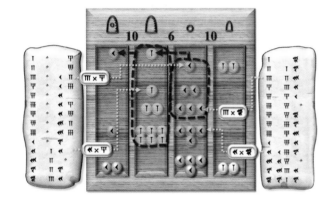

𒐏𒐕 × 𒐖𒐗 = 𒐏 𒐏𒐕

PRACTICE: Multiply 𒐏𒐕 by 𒐖𒐗.

ANSWER: 𒐏𒐕 𒐖𒐗

THE DIAGONAL AND THE SOUL

Multiplying a Digit by a Number

Can you use mathematics to prove that the soul is eternal? Socrates, an ancient Greek philosopher, seemed to think so. In a discussion with Meno, a young, well-to-do visitor to Athens, the two try to determine the nature of virtue. In part of their conversation, Socrates uses the mathematical properties of the diagonal of a square to prove a point.

Socrates takes aside a slave boy who Meno admits knows no mathematics. Socrates draws a square on the floor and challenges the boy to find a square with twice the area. The boy's first attempt consists of doubling the sides of the square. Socrates, however, leads him down the path of reasoning that the new square consists of four copies of the old square and hence has four times the area.

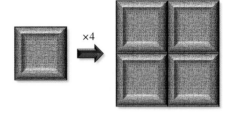

Twice the side length makes a square four times as large.

The slave tries another false guess, at which point Socrates adds diagonals to the square. With leading questions, the boy eventually realizes that a square made from four diagonal sections uses halves of each small square; hence, it's half the area of the larger square. Since the larger square is four times as large as the original square, the new square must be half of this, which is twice the area of the original square. This new square is exactly the one they were looking for.

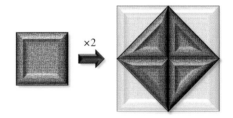

The square made by the diagonals has twice the area.

Socrates points out that the child has just solved a complex geometry problem on his own. The boy somehow knew things that he was never taught. Where could he have learned these? If not in this life, the slave must have learned it before he was born; hence, the soul must exist independently from our earthly existence.

While the above argument doesn't fully convince the modern mind, Socrates is at least partially correct. The human mind is prewired to understand certain concepts like those dealing with quantities and geometry.

As a result, many ancient cultures, like the Vedic priests in India, knew the very mathematical relation Socrates described.

The ancient Greeks focused on geometrical relationships; the Babylonians, like us, focused on the arithmetical. Consider a 1-by-1 square. This square clearly has an area of 1×1, which is 1. The diagonal, as we've just seen, makes a square of twice the area, 2. Since the area of a square is the length of the side multiplied by itself, the side of our square is the number that when squared is 2. In modern language, it's the square root of 2. The Babylonians were aware of this fact. They even knew the Pythagorean Theorem, a generalization of this notion.

Consider the following illustration of an actual Babylonian tablet. It shows a square drawn in much the same way as Socrates's figure.

The diagonal of a square tablet.

The side of the square is labeled with the length 《, which of course is 30. Along the diagonal is the particularly large number ˥ 《ᵀᵀᵀ 《《ᵀ ‹. It's actually not a large number. Remember that the Babylonians didn't write decimal points. If they had, they would have written the number as ˥.《ᵀᵀᵀ 《《ᵀ ‹. This can be translated into our number system as follows:

$$1 + {}^{24}\!/_{60} + {}^{51}\!/_{3600} + {}^{10}\!/_{216000} \approx 1.41 \ldots$$

This is an excellent approximation of the square root of 2, off by less than one part in a million. The number below it is ᵀᵀᵀ 《ᵀᵀ 《《ᵀᵀ, exactly what you get when you multiply 30 by the approximation. Remember that if the side of the square were 1, the diagonal would be root 2. A square with a side of 30 is thirty times larger, so we would expect the diagonal to be 30 times the square root of 2, exactly what the Babylonian tablet calculates. Let's use the Babylonian tablet to estimate the diagonal of a square.

EXAMPLE: Estimate the length of a diagonal of a square whose side is ᵀᵀᵀ.

We need to calculate the value of ᵀᵀᵀ multiplied by our estimate of the square root of 2. Because ᵀᵀᵀ can be thought of as a base-sixty digit, this is roughly equivalent to the way we would multiply a digit like 7 by a number such as 256. We'd first multiply 7 by 6, getting 42. We'd write down the 2 and carry the 4. Then we would proceed to the next number, multiplying 7 by the 5 of 256. This gives 35. The 5 needs to be added to the carry of 4, giving 9. The 3 of the 35 needs to be carried to the next multiplication. Finally, we'd multiply 7 by the 2 of 256, giving 14. Added to the carry of 3, gives 17. The following diagram illustrates the above procedure and represents a mix of modern paper methods with counting-board procedures.

7×256 done as arithmetic and with calculi.

Notice that the methods don't differ greatly. When we multiplied 7 by 6, both methods put the 2 of the 42 down below and put the carry of 4 above in the next column over. In the modern method, some of the adding is done in one's head, such as when we add the carry of 4 to the 5 of the $7 \times 5 = 35$. On the counting board, that addition can be left off until the end.

To begin our multiplication of 𒁹.𒐏𒐕 𒐏𒐖𒁹 𒌋 by 𒐏𒐊, we first need to look at our multiplication tablets for 𒐏 and 𒐊.

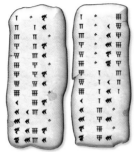

We now start with the rightmost digit, 𒌋, of 𒁹.𒐏𒐕 𒐏𒐖𒁹 𒌋 and multiply it by 𒐏𒐊.

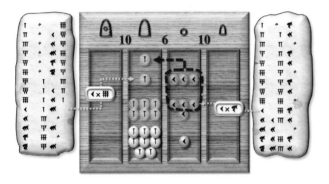

We can now enter this value of 𒐊 𒌋 on another counting board. Note that although our above board has four columns, we can think of our answer as a two-digit base-sixty number. The two rightmost digits give the regular number, in this case 𒌋. The two on the left give the carry, 𒐊. The number gets placed on the board on the far right with the carry placed above.

We now need to multiply 𒐏𒐊 by the next digit of 𒁹. 𒐏𒐕 𒐏𒐖𒁹 𒌋, which is 𒐏𒐖𒁹.

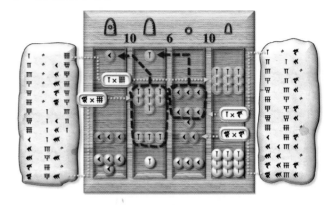

We now enter the product we just obtained, 𒐏𒐖𒁹 𒐏𒐊, on the second counting board. The 𒐏𒐖𒁹 is the carry so that it gets entered diagonally above to the left just as before. I've circled the new calculi in green to highlight them.

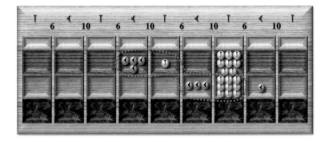

We now repeat, going to the next digit of 𒁹. 𒐏𒐕 𒐏𒐖𒁹 𒌋, which is 𒐏𒐕.

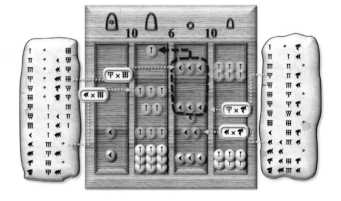

This gives us the answer 𒌋𒐊 𒐏𒐊, which we once again enter on the second counting board.

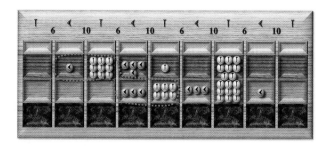

Usually we'd need to repeat the process one more time; however, the next multiplication is ⊺×⧊⧉, and the fact that there is no times-1 table, we can easily assume that the Babylonians knew that this is ⧊⧉. When we place this on the second table, we get the following:

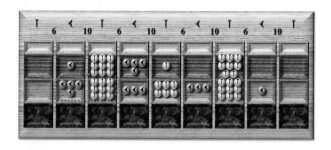

Finally, we add the numbers in the two columns, carrying when necessary.

Hence, the answer is ⊺ ⧉ ⟨⧉ ⧊⧉ ⟨. Notice that there is no decimal point. As I've said before, the Babylonians didn't use a point to denote where the whole number part of the answer ends. It really isn't necessary. The whole-number part of ⊺ ⧉ ⟨⧉ ⧊⧉ ⟨ could be ⊺, ⊺ ⧉, or ⊺ ⧉ ⟨⧉. These are 1, 69, and 4164, respectively. Remember that the problem is asking us to find the diagonal of a square that has a side of 49. The diagonal should thus be a number a little bigger than 49. The answer obviously has a

whole part of 69. The remaining digits, ⟨⧉ ⧊⧉, and ⟨, are the fractional part of the number.

Now try this problem to find the diagonal of a square with side ⟨⧉. You will need the following multiplication tables. The whole set of tables can be found at the front of the book. Check your answer as you go with the one in the back of the book. Don't worry if you make mistakes. Remember, my goal is to show you how difficult this math is compared to Egyptian methods. Your difficulties actually prove my point.

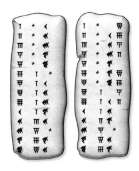

PRACTICE: Find the length of the diagonal of a square whose side is ⟨⧉. Use common sense to place the decimal point.

ANSWER: ⟨⧊.⊺⊺ ⟨⧉ ⧊

The above method can be used to multiply a Babylonian digit by a number. Use the following tables to solve the following problem:

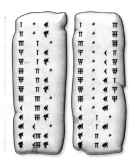

PRACTICE: Multiply ⟨⊺⊺⊺ by ⟨⧊ ⧊⧊.

ANSWER: ⧊ ⧊⊺⊺⊺ ⧊

THE IGNORANT, LIARS, AND THE INSANE

Multiplication

Time is running out. I check and recheck every step, and I can't figure out what I did wrong. Out of the best intentions, I promised my math history class an answer key to the Babylonian multiplication homework and class is about to start in ten minutes. The ridiculous thing is that I already know what the answer is since I cheated. I solved the problem using modern methods and then converted the answer back into Babylonian giving the correct solution. Of course, as I try to obtain the answer using Babylonian methods, it's not what I keep getting.

I've tried many of the tricks I've learned over the years. One useful one is to subtract the correct answer from the one you get. It's often off by a small multiple of a Sumerian calculus. If you understand the system well enough, it tells you which column the mistake is in. For example, if it's off by 1200, which is 600×2, then it's off by the 600-value calculi, which is the second ⟨ column from the right.

After looking for the mistake for almost half an hour, I finally catch the error. I put one entry in the wrong column. I quickly make changes to my paper and rush off to the copy machine, knowing full well that I'll be a few minutes late to class.

At the beginning of this chapter I promised you pain. You may have thought that the last section was a bit tedious, but it was about only a single digit times a number. We'll now multiply two Babylonian numbers together. I tried to warn you to stop reading. Since you're still with me, I can assume that you have no respect for my opinion. Don't worry. I'm not offended. I've got kids and I'm used to it.

EXAMPLE: Multiply ⌐ ⦉⦉⊞ ⦉⌐ by ⫫ ⦉⫸ ⟨⊞.

Let's first look at one of our modern multiplications to gain an appreciation of the procedure.

$$
\begin{array}{r}
34 \\
\times 12 \\
\hline
68 \\
34 \\
\hline
408
\end{array}
$$

The 68 above is the result of the digit multiplication 2×34, similar to the procedures we did in the last section. The 34 below the 68 is the solution to the multiplication 1×34. It's shifted to the left one position because the 1 is actually the 10 in the number $12 = 10 + 2$. The same thing happens in base-sixty calculations; however, we need to shift our products two columns to make the alternating 10 and 6 bases act like base 60. If there were another digit in front of the 12, we would put its product with 34 on the next line shifted over one more.

The Babylonian number ⌐ ⦉⦉⊞ ⦉⌐ is a three-digit base-sixty number. This means we will need three rows to enter in our digit multiplications. Each row consists of two sub-rows, one for the number and the other for the carry. At the top, as before, we'll put in a carry row for the final addition, and at the bottom we'll add a row for the simplified answer. But first we will multiply the digit ⦉⌐ by ⫫ ⦉⫸ ⟨⊞. Here are the three separate digit multiplications followed by the first-row entry.

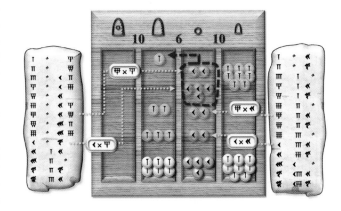

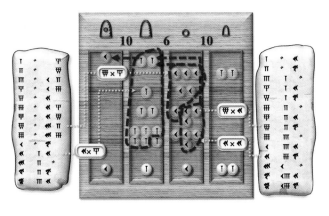

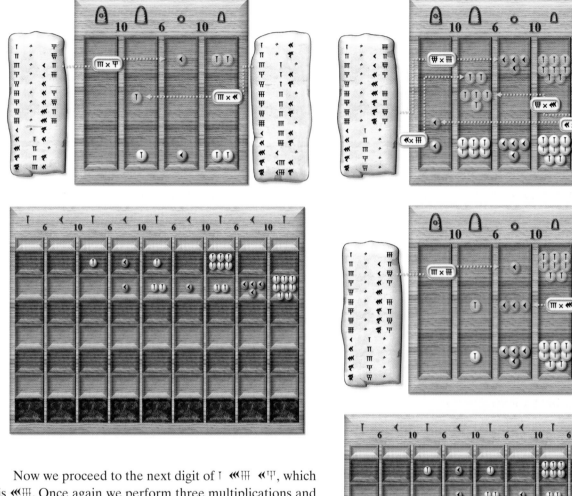

Now we proceed to the next digit of Ⅰ ⧗Ⅲ ⧗Ⅱ, which is ⧗Ⅲ. Once again we perform three multiplications and place them on the large board.

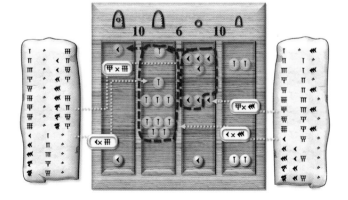

Hopefully you're tired of this and will appreciate that the next multiplication is by Ⅰ. When computing Ⅰ × Ⅲ ⧗Ⅲ ⦉Ⅲ we just put the calculi of the number down on the second board. It's shifted to the left another two squares.

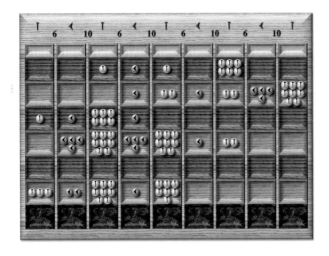

All that's left to do is add these calculi, giving us the final answer.

So the solution is 𒌋 𒌍𒌋 𒌍𒐖 𒌍 𒐕𒌋. That's a lot of work for one multiplication. If you enjoy pain and feel the need to try one, have a go at this.

PRACTICE: Multiply 𒐕𒐖 𒌏𒐽 by 𒐖 𒌍𒐕 𒐕𒌋.

ANSWER: 𒐖 𒌏𒐖 𒌏𒐽 𒐕𒐕 𒌏𒌋

If you had trouble with that one, you can take some solace in the fact that I had to do it twice to get the right answer, and I've been doing this for years.

In the example worked out above, the multiplication consisted of nine digit multiplications, which contained a total of twenty table references. Most digit multiplications required simplifications. The final board, which contained eighty-four calculi spanning nine columns, also needed simplification. The work took up more than an entire page. It would have been bigger if I hadn't lazily decided to make one of the digits 1, in order to reduce the number of diagrams I needed to draw.

Let's now do the exact same problem using Egyptian math. If we convert the Babylonian numbers being multiplied into our system, they become 5,784 and 12,497. Here are the numbers multiplied using the Egyptian method. It took me about two minutes. I could have done it faster, but I wanted the solution to consist of as few lines as possible. The multiplication is not standard, but it uses simple tricks accessible to anyone who has practiced Egyptian math. Below I make 5000 in the obvious way. I know I can use the 500 and half of it, 250, to get to 5750. I'm now 34 away. A tenth of the 250 gets me to 5775, leaving only 9 to go. I use the 5 I already have and make a 4 by doubling the starting 1 twice—and I'm done.

	5,784	?	
1)	1	12,497	
2)	10	124,970	(1) × 10
3)	5	62,485 ✓	(2) ÷ 2
4)	50	624,850	(3) × 10
5)	500	6,248,500 ✓	(4) × 10
6)	5000	62,485,000 ✓	(5) × 10
7)	250	3,124,250 ✓	(5) ÷ 2
8)	25	312,425 ✓	(7) ÷ 10
9)	2	24,994	(1) × 2
10)	4	49,988 ✓	(9) × 2
		72,282,648	

We get the same answer with a miniscule fraction of the work.

It seems that every time I pick up a book on the history of mathematics, I read about how awkward the Egyptian system is in comparison to the "enlightened" Babylonian one. We have to assume that these authors have never actually done a calculation in either and have no clue what they're talking about. We make this assumption to be generous because otherwise we would need to assume that they are ignorant, liars, or just plain insane.

8
JUDGMENT DAY

GOING FORTH BY DAY

Initial Comparisons of Egyptian and Modern Methods

We generally consider the grave as our final resting place. For an Egyptian, being placed in a tomb was merely the start of a great journey; the goal was to reach the land of the setting sun. The *ka*, a form of the deceased's soul, had a head start since all tombs were built on the western side of the Nile for that very reason. To begin the journey, the ka would open the false door within the tomb. To the living, the door seemed merely painted on the wall, but to the dead, it opened to the underworld. On the soul's way westward, a number of obstacles could slow down its progress, like gatekeepers and other monsters. Those wealthy enough in life could afford a copy of the papyrus called *The Book of Going Forth by Day*, which we now call *The Egyptian Book of the Dead*. This user's guide to the afterlife contained spells designed to easily bypass these barriers. It was often buried with the deceased so it would be on hand for convenient use by the departed.

Eventually the soul would end up in the Hall of Osiris, the benevolent ruler of the underworld. On its arrival, a horde of forty-two monsters, many with terrifying names, would each demand that the deceased take an oath of innocence. To the Swallower of Shadows, the ka would have to swear that it had never murdered. To He Whose Eyes Are in Flames, it would deny ever having stolen property of the gods. To the other forty monsters, it would deny crimes like adultery, acts of violence, and cursing gods. The Egyptians valued law and order, and hence many smaller crimes appeared in the list of damnable offenses. So crimes like altering weights and measures, blocking water from flowing, and engaging in wrongful property

disputes also had to be denied. It's hard to imagine how anyone could achieve Egyptian salvation, because they had to deny ever lying, talking too much, getting angry, and feeling sullen.

After satisfying all forty-two judges, the true judgment began. The heart of the deceased was placed on one side of a scale. It was the only organ left in the mummified body and was considered the seat of the spirit. Even if the defendants lied to the judges, their hearts contained the true character of their being. On the other side of the scale was a feather of the goddess Ma'at, the goddess of truth, justice, and order. If the scale balanced, meaning the soul was pure with Ma'at, the ka was admitted into the Field of Reeds for an eternity of peace and prosperity. If the

Anubis and a hungry monster weigh a heart.

scale failed to balance, the heart would be consumed by a monster that was part crocodile, lion, and hippopotamus.

It's now time to judge Egyptian mathematics, not by itself, but in comparison to modern methods. In this chapter, we'll evaluate the systems on a number of criteria. In the rest of this section I'll recap some arguments already made, and later I'll make some new arguments. I need to make it clear that this is an imbalanced comparison. Egyptian mathematics consists of one unified system. Modern mathematics is a compilation of three systems: the first is decimal, in which numbers are represented in much the same way the Babylonians did, and the second are third are composed of fractions. The two systems consist of standard fractions, like $\frac{1}{3}$, and mixed fractions, like $2\frac{1}{3}$.

Approximation Value

Egyptian fractions form a good approximation of a number when properly simplified. The value $7\,\overline{2}\,\overline{9}\,\overline{36}$ is a little bit more than seven and a half. Decimal notation has similar properties. The number 8.52324 is about $8\frac{1}{2}$. If we want better approximations of these numbers, we can include extra fractions for more accuracy. So the above two numbers can be approximated as $7\,\overline{2}\,\overline{9}$ and 8.52. We can trade off accuracy for simplicity as we see fit. Fractions, on the other hand, have little approximation value. The fraction $\frac{8446}{347}$ is difficult to estimate even as a whole number. To help rectify this situation, "mixed fractions" are used, with which $\frac{8446}{347}$ becomes $24\frac{118}{347}$. This new form tells us that the fraction is about $24\frac{1}{3}$, although it's not clear if it's above or below this value. Even though the mixed format tells us the whole component of a number, the difficult-to-estimate fractions still remain. Hence Egyptian fractions and decimal notation both score well in this category, whereas fractions fail miserably.

Time to Learn: Whole Numbers

The Egyptian method for whole numbers is trivial. If you can add two numbers, you can already multiply by 2, and hence you can double. If you can double, you can multiply. If you can multiply, then you already know the basis of division. Most of my students can learn to do multiplications and divisions with remainders in a matter of minutes.

Whole-number multiplication in our system is a bit tricky. You first need to memorize the multiplication tables. The table is ten digits by ten digits, and hence 100 entries. You could argue that multiplication by 0 and 1 require no memorization, but even if we get rid of these numbers, there are still sixty-four multiplications to memorize. Because the human brain doesn't register these as mathematics, they get stored in the verbal part of the brain. They're all similar, so the mind gets them confused. Since they follow no obvious mental pattern, memorizing them is boring, and the brain by and large ignores them. Hence this process usually takes months to learn and be error free.

When you perform multidigit multiplications, you need to do a series of digit multiplications. The number of such multiplications is often quite large and is equal to the product of the number of digits of the two numbers. For example, a five-digit number times a four-digit number, like $38,430 \times 2,259$, has 5×4, or 20, different digit multiplications to be made. The result of each has to be precisely placed on a grid, just as we placed the results of the Babylonian digit multiplications on a counting board. Some will have carries, which must also be placed. The final result gets totaled. Mastering the whole process takes years.

Today's students need to learn whole-number division. Unlike the Egyptian system, there is little resemblance to whole-number multiplication. You need to guess repeatedly how many times a number goes into another, multiply the guess by the divisor, and then subtract. As with multiplication, it's important that the result of each operation is carefully placed on a grid so that digits of the same value line up. This system takes a while, especially since it requires students to make good guesses to do it properly. Clearly, in terms of whole numbers, Egyptian mathematics is much easier to learn since we're comparing a learning time of ten minutes to six months.

Time to Learn: Fractions

Working with Egyptian fractions is not hard, but it's also not always easy. Simplifications and working with parts takes some care. The problem with evaluating the time it takes to work with fractions is that the issue is vague, because it largely depends on the skill of the mathematician. Anyone can begin to work with Egyptian fractions right away; however, it takes some sophistication to do it well. I'm going to assume that you made it through the first five chapters of this book in anywhere from a week to a month, which is not that long compared to the time it takes to learn most things in mathematics.

Decimals score well in terms of ease of learning. Fractional operations work just like their whole-number counterparts. For example, $8.32 + 1.25$ is 9.57, since $832 + 125$ is 957. You simply need a few rules to determine where decimal points belong. The same thing also happens for subtraction, multiplication, and division.

Standard fractions are not so simple. Consider the addition $7/12 + 4/15$. Before they can be added we need to find a "common denominator." An easy way to find one is to multiply the denominators 12 and 15 to get 180. This method makes the worst denominator possible, since the number is so large. Both the numerator and denominators grow, making the fraction more difficult to work with and harder to interpret.

We could find what's called the *least common denominator* (LCD), which is the same as the least common multiple. In the case of 12 and 15, the LCD is 60. One way of finding the LCD involves factoring both denominators into primes. The notion of prime factors and how to find them requires a fairly large skill set that takes a while to learn. Even if you're knowledgeable, the process is impractical. If I show you 35, you might immediately recognize that it is 5×7, the prime factorization of 35. If, on the other hand, I showed you 5339, would you recognize that it is 281×19? Factoring numbers into primes is so difficult that a common modern encryption code is based on it. In other words, there's a secret code that can be broken if you can factor a known number into two primes. Even the most powerful of computers can't find the factorizations in a reasonable amount of time, which is why the codes are essentially impossible to break. If supercomputers can't simplify most fractions, how can we?

All we've done so far is talk about finding a common denominator, and there are other steps to adding fractions. Our above sum $7/12 + 4/15$ can now be written as $35/60 + 16/60 = 51/60$. We now face the problem that our sum isn't simplified, a difficulty we often face in Egyptian mathematics. In order to reduce a fraction, we need to find common factors, a process that we've just shown to be difficult. In our fraction, both the 51 and the 60 share a common factor of 3 reducing to $17/20$.

Subtraction of fractional values shares exactly the same problems as addition. Multiplication and division don't have the common denominator problem but do have the simplification difficulty. Clearly, fractional operations take a long time to master.

I won't even seriously discuss the time it takes to do operations on mixed fractions. Most mathematicians convert these into standard fractions before doing any operation because of the difficulty involved. This just adds more steps to master to the already-complicated fractional operations.

Time to Do

If you haven't blocked out your memories of doing Babylonian multidigit multiplications, you realize that these can take an excessive amount of time. Practice can shorten it, and the same is true for Egyptian mathematics. Fractions seem easy to work with, but the problem of calculating prime factorizations for the denominators can't be ignored. To be fair, picking parts for Egyptian fractions can run into the same difficulties; however, I've timed myself doing operations in both Egyptian and modern mathematics and have found that I can usually work faster in the Egyptian system. I will admit that there

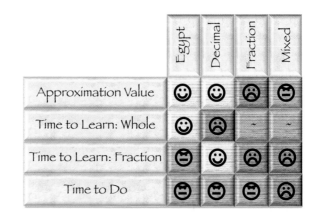

Best system according to particular criteria.

are times when I get stuck while doing an Egyptian computation. Modern methods generally take longer but are more consistent in terms of time. I'll call this one a draw, even though I personally would call this one a win by a nose for Egypt. Here's a table tracking scores in which the happier the face, the better the system rates.

You might notice that fractions and mixed numbers do badly in the comparisons. The decimal system is almost holding its own. However, especially in the "Time to Learn" categories, you can't look at decimal and fraction values separately because you need to learn all of them. At this point it seems that if we just learned decimals, the Egyptian system might not win; however, as we shall see, there is something seriously wrong with the decimal system.

THE ROOT IS ALL EVIL

Babylonian Square Roots and Division

Two and a half thousand years ago the mystic and philosopher Pythagoras was fascinated by the ratios of music. The harmony made by plucking two lyre strings of equal tension was directly related to the ratios of the string lengths. For example, a string of length 6 plucked simultaneously with a string of length 4 produces a harmony called the musical fifth since they are in ratio 3:2. This can be calculated by thinking of them as fractions. Just as $\frac{6}{4} = \frac{3}{2}$, so does 6:4 = 3:2.

Attempts were made to systematically organize the musical scale according to mathematical law. Pythagoreans realized that the note musically in the middle of two other notes had a string length that is the *geometric mean* of the other two lengths. Taking a modern perspective, the geometric mean is calculated by multiplying the string lengths and then taking the square root. For example, consider strings of length 4 and 9. Multiply them to get 36 and then take the square root of 36 to get 6. We can interpret this as thinking of 6 as the half-way point between 4 and 9 in a harmonic sense. The musical distance between 4 and 6 is the same as the distance between 6 and 9. This means that the harmony made by plucking the 4 and 6, which we've seen above plays a fifth, is the same as the harmony made by plucking the 6 and 9. Mathematically,

this can be confirmed by realizing that 9:6 and 6:4 both equal 3:2, which is the musical fifth.

It is possible that by considering proportional string lengths as sides of proportional triangles, the Pythagoreans came up with the Pythagorean Theorem. Many people credit them with the discovery of this theorem. They may well have proved it, but they certainly didn't discover it, since it was well known to the Babylonians long before. It's perhaps no surprise that Pythagoras spent a few years as a slave of the mathematician-astronomers of Babylon (magi), after being captured in Egypt by the conquering Persian Empire.

Consider two lyre strings of lengths 5 and 8. If there was a string directly between the two, it would have a length $\sqrt{40}$ since 40 = 5 × 8. Unfortunately, we never learned what number when multiplied by itself is 40. We couldn't have learned this, since the square root of 40 is irrational. That means that there is no finite way to write it as a series of additions, subtractions, multiplications, or divisions of whole numbers. We can merely estimate such numbers.

The Babylonians devised a reasonable way to make these estimates quickly. Since 40 is between the two squares 36 and 49, the answer is between 6 and 7. So it's going to be 6 plus a fraction. The trick is to write 40 as a square plus a remainder. In this case it's $6^2 + 4$. The extra fraction is the 4 divided by twice the 6, or $\frac{4}{12}$. This simplifies to $\frac{1}{3}$ so the answer is $6\frac{1}{3}$. When rounded to five digits, it is 6.3333. The actual value of $\sqrt{40}$ when rounded is 6.3246, which means that our approximation is not bad considering how fast we came up with it. It has an error of about one part in a thousand. The ≈ symbol below means "is close to" and is used instead of = in estimates.

EXAMPLE: Estimate the square root of 67.

SOLUTION: $67 = 64 + 3 = 8^2 + 3$ So $\sqrt{67} \approx 8\frac{3}{2 \times 8} = 8\frac{3}{16}$.

PRACTICE: Estimate the square root of 102.

ANSWER: $10\frac{1}{10}$.

The Babylonians didn't solve it exactly this way. They couldn't, because they didn't use fractions. They had to do the divisions in their base-sixty decimal system. When we got $\frac{4}{12}$ above in our estimation of the square root of

30, the Babylonians would have needed to divide 4 by 12. Instead, they would have mysteriously multiplied 4 by 5. As strange as this may seem, we need to remember that the Babylonians never wrote decimal points.

Let's look at a similar case in a base we're more familiar with. I could calculate $18 \div 2$ by simply dividing by 2 to get 9. But what happens if I multiply 18 by 0.5? I get the same answer. That's because dividing by 2 is always the same as multiplying by 0.5. In fact, we can think of 0.50 as 50%, where it more clearly represents a half. Similarly, instead of dividing by 4 we could multiply by 0.25, which is 25%. The numbers 2 and 0.5 are called *reciprocals* because they multiply to 1. Similarly, 4 and 0.25 are reciprocals since $4 \times 0.25 = 1$. It's well known to mathematicians that dividing by a number is equivalent to multiplying by its reciprocal.

So let's go back to our example of how a Babylonian would calculate $6 \frac{4}{12}$. The fractional part, $\frac{4}{12}$, is computed by multiplying 4 by the reciprocal of 12. At this point they would pull out a table that would tell them that the reciprocal was ⌀.𒅉. You can convince yourself of this by thinking of ⌀.𒅉 as $\frac{5}{60}$. If you multiply this fraction by 12, you get 1.

You can actually use time to calculate some simple reciprocals. Above we're looking for a representation of $\frac{1}{12}$. Note that 5 minutes is $\frac{1}{12}$ of an hour, but recall that minutes are really just Babylonian decimals. Since 5 minutes is ⌀.𒅉 hours, the reciprocal of 12 is ⌀.𒅉. Similarly, 15 minutes is a quarter of an hour, so the reciprocal of 4 is ⌀.◁𒅉.

Let's get back to the computation of $6 \frac{4}{12}$. We need to divide 4 by 12 using Babylonian methods. Just as dividing

by 2 is basically the same as multiplying by 0.5, the Babylonians would pull out their times-𒅉 table to divide by 12 and look up 𒅉 × 𒅉 to determine $\frac{4}{12}$. They would get the answer ◅◅, and hence 6 is ⫶ . ◅◅ in Babylonian. Of course there would be no decimal point in their notation.

EXAMPLE: Divide ◅◅◅𒅈 by ◁𒐈 knowing that the reciprocal of ◁𒐈 is 𒅉.

SOLUTION: We simply need to multiply ◅◅◅𒅈 by 𒅉 inserting the proper decimal point.

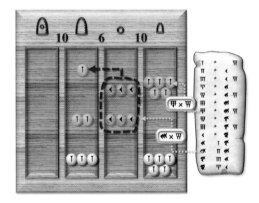

𒅉 × ◅◅◅𒅈 is ⫶⫶⫶ 𒅉.

By estimating ◅◅◅𒅈 ÷ ◁𒐈 we know the answer is about ⫶⫶⫶, so the decimal point goes after the ⫶⫶⫶, giving a final answer of ⫶⫶⫶.𒅉.

PRACTICE: Divide 𒐊𒅈 by ◁𒐈 knowing that the reciprocal of ◁𒐈 is 𒅉.

ANSWER: ⫶⫶⫶.◅◅◅𒅉

When we found the estimate for the square root of 67 above, we got the answer 8 . The Babylonians now needed to divide by 16, which means they had to multiply by the reciprocal of 16. In Babylonian, the reciprocal of ◁𒅈 is .⫶⫶⫶ 𒐊𒅉. We could do a standard multiplication of ◁𒅈 by .⫶⫶⫶ 𒐊𒅉, which as we've seen is tedious. The Babylonians apparently knew it was awkward because they made up a multiplication table for ⫶⫶⫶ 𒐊𒅉 made especially for division by ◁𒅈.

The times-5 table is also the divide-by-12 table.

The times-𒐗 𒌋𒌋 table is the divide-by-𒌋𒐈 table.

So to calculate 𒐗 divided by 𒌋𒐈, we treat the above table as a divide-by-𒌋𒐈 table rather than a times-𒐗 𒌋𒌋 table. The third row tells us that the answer is 𒌋𒐊 𒌋𒌋. The division $3/16$ is small, but not too small. The starting 𒌋𒐊 is reasonably assumed to be $11/60$, which is close to $3/16$. Hence the answer is . 𒌋𒐊 𒌋𒌋.

PRACTICE: Use the above table to divide 𒌋 by 𒌋𒐈.

ANSWER: . 𒌍𒌋𒐈 𒌍

We could do many more divisions that become more and more complicated just as we did with multiplication; however, I have no wish to get bogged down with Babylonian details. Instead, let's consider the following problem.

EXERCISE: Estimate the square root of 52.

SOLUTION: $52 = 49 + 3 = 7^2 + 3$. So $\sqrt{52} \approx 7\frac{3}{2 \times 7} = 7\frac{3}{14}$.

To simplify this in Babylonian, we simply need to find the multiplication table for the reciprocal of 14. Unfortunately there is none. There are none for 7, 11, 13, 14, 17, 19, 21, 23, and an infinite number more.

Recall that in the last section I said that in our comparison the decimal system seemed to hold up well compared to the Egyptian system. But there is a serious problem with decimal and all other similar systems. There are certain things that you can't do. The Babylonians dealt with it by avoiding the problem. You simply don't divide by certain numbers, at least not in their textbooks. Imagine a system in which it's basically impossible to divide by 7, and you begin to see the magnitude of the problem.

FIBONACCI'S GREED

When Can You Divide by a Number?

Eight hundred years ago Guilielmo wanted to give his young son an edge. So he summoned his son, Leonardo, to join him in the North African city of Bougie. His intent was to provide his son with an education that was unavailable in medieval Europe. The Crusades had been raging for a hundred years, with European soldiers continually gaining and losing territory in Islamic lands. Although the violence accomplished little, Arabic goods began to pour into relatively primitive Christian Europe. Initially aiding in the Crusades, the Republic of Pisa realized that their powerful navy could be used to dominate trade in the Mediterranean and gain considerable profit from the goods. Through both violence and diplomacy, they gained numerous trade outlets. Pisan men, like Guilielmo, were needed as diplomats throughout the region.

Guilielmo apparently noticed the fast and effective accounting of the Arabic merchants, so he summoned his son and arranged for learned men to teach him. Leonardo, now known as Fibonacci, learned all that his Arabic teachers knew about mathematics. He learned the Hindu methods of computations, Greek geometry, Babylonian base-sixty computations, and the budding algebraic methods of the Arabs. He also learned unit fractions, which were basically fractions of the ancient Egyptians that had somehow survived the two and a half thousand years since the time of Ahmose.

He discussed a few methods for expressing division in terms of unit fractions. One strategy he discussed was called a "greedy" algorithm. Mathematicians have various ways of solving problems. A set of specific instructions that explains how to solve one in a particular way is called an *algorithm*. A greedy algorithm is one that tries to do the best as soon as it can without caring about the future consequences of its choices.

Consider a mouse moving through a maze. Its goal is to acquire as much cheese as possible, but it will go after only cheese it sees. Our greedy mouse enters the maze pictured below and grabs the first piece of cheese. It then looks left and right seeing two and one pieces respectively. Only caring about the here and now, the mouse runs to the two pieces without looking around the corner of the side containing the one piece. It then runs to the end of the maze, eats the two remaining pieces, and stops because it can't see any more. However, if our mouse hadn't been so greedy in the short term, it would have noticed the huge pile of cheese around the right corner.

Greedy algorithms sometimes work and sometimes they don't. Our greedy mouse was able to get cheese using his shortsighted method, although he didn't get all that he could have. We need to consider the question, what happens if you "greedily" divide with the largest pieces possible?

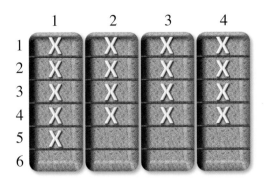

Four loaves cut in sixths makes 6 rows of 4, more than enough for the 17 workers.

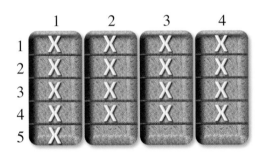

Four loaves cut in fifths make five rows of four, giving 20 slices, one for each worker and three left over.

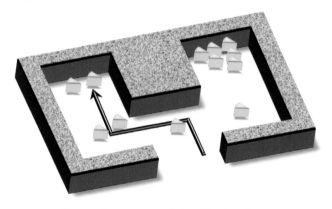

The greedy mouse goes left, getting less cheese.

Consider dividing four loaves of bread between 17 workers. We could use slices of size $\overline{6}$. If we divided each of the four loaves in sixths, we'd get 24 pieces. We can think of this as 6 rows of 4, giving 24 slices, more than enough for the 17 workers. In fact there are 24−17, or 7, slices left.

Are we being greedy enough? Is it possible that we could cut bigger slices? If we cut slices of size $\overline{5}$, we'd have five rows of four, giving 20 slices. After the workers each get one, there would be 20−17, or 3, slices left.

Visually it should be clear from the above two diagrams that sixths is not the greediest cut. In the first illustration of loaves, the bottom row is unused. When we cut the loaves in fifths, we essentially eliminate the superfluous row. So instead of getting 7 leftover slices, we end up with 3. The number of slices drops by 4, the number of slices in the eliminated row.

Part of the four loaves has been split evenly between 17 workers, with each getting a fifth of a loaf. Arithmetically, we've shown that $4/17$ is equal to $\overline{5}$ and a small remainder consisting of three slices. Let's eliminate the unused slices by dividing them up in a greedy fashion, treating them as whole loaves.

Starting with three mini-loaves, we need to cut them up so that there are enough slices for each of the workers. In essence we're asking, what do we multiply 3 by to get a number just above or equal to 17? The answer is 6, since 3×6 is 18. The cut now looks like the figure at the top of next page.

The new mini-slices are fifths cut in sixths. So each piece is size $\overline{30}$ because 5×6 is 30. So now when we divide the four loaves between 17 workers, each worker gets the original slice of size $\overline{5}$ and an additional one of size $\overline{30}$, with one last mini-slice remaining. Arithmetically this means that $4/17$ is $\overline{5}\ \overline{30}$ with a miniscule remainder.

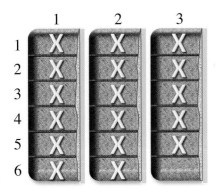

The three slices sliced in sixths leaves one piece left over.

Now we are left with one piece of size $\overline{30}$ to distribute between the 17 workers. When we cut it into 17 pieces, each piece is size $\overline{510}$ because 30×17 is 510. There is now nothing left to distribute. Each worker has one piece each of size $\overline{5}$, $\overline{30}$, and $\overline{510}$. Since this is the result of four loaves divided evenly between 17 people, we know $\tfrac{4}{17}$ is $\overline{5}\ \overline{30}\ \overline{510}$.

This method of performing divisions using sliced bread is nothing new to us. Early in this book we saw that this method can produce solutions, but now I want to pose a more theoretical question. *Must* it produce solutions? The answer to this question is not as obvious as you might think. Consider an attempt to find the decimal representation of $\tfrac{4}{17}$ using long division. Each step in the calculation is clearly defined. We never get stuck, so the algorithm appears to work just fine. The problem is, it never ends. The division goes on forever, and as a result, the decimal expansion of $\tfrac{4}{17}$ does too.

$$
\begin{array}{r}
0.2352\ldots \\
17\,\overline{)\,4.0} \\
3.4 \\
\hline
60 \\
51 \\
\hline
90 \\
85 \\
\hline
50 \\
34 \\
\hline
160 \\
\ldots
\end{array}
$$

The Egyptian method worked because the division process terminated. At some point the last piece of bread was handed out to the workers, ending the algorithm. We now need to ask, will this always happen? This is an important question. I've criticized the decimal system because some fractions can't be expressed in a finite form. However, to be fair, we need to apply the same criterion for Egyptian fractions. Can all fractions, hence all divisions, be expressed with a finite number of Egyptian fractions? We've seen that the sliced-bread method produces solutions for some problems, like $\tfrac{4}{17}$. Must it work for all?

Let's look at another example with this question in mind. Consider $\tfrac{5}{13}$ computed using a greedy application of sliced bread. We first cut the five loaves into 3 slices each. This produces 15 slices. We hand out 1 slice to each of the 13 workers, leaving 2 slices. We then reapply the method to the 2 pieces, treating them as whole loaves. The important thing to notice is that the number of loaves dropped from five in the first application to two in the second.

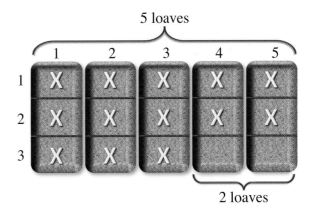

The number of loaves drops from five to two.

Is it possible that the number of loaves could have stayed at five, or even have grown larger? The answer is no, not if you're greedy. Consider what would have happened if we had sliced the five loaves into 4 pieces, a nongreedy cut.

5 loaves

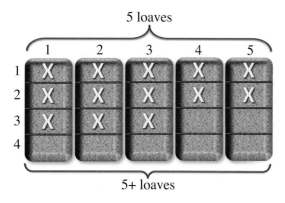

5+ loaves

A non-greedy cut leaves too many loaves.

In the above cut, we have 7 slices left. This means that the whole bottom row is unused; hence, we didn't need as many rows. We could have cut our loaves into fewer slices, creating fewer rows; hence, our cut was not greedy. If your cut is greedy, every row will be used and only the bottom row will have unused slices. The total number of slices, used and unused, in the bottom row is exactly the number of initial loaves. Since at least one must be used in a greedy cut, the number of loaves must go down.

The same thing happens when we reapply the process to the 2 remaining slices. We cut them in sevenths, creating 14 slices. We distribute 13 of them, leaving 1 slice. The number of loaves drops from two to one. The bottom row

2 loaves

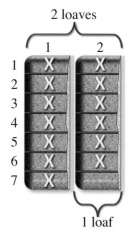

1 loaf

Using one slice on the bottom drops the number of loaves from two to one.

has two slices, the same as the number of loaves. As in all cases, at least one is given out to the workers, leaving fewer loaves.

The last loaf can be cut in 13 and distributed. The initial cut was in thirds, which was then cut in sevenths and then in thirteenths. This means that $5/13$ is $\overline{3}\ \overline{21}\ \overline{273}$. What's more important is that we have an answer. We started with five loaves and each time the number of loaves dropped. Eventually we must run out of loaves, and the process will terminate, generating a finite answer. We can apply this logic to any division. Hence, every fraction can be expressed in Egyptian fractions.

THE IMPOTENT DECIMAL

Decimals Don't Do Fractions

Perhaps you are not impressed that any fraction can be written as an Egyptian fraction. Perhaps you think that being able to write a complete number is the least a number system should be able to do. Perhaps you're right, but did you know that the decimal system can't? It's something we tend to forget due to our overfamiliarity with our version of mathematics. To be sure, some fractions work. For example, $7/8$ is 0.875 but others, like $5/7 = 0.714285714285\ldots$, never end.

Some of you are thinking that $5/7$ becomes a repeating decimal and can be expressed as $0.\overline{714285}$, where the overbar means "repeating" (not the Egyptian fraction). This is because your mind has been corrupted by the anti-Egyptian propaganda of your grade-school math teachers. I can see that your brainwashing has been fairly thorough, so I'll need to engage in some extensive deprogramming.

Consider the following addition:

$$\begin{array}{r} 137 \\ + 245 \\ \hline 382 \end{array}$$

The two middle, numbers 3 and 4, add up to 7, but we see an 8 below them. This is of course because there is a carry when we add the 7 and 5 to their right. This shows us that when we add decimals, we need to start on the rightmost digit. Now consider trying to add the following repeating decimals:

$$1.474747\ldots$$
$$+\ 2.186186\ldots$$
$$\overline{?????????}$$

In order to begin you'll need to go to the rightmost digit, which unfortunately doesn't exist because each number stretches to the right infinitely.

Multiplication is even worse. In the following multiplication we need to multiply each digit of one number by each of the other.

$$
\begin{array}{r}
137 \\
\times\ 245 \\
\hline
685 \\
5480 \\
27400 \\
\hline
33565
\end{array}
$$

So to start we need to multiply the rightmost digit of the second number, 5, by each of the digits of the first number: 1, 3, and 7. We then repeat for each of the other digits of the second number, 4 and 2. Since each of our two original numbers is three digits long, we need to perform three digit multiplications for each of three digits for a total of 3×3, or 9, multiplications. Then we need to add three numbers that are at least three digits long. When we try to do the same operations on two repeating decimals, we need to multiply an infinite number of digits by an infinite number of digits for a total of *infinity × infinity* or *infinity*^{2} multiplications. We then need to add an infinite number of infinitely long numbers.

If you wish to try, go ahead. I'll wait. I'm sure this book will last until you're forced to stop due to a near terminal case of writer's cramp. Of course, in order to begin you'll still need to find the rightmost digits, which don't actually exist. I can see that you are stubborn and are thinking that just because you can't do mathematical operations on them doesn't mean they're not numbers. I abhor draconian measures, but your obstinacy leaves me no choice. In order to make a point, I'm going to force you to swallow a pile of rocks, and when you complain I'll reply that just because you can't digest them doesn't mean they're not food! Numbers are mathematical objects. So if you can't perform math on them, they are not numbers.

Now that we know there's a difficulty with decimals, we need to determine how big a problem it is. When does this problem occur and how often? Unfortunately it's far too common, and to fully understand this we need to know something about base mathematics.

Consider the fraction $\frac{7}{8}$, which equals 0.875. We can turn 0.875 into a whole number by multiplying by 1000, giving $0.875 \times 1000 = 875$. Multiplying by 1000 just moves the decimal point three times, one for each zero. If we have more digits behind the decimal point, we simply need more zeros. Multiplying by 1000 is the same as multiplying by 10 three times. So we get $0.875 \times 10 \times 10 \times 10$. Since each 10 is just 2×5, we get $0.875 \times 2 \times 2 \times 2 \times 5 \times 5 \times 5$. We get one 2 and one 5 for each 0 in the 1000. In order to understand what any of this has to do with nonterminating decimals, we need to see how this works in terms of fractions.

Remember that 0.875 is $\frac{7}{8}$, so if 0.875×1000 is a whole number, then so is $\frac{7}{8} \times 1000$. We need to understand what it is about $\frac{7}{8}$ that makes it an integer when multiplying by 1000. We can rewrite $\frac{7}{8} \times 1000$ as $\frac{7}{8} \times 2 \times 2 \times 2 \times 5 \times 5 \times 5$. This becomes

$$\frac{7 \times 2 \times 2 \times 2 \times 5 \times 5 \times 5}{8}$$

Now the 8 on the bottom can also be factored just as we've done to the top giving

$$\frac{7 \times 2 \times 2 \times 2 \times 5 \times 5 \times 5}{2 \times 2 \times 2}$$

We can simplify the expression by canceling an equal number of 2s on the top and bottom giving

$$\frac{7 \times 2 \times 2 \times 2 \times 5 \times 5 \times 5}{2 \times 2 \times 2} = 7 \times 5 \times 5 \times 5$$

Note that the three 2s from the 8 completely cancelled. This is why multiplying $\frac{7}{8}$ by 1000 produces a whole number.

Similarly, when we divide 11 by 50 we get

$$\frac{11}{50} = \frac{11}{2 \times 5 \times 5}$$

In the bottom of the fraction we have one 2 and two 5s. So we are going to need to multiply by 100, which is $2 \times 2 \times 5 \times 5$ in order to get enough 5s so that both cancel. Note that this means that $\frac{11}{50}$ has exactly two digits after the decimal point, since multiplying by 100 shifts the decimal point two places, leaving only a whole number. In fact $^{11}\!/_{50}$ is 0.22, so the multiplication of 0.22×100 is 22, a whole number, just as we predicted.

Now what happens if we try to turn $^3\!/_{14}$ into a whole number by multiplying by a power of 10? We start with

$$^3\!/_{14} = {}^3\!/_{2 \times 7}$$

If we multiply by 10, we add a 2 and a 5 to the top of the fraction. The 2 on the top quickly cancels with the 2 on the bottom.

$$^{3 \times 2 \times 5}\!/_{2 \times 7} = {}^{3 \times 5}\!/_{7}$$

The problem is there's still a 7 on the bottom of the fraction, so it's not a whole number. We can try multiplying by 100 to get

$$^{3 \times 2 \times 2 \times 5 \times 5}\!/_{2 \times 7} = {}^{3 \times 2 \times 5 \times 5}\!/_{7}$$

But there's still a 7 on the bottom. Hopefully you realize that no amount of 2s or 5s on the top will cancel the 7 on the bottom. In other words, it's impossible to multiply $^3\!/_{14}$ by any power of 10 and get a whole number. So $^3\!/_{14}$ in decimal notation has to go on forever.

Let's recap the argument because it's a little convoluted. If a decimal doesn't go on forever, like 0.2463, then there must be a power of 10, like 10,000, by which we can multiply it to get a whole number. For example, 0.2463×10000 is 2463. However, no power of 10 multiplied by $^3\!/_{14}$ will ever make it a whole number because all the 2s and 5s in the world won't cancel a 7. Therefore $^3\!/_{14}$ can't terminate.

What is it about 14 that causes the above problem? It's the prime factor 7, and its crime is not being 2 or 5. A prime factor is a number that can't be broken down farther by factoring. So 14 is not prime because it can be written as 2×7, and both the 2 and 7 are primes because they can't be factored into smaller pieces. There are many primes. The list of primes begins 2, 3, 5, 7, 11, 13, 17, 19, and so on. Mathematicians can prove that this list goes on forever; hence, all but two of an infinite list of primes will cause problems in division.

How many numbers have only 2s and 5s in them? That depends upon how large the numbers you're looking at are. In the first ten numbers we have 2, 4, 5, 8, and 10 with

only 2s and 5s, about half. But in the first million numbers there are only about 90, less than one hundredth of one percent. This proportion drops closer and closer to zero the larger we allow our numbers to get. Hence, most numbers cause problems when divided in decimals.

This is not just a problem with our base-ten system but also with any base system. The Mesopotamians used base 60. Since 60 is $2 \times 2 \times 3 \times 5$, they could divide by numbers with only 2s, 3s, and 5s without fear. The extra 3 gives a little more latitude, but three primes out of an infinite number is still virtually nothing especially when the numbers get large. In fact, any base when factored will have only a finite number of primes, leaving an infinite number. The more primes we try to include, the more difficult the base is to work with. Base ten has a 10-by-10 multiplication table with 100 entries to be memorized. If we instead picked base 210, because 210 is $2 \times 3 \times 5 \times 7$, with four different primes, we'd need to memorize a 210-by-210 table with 44,100 multiplications. Even if we spent the time required to master this base, we would still run into problems with the prime factors of 11, 13, 17, and so on. In the decimal system, all we can do is delay the problem, not avoid it.

So why do we use the decimal system if it's inherently flawed? The answer is simple. The decimal system is not a number system at all but rather a system of arbitrarily good approximations of numbers. For example, when we divide 3 by 11 we get $^3\!/_{11} = 0.27272727\dots$. We could write the 27s forever, but we will eventually get tired and write "…". The three dots at the end of our "number" literally mean "I'm bored and I hope you get the idea since I'm stopping now." Although the decimal system cannot express $^3\!/_{11}$ in any usable way, it can let us approximate the fraction as close as we'd like. So we can say it's roughly 0.27 and be off by at most 0.01. Or we could say it's roughly 0.27272 and be off by at most 0.00001.

The decimal system is all about being annoyed. If you calculate the interest you get on a savings account and get $243.82453, do you really care about the $^3\!/_{1000}$ of a penny on the end? I personally find whole pennies more trouble than they're worth. We get irritated and say, "I've got $243.82." Remember the π we all learned in grade school when doing circle problems? Its decimal representation goes on forever, never repeating, yet we all memorized 3.14. Our approximation of π is a testament to how quickly we get annoyed. Yet for most practical applications it makes little difference, and even if it did, we could get our calculator to give us a few more digits, delaying our irritation and giving us the required accuracy.

RULES ARE FOR DEVIANTS

ALGORITHM VERSUS INTUITION

Imagine you're sitting at home watching television. A special report interrupts your favorite show and the newscaster announces that due to a technicality in legislative procedure, a judge has ruled that the law prohibiting homicide is invalid. It's the weekend and the lawmakers are rushing from their homes to the capital in order to re-enact the law properly, but until they do, murder is legal. The journalist says that they should change the law in one hour.

What would you do? Would you go out and kill someone? I'm going to assume not. Sure, we would smirk as we thought of a few people who may or may not have it coming, but in the end, no one would die by our hands. So if we don't need a law to refrain from murder, why do we require such rules? The answer is simple. Not everyone is as moral as you or I. The world is full of deviants, and we need the laws to force them to exhibit proper behavior.

Now consider the following addition:

$$\tfrac{2}{4} + \tfrac{3}{4} = \tfrac{5}{4}$$

When adding the left-hand side, the typical student will go through the following reasoning. They will first note that the denominators are the same, and then they will apply the rule that the sum of the fractions is the sum of the numerators over the common denominator. This gives a numerator of $2 + 3 = 5$ and a denominator of 4, producing $\tfrac{5}{4}$, the answer.

Now I want you to read the following three lines out loud.

Two dogs plus three dogs is five dogs.
Two dollars plus three dollars is five dollars.

Two quarters plus three quarters is five quarters.

I assume you found each of the statements obvious. You didn't need any complicated rule to believe they were true. Now read the addition of fractions, interpreting fractions like $\tfrac{2}{4}$ as *two quarters*.

$$\tfrac{2}{4} + \tfrac{3}{4} = \tfrac{5}{4}$$

Notice that word for word it's almost exactly the same as "two quarters plus three quarters is five quarters." The only difference is that the word "is" is replaced with "equals." If you met someone who didn't know that two dogs plus three dogs is five dogs, you would probably consider them to be not very bright. However, if you met someone who didn't know how to add $\tfrac{2}{4}$ and $\tfrac{3}{4}$, you might excuse their ignorance by thinking that everyone forgets the rules of math now and then.

Why do you need a rule to add fractions but not dogs? I'm sorry to inform you, but you're a mathematical deviant. You need rules to force you to do what's right. Just like the psychopath who doesn't think about morality when committing horrible acts, you refuse to think about right and wrong when doing mathematics. Fortunately your teachers forced you to obey the laws of math, and there is no reason to incarcerate you; however, I am disappointed.

There's a serious problem with rules. Because they don't need to be understood, you can't possibly apply your knowledge to new situations. Consider someone who knows how to add fractions. If they see something like $2x + 3x$, they would immediately ask the teacher for the rule for adding such things. Of course two x's and three x's is five x's since two of anything and three more are five of those things. So a nondeviant would immediately know that the answer is $5x$, even if they had never seen such a problem before. Likewise, they would know that 2^n and 5×2^n is 6×2^n because one 2^n and five more are six 2^n's.

Not only do deviants not know when to apply a rule, the often don't know when not to. When they see $2x + 3y$, they might be tempted to write $5xy$. Yet the same person would scoff if they heard someone say that two cats plus three birds is five catbirds.

Think about decimal long multiplication and division. The procedures and steps are defined and unambiguous. You simply obey the rules and obtain an answer. Do you know why long division works? Could you apply its ideas to divide the polynomial $x^4 + 3x^3 + 7x^2 - 5x + 9$ by $x^2 + 5x - 1$? Could you adapt the ideas to the Euclidean algorithm to "divide" the lines _____ by _____? I'm guessing not and that's because most people don't think about the mathematics they do.

Egyptian computation, on the other hand, forces you to think. We've seen again and again that simple doubling is usually not the best way to solve a problem. Shunning straightforward algorithmic approaches forces you to think about what you have, what you need, and what you must to do to get it.

Once you start thinking, wonderful things start to happen. I once assigned the following problem taken from Ahmose's papyrus. In what follows, a *sha'ty* is a unit of value and a *deben* is a unit of weight.

A *deben* is a ring whose weight is used as a standard.

EXAMPLE: A bag of precious metals contains an equal amount of gold, silver, and lead. The total value of the metals is 84 sha'ty. If a deben of gold is 12 sha'ty, a deben of silver is 6, and a deben of lead is 3, how much do we have of each?

The papyrus finds the solution by adding 12, 6, and 3 to get 21 and then divides 84 by 21 to get 4, the deben of each. It finally multiplies 4 by each of the prices to get the value of each precious metal. However, when I asked a student to put his solution on the board, this is what he wrote:

| Deben | Sha'ty of | | | |
	Gold	Silver	Lead	Total
1	12	6	3	21
2	24	12	6	42
4	48	24	12	84 ✓
	48	24	12	

This was unlike anything I had taught. There are only two columns in an Egyptian computation. It's vaguely reminiscent of changing the parts in a division, but I hadn't even reached that portion of the material when the student put this solution up. The student's understanding of the method enabled his to invent his own variations. He had made the method his own and applied it in unique ways to new problems.

If you understand Egyptian methods, there's nothing complicated about the above solution and you might not be impressed. But neither is there anything complicated in the notion that two of something and three more are five of those things. The trick to effective mathematics is seeing the inherent simplicity in things that others only see as an arbitrary set of rules.

Rules enable us to abdicate thought, but it is thought that enables us to solve the problems that the rules supposedly address. Mathematics is about things. We live in a world of things and know how they work. The so-called rules of mathematics are just symbolic representations of what we already know. When you understand this, the rules become clear and how to apply them to the world becomes transparent.

PTOLEMY'S ANGST

Claudius Ptolemy, a scholar in Alexandria, Egypt, simply wasn't satisfied. Greek astronomers had had mathematical models of celestial objects for centuries; in particular,

Hipparchus had made models for the sun and moon. However, it was known that the Greek models only roughly estimated the positions of the objects of the sky. Greeks tended to care only about the abstract theoretical models and dismissed the hard number crunching required to make accurate predictions. The few Greeks who desired specific numbers relied heavily on Babylonian measurements that had been made long ago.

The Babylonians had their own models of the celestial sphere long before the Greeks, and oddly enough, those were far more accurate. The Babylonians cared more about getting a good answer than formulating an intellectual paradigm for the workings of the universe. For example, they estimated the time between something called *planetary retrogressions* by fitting curves with parabolas, but the retrogressions are periodic, meaning they repeat, and parabolas don't; hence, the Babylonians were forced to slice up the parabolas and join them.

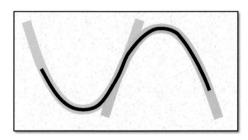

A modern curve in black fitted by a green and blue parabola.

Apparently the Greeks did not think that the universe should be constructed piecemeal but rather should be the result of a unified theory. The Greeks considered circles with their symmetry to be perfect objects. A perfect universe could not help to be built from anything but from perfect objects, so they concluded that stellar motion had to be circular. However, simple observation makes it clear that the motion of the sun, moon, and planets is not quite circular. Astronomers like Hipparchus dealt with this by putting spinning circles on spinning circles. While this motion is difficult to explain in a static book with no

animation, anyone who has played with a Spirograph should understand the basic concept. One wheel travels around another wheel. The center of the wheel moves in a circular path, but an off-center point wobbles.

The red dot on the purple wheel traces out a wobbly path around the blue circle.

Such models apparently satisfied many ancient Greeks because it explained how the types of motions in the sky could be explained through a "pure" theory. Ptolemy, however, wanted the models to make accurate predictions, not just exist as intellectual curiosities. So he made lists of astronomical observations, refined the current models by adding circles, and applied the methods to the motions of the visible planets. All of this required great accuracy in mathematics. Many Greeks at this time used Egyptian fractions, but Ptolemy made the unfortunate statement that he would use base-sixty computation because Egyptian fractions were "embarrassing."

Hopefully by now you've learned to appreciate Egyptian mathematics and find Ptolemy's remarks dismissive. In order to confront Ptolemy's statements, we must understand his needs. What is the best tool? If you have a box of nails, it might be a hammer, but if all you've got is screws, it might be a screwdriver. You can't judge a tool out of context. You need to know what you want it to do and what aspects of the job you consider important. By understanding what Ptolemy used his numbers for, you can begin to understand why he preferred decimal systems.

The main flaw with decimal notation is that you will almost always need to round. As we've seen, many simple fractions can't be expressed in decimals. For Ptolemy to prefer decimals, that liability couldn't have been an important concern, and it wasn't. Ptolemy had to round because he measured the position of the stars. Physical instruments are used to measure the angular position of objects in the night sky. They work in much the same way rulers do, using lines on a device to take a reading and such measurements are always approximations.

Consider the following hypothetical device pictured below. It's a tool to measure the angular altitude of a star. The device works by turning it toward a star and pointing the stick directly at the one to be measured. You then take a reading based on which tick mark the stick coincides with. If the stick coincided with the lowest tick mark, it would be horizontal. Such a star would appear to be on the horizon, and its altitude would be 0 degrees above the horizon. The highest tick mark points straight up and points to an altitude of 90 degrees. In general, the tick marks represent increments of 10 degrees. Counting from 0, the stick covers the third tick mark, so we might say the star has an altitude of 30 degrees.

A device measuring an altitude of a little more than 30 degrees.

There's a problem with this. Note that the stick clearly appears closer to the 40-degree mark than to the 20-degree mark. This suggests that the measurement, while clearly being less than 40, is more than 30. How can you determine exactly how much more? You could try to add

more tick marks. This would give you a better answer, but ultimately you will still need to approximate. Sooner or later, your imperfect eyesight, the ability to point the stick, and the precision of the parts of the machine would make it impossible to obtain an exact answer.

Remember that one of the main advantages of Egyptian math is that it provides exact answers. Imagine I ask you what 3×4 is, and you proclaim that the answer is 12. I then tell you that you're wrong; the answer is in fact 12.6. How could this be? I measured the 4 wrong. It should have been 4.2. My point is that you can't get an exact answer using inexact numbers. Worrying about the rounding of decimal in these cases is a bit like worrying about getting dirty after you've been doused with mud.

Ptolemy also refined the trigonometric tables for the chord, the Greek equivalent of the sine. His mathematical model of the solar system used chords to compute the locations of the planets, but he knew that his values for various most chords were wrong. They had to be, because most are irrational. Like π or the square root of 2, their decimal expansions not only go on forever, they never repeat. Hence his chord values had to be approximations. Once again there was no need for the exact computations of the Egyptian system.

Tables also favor decimals in another way. Humans like patterns. If I were to give chord values of 2.0, 2.1, and 2.2, you would expect the next entry to be 2.3. The uniform increase in value seems natural, but Egyptian mathematics simply refuses to do this well. A table that gives 2.0 to 2.9 both in decimal and Egyptian math is at the top of the next page.

Note that while the Egyptian mathematics is perfectly capable of expressing these values, the steps seem unnatural. Patterns that arise naturally using these numbers in decimals would not occur in Egyptian, or at least not be transparent.

The final reason Ptolemy preferred decimals over Egyptian math was cultural. Greeks learned most of their astronomy from the Babylonians. For many centuries, the Babylonians, and then the Greeks, had been making stellar observations and recording them using the Babylonian system. Imagine you knew that the location of a star was 𒐖𒐋 . ◄𒐊 𒐏𒐊 degrees. Think of how difficult it would

Decimal	Egyptian
2.0	2
2.1	2 $\overline{10}$
2.2	2 $\overline{5}$
2.3	2 $\overline{5}$ $\overline{10}$
2.4	2 $\overline{3}$ $\overline{15}$
2.5	2 $\overline{2}$
2.6	2 $\overline{2}$ $\overline{10}$
2.7	2 $\overline{\overline{3}}$ $\overline{30}$
2.8	2 $\overline{\overline{3}}$ $\overline{10}$ $\overline{30}$
2.9	2 $\overline{\overline{3}}$ $\overline{5}$ $\overline{30}$

be to translate this number into Egyptian. The fractional part is $(18 \times 60 + 35)/(60 \times 60)$. You would need to do this division in Egyptian to express it in unit fractions. Now imagine the work it would require to translate pages and pages of stellar observations. Even if you took the time, your peers probably would not, making your work useless to them and, as a result, unappreciated.

So Ptolemy perhaps made the correct choice, but for the wrong reasons. He didn't fully understand the advantages and disadvantages of both systems. If he had, he probably would have converted the base-sixty of the Babylonians into the base-ten that the Greeks used. Ptolemy expressed the whole number part of a value in base ten, but he wrote the fractional part in base sixty. This is terrible. One of the main advantages of the base system is that multiplying fractional values is just like multiplying whole numbers with the exception of placing a decimal point. So 283×643 is just as easy as 2.83×64.3. However, for Ptolemy, multiplying two whole numbers has different rules than multiplying two fractional values. Even worse, a whole part times a fractional part has its own convoluted procedures.

We need to remember that Ptolemy was not actually a mathematician, but he was an excellent astronomer. The model he used for the solar system had been invented by earlier Greeks, possibly by Hipparchus. The same holds true for his table of chords. He carried out the same computations but rounded at smaller numbers and added more entries for more angles. He made more stellar observations and did them with great accuracy. Finally, he used these improvements to enable the mathematical model of the solar system to give better predictions to the locations of the sun, moon, and planets. As an astronomer, he based his low opinion of Egyptian mathematics on astronomical needs, not mathematical.

We now need to ask ourselves, do we have a good reason, like Ptolemy, to avoid the Egyptian system? The answer is yes. Up to this point, I've completely ignored fractions in my analysis. There's an excellent reason for this. Fractions are evil. You may not think so. Perhaps you found working with them fairly easy, but that's because the examples you were given in school were made to be easy. Fractions grow as you use them. Every time you add, subtract, or multiply fractions, the denominators add. So if I add fractions with denominators of size four and six digits, the resulting fraction will often have a denominator of the size of ten digits.

$$\tfrac{241}{8,327} + \tfrac{83,132}{398,723} = \tfrac{788,332,407}{3,320,166,421}$$

Fractions with denominators of four and six digits tend to produce a denominator of a size equal to $4 + 6$.

Division also increases the size of the denominator, except it does so by the size of the numerator. This means if we add, subtract, multiply, or divide by five fractions with three digits in their parts, the denominator will have about fifteen digits. This in turn means that the fraction will be in parts of quadrillionths, and such numbers are almost incomprehensible to the human mind. Egyptian fractions and decimals have similar problems, but their easy approximations allow us to ignore most of the details and still have a good sense of the value of a number.

Fractions are inferior in almost every way. So why do we use them? One obvious reason is that fractions give exact answers, whereas decimals do not. But Egyptian fractions also give exact answers. This is the power of the Egyptian system. It's a compromise. We have two needs: one is for exact answers, and the other is for quick approximations. Fractions and decimal each satisfy one of these needs but fail the other utterly. Egyptian fractions do both reasonably well.

This is like needing a sledgehammer and a saw and then realizing that an ax can do both. The sharp side can cut wood, and the flat side can be used as a hammer. The advantage to the compromise is that you don't have to buy two tools. Similarly, if you know Egyptian mathematics, you don't need to memorize the methods of both decimals and fractions. There's a problem with the above analysis. In terms of our analogy, we have to buy a saw. We still need a bludgeoning tool, so we can't buy just one implement. Despite all the problems inherent in fractions, we need to teach it in modern society. Oddly enough, we don't need it to deal with numbers.

Modern science has grown hand in hand with algebra. Algebra lets us generalize our computations. For example, a rectangle with base 3 and height 4 has an area of 3×4, or 12. Algebra lets us find the area of any rectangle. We do this by using variables and being vague about the values. In some sense, we don't know the base of the rectangle, but we give it a name, say B. Similarly we can call the height, H. This means the area is $B \times H$. How can I divide B and H? How can you divide any two numbers whose values are unknown? The answer is simple. I divide B by H by "dividing B by H." In other words, I simply write B/H. Notice that this is a fraction.

If you want to know how to add, subtract, multiply, or divide by such objects, you need to know how to work with fractions. Every engineer who builds the technology we rely on, every programmer who tell computers what to do, every physicist and chemist who model the real world in numbers needs to know fractions. They need to be taught them in terms of numbers in school before they can abstract them in algebra later. If we stop teaching fractions, science and technology will screech to a halt, and the modern world as we know it will cease to exist.

Now that we're forced to learn fractions, a system that gives exact answers, we need one that gives good approximations. We can choose between the decimal or the Egyptian; however, Egyptian fractions no longer have the same appeal they once did. The system's ability to give exact answers is almost redundant with the fractions we

already know. The "compromise" choice is no longer as obvious because we have to learn two systems anyway.

I CHING AND THE DIGITAL AGE

Three thousand years ago in China, King Wen was imprisoned. During this time it was said that he put his finishing touches on the sixty-four symbols that form the *Book of Changes*, more commonly known as *I Ching*. Various incarnations of the book were considered ancient, even in the time of King Wen. Tradition has it that the inventor of the linear signs that make up the sixty-four symbols was Fu Hsi, who is so ancient that he's considered the inventor of cooking.

The two linear symbols are the broken and unbroken lines, which represent "no" and "yes," and were used in oracles. Presumably, someone would do something like flip a coin. A head or something equivalent would be written as a horizontal line indicating the answer to a question was yes. Likewise, a tail indicated two shorter horizontal line segments, indicating no.

Heads means yes, tails means no.

After some time, the answer of yes or no probably seemed insufficient. The diviners probably moved to *trigrams*, that is, figures composed of three yes/no line segments stacked vertically. Eventually they settled on *hexagrams*, two stacked trigrams, and hence six linear symbols.

The version of the *Book of Changes* I used instructed me to ask a question. I was then to throw three coins six times. If in the first throw, most of the three coins were heads, I was to draw the yes line, and if most were tails, I was to draw the two lines of the no. I then was instructed to repeat the process five more times, putting each new linear symbol above the previous one. When the six rows were completed, I had to look up the drawn figure on a table that in turn directed me to a description of the meaning of the hexagram I'd just created.

The *I Ching* symbol for opposition.

Following the example of Carl Jung, the author of the foreword of my edition of the *Book of Changes*, I asked about the future of this book, *Count Like an Egyptian*. I was none too pleased when I saw that the hexagram was named "opposition." It was easy to see why the creators of this symbol named it that. The top three lines form the trigram *the clinging fire* and the bottom three form *the joyous lake*. Nothing says opposition like fire and water. I was relieved to see that the text was not as ominous.

THE IMAGE

Above, fire; below, the lake:
The image of Opposition.
Thus amid all fellowship
The superior man retains his individuality.

THE JUDGMENT

Opposition. In small matters,
good fortune.
(I Ching, Book of Changes)

The image seems to imply that it's a symbol of individuality. My book is certainly unusual, and I can hardly be upset about having my "uniqueness" making me a "superior man." The "opposition" still concerns me, but that it ends with "good fortune" won me over.

What's interesting about *I Ching* to a mathematician is that its symbols are essentially binary numbers. We can think of the yesses as 1 and the nos as 0. Going from top to bottom, my hexagram of opposition becomes the binary 101011_2.

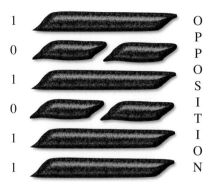

"Opposition" is the binary number 101011_2.

We already know binary because it's just a fancy word for base two. Recall that in base two, the lowest-valued calculus has a value of 1 and they increase by a factor of 2, so the objects have a value of 1, 2, 4, 8, and so on. We can view binary as numbers expressed in the following calculi:

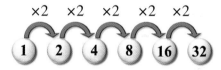

The calculi of binary.

A 1 in binary represents the notion of having "one" of the appropriate calculi, and 0 represents not having it. So our number for opposition can be viewed as the following:

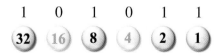

101011_2 is 32, 8, 2, and 1, which total 43.

I've "grayed out" the missing calculi. The calculi that remain are the 32, 8, 2, and 1, which total 43, the value of 101011_2.

EXAMPLE: What is 011101_2 in base ten?

SOLUTION: $(0 \times 32)+(1 \times 16)+(1 \times 8)+(1 \times 4)+(0 \times 2)+(1 \times 1)$ $= 29_{10}$.

PRACTICE: What is 1000110_2 in base ten?

ANSWER: 70_{10}

You could argue that the *I Ching*'s symbols are not numbers. They're not used as, or ever intended to be, numbers. But this is exactly why they're so appropriate as an example of binary. In order to understand why, we need to know something about the digital revolution.

It has been long known that machines could do basic mathematics. Blaise Pascal invented the first digital calculator hundreds of years ago. These original devices were clunky, awkward, and never popular. During the twentieth century it was realized that high and low voltages could represent the 1s and 0s of binary. This means that electronic machines could literally store numbers within their wires. Let's assume we need to a machine to contain the number 101_2. Since there are three digits, we need three wires. The first digit is a 1. To represent this, we flick a switch, sending electricity down the first wire. The second digit is a 0, so we leave the switch corresponding to the second wire in the off position. Finally, the last digit is another 1, so we allow electricity to flow down the third wire.

The following diagram represents the number 101_2 we discussed above. Yellow represents the wires with electricity flowing through them. Similarly the blue represents the cold wire. Note that 101_2 can be represented in calculi. In base two, the first three calculi have values of 1, 2, and 4. The first 1 in 101_2 is the highest digit and represents a 4-valued calculus. The 0 in the middle means we have no 2-valued calculus. Finally, the last 1 means we have one 1-valued calculus. Note that the calculi correspond to the three wires. Since the top and bottom wires are electrified, we have a 4 and a 1 calculus, respectively. The cold middle wire means we have no 2 calculus.

The number 6 is represented by the binary number 110_2. In this case, the top two wires would be coursing with electricity since the first two digits are both 1. This means that we have one 4 and one 2 calculus. Similarly, the bottom wire would be blue, with no electricity, to represent the 0 in the number and the missing 1 calculus.

As we've seen, we can represent numbers in an electronic device. We can also represent operations like + and × using electronic "gates." These gates manipulate the electricity that flows through them. For example a NOT gate switches the voltage from hot to cold and vice versa. The following diagram has a computer circuit that adds 1 to a number from 0 to 3. In order to stay on topic, I'm not going to explain how or why this works. The only thing I want you to see is that the number $10_2 = 2_{10}$ flows into the wires on the left and the number $011_2 = 3_{10}$ flows out on the right. In some sense, the circuit added 1 to 2 and got 3.

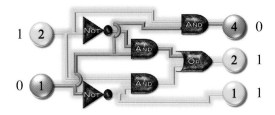

The NOT, AND, and OR gates form an add-1 circuit.

So we can see that computers can express numbers and do math, but how does a computer know what to do? The answer is trivial: it has an instruction that is expressed in binary. In my old programming days, I used to use a language called 6502, named after the chip the language ran on. It was the language the computer "spoke," so no translation was required. We used this language because it was lightning fast. One of the most commonly used instructions was Load the Accumulator, abbreviated as LDA. It simply took a number from the computer's memory and put it in a special place called the *accumulator*, where most of the math was done. One form of the command was expressed by the computer as 10101101_2, which is 173_{10}.

The number $101_2 = 5_{10}$ is represented by high voltage
in the 4 and 1 positions.

Notice that it's just a number, but in reality, it's a command. How did the computer know which command to execute? There was something called the *program counter*, another number, which told the computer where the next command was located. How did the computer know what to put in the accumulator? There was a number after the LDA, which told it where to get the number from. Everything in a computer is a binary number. All text, music, programs, and images are just binary numbers. The computer "distinguishes" between these by the context in which it encounters them. So the *I Ching* symbol is a number in this sense, just as everything else is.

Computation in binary is trivial. Two calculi of the same type are traded up for the next larger one. This means that the only digits are 1 and 0. As a result, multiplication is simple. You simply need to remember that 0 times anything is 0, and 1 times anything is itself. Consider the following multiplication of 1101 by 1011. Note that I put in 0 calculi when there are none, so the relationship between calculi and binary is transparent.

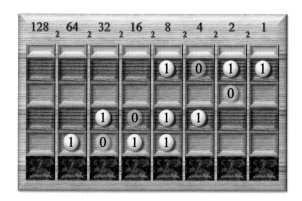

We take the digits of 1101 and multiply each of them by 1011. The leftmost digit of 1101 is 1. When we multiply it by 1011, we get 1011, of course. It appears in the top row. The next digit of 1101 is 0, and when it is multiplied by 1011, we get 0. This is the second row, which is shifted one place to the left. The next two digits of 1101 are both 1, which each make 1101 when multiplied by 1101. Hence the last two rows are 1101, shifted appropriately. Essentially all we do is rewrite 1101 on our counting board over and over, putting in a 0 when the corresponding digit is 0.

We can now add the calculi on our counting board. Binary is just base two, so two calculi in one column get exchanged for one in the next column. After doing the calculation we get the answer of 10001111_2.

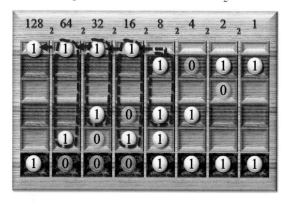

PRACTICE: Multiply 1110_2 by 101_2 using binary calculi.

ANSWER: 1000110_2

I actually don't care that much about the answer. Look again at the first counting-board diagram. The first row has a calculus in the 8, 2, and 1 columns. This makes the number $8 + 2 + 1$, which is 11, the value of 1101_2. Similarly the third and fourth rows are $32 + 8 + 4 = 44$ and $64 + 16 + 8 = 88$. These are both multiples of 11. This should come as no surprise because the rows are just "shifted" 1011_2's. Shifting a number is like multiplying by the base. When we take 83 and shift it one, giving 830, we've multiplied by 10. When we shift 1011_2 over twice to get the third row of 101100_2, we've multiplied by 2 twice, giving us 4×11. Similarly, the last row is shifted again, giving 8×11. So each row is double the one before it, if you ignore the 0s that occur from the multiplication by 0.

Now look at the Egyptian multiplication of 13 by 11, the same two numbers, now written in base ten.

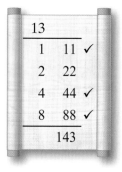

13×11.

Notice the checked off numbers of 11, 44, and 88. These were exactly the numbers we added in our binary multiplication. This is not a coincidence.

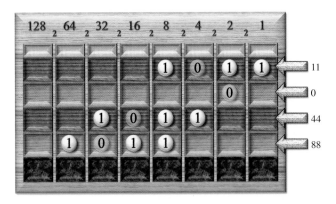

The same three numbers come up and in the same order as in the Egyptian multiplication.

We've already seen that the rows on the counting board double. They start with 1011_2, which is 11. All we need to convince ourselves of is that the 0s, which correspond to unchecked numbers, coincide.

There's a surprisingly easy way to use Egyptian multiplication to turn a number into binary. Notice the first column in the Egyptian computation. From it we can deduce that 13 is $8 + 4 + 1$. Note that these are binary calculi. We can express this using a "grayed out" 2 to represent a zero as follows:

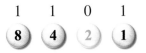

We have the 8, 4, and 1 checked off, so we have a 1 in the 8, 4, and 1 positions. The 2 isn't checked, so we have a 0 there. These correspond to the checkmarks read from bottom to top. We see check, check, no check, check, so we write 1101 as its binary value.

EXAMPLE: Use a one-column Egyptian-style table to convert 37 into binary.

SOLUTION: Double to get 37 and read the checks as 1s going from bottom to top getting 100101_2.

37		
1 ✔		1
2		0
4 ✔		1
8		0
16		0
32 ✔		1

PRACTICE: Use a one-column Egyptian-style table to convert 55 into binary.

ANSWER: 110111_2

COUP DE GRAS

"Horus Eye" Fractions and Binary

I once heard a story so long ago I don't remember where I heard it or even if it was fictional. To the best of my recollection, it involved the meeting of two cultures, and it went something like this. A European man and a Chinese man were talking about food. The European was surprised that the people in China ate with chopsticks. Their culture was so rich, yet they ate with such primitive tools. He asked why they didn't eat with more sophisticated implements like knives and forks.

To this the Chinese man replied that they once did but that they had evolved beyond them. He pointed out that you can tell how sophisticated a culture is by how well prepared their food is before they eat. A wild animal tears flesh from a carcass, a primitive eats cooked meat off a bone, while the Europeans stab and slice their meat with forks and knives. On the other hand, Chinese food is completely prepared beforehand. All that is needed is to pick it up and put it in one's mouth; therefore, chopsticks suffice. Since Chinese food is better prepared for eating, the Chinese are the more advanced peoples.

My point in telling this story is that whether or not you think chopsticks are more advanced than knives or forks, the Chinese made a choice, and that's what's important.

Just because someone reads a book rather than watches TV or walks a few blocks rather than drives a car doesn't mean that person is primitive. Primitiveness would imply ignorance of technology, and if you consciously choose not to use something, you are not ignorant of it.

Many people see the Egyptians as ignorant. They claim that their "awkward" mathematics was a product of their lack of knowledge of more advanced systems. But if they were aware of a decimal-like system and chose theirs over it, this claim falls to pieces.

In one of the problems in the Rhind Mathematical Papyrus, Ahmose takes an answer of $\overline{5}\,\overline{10}$ and converts the units. Initially it's in hekat, and Ahmose converts it into *ro*, a unit roughly equal to a tablespoon, by multiplying by 320 as follows:

$$
\begin{array}{r|r}
\multicolumn{2}{l}{\overline{5}\,\overline{10}} \\
\hline
1 & 320 \\
\overline{10} & 32 \checkmark \\
\overline{5} & 64 \checkmark \\
\hline
& 96
\end{array}
$$

$\overline{5}\,\overline{10}$ hekat is 96 ro.

Then Ahmose does something strange. He converts it back into hekat. He does this only by dividing by 2. He uses no tricks or shortcuts, and he leaves a remainder after he gets to $\overline{64}$.

$$
\begin{array}{r|r}
? & 96 \\
\hline
1 & 320 \\
\overline{2} & 160 \\
\overline{4} & 80 \checkmark \\
\overline{8} & 40 \\
\overline{16} & 20 \\
\overline{32} & 10 \checkmark \\
\overline{64} & 5 \checkmark \\
\hline
\multicolumn{2}{l}{\overline{4}\,\overline{32}\,\overline{64}\ \text{rem. 1}}
\end{array}
$$

Ninety-six ro is $\overline{4}\,\overline{32}\,\overline{64}$ hekat and 1 ro.

The answer given is $\overline{4}\,\overline{32}\,\overline{64}$ hekat and 1 ro. The fractions are not given in their usual form, but in an unusual form of hieratic script. Special symbols are given for each of $\overline{2}, \overline{4}, \overline{8}, \overline{16}, \overline{32},$ and $\overline{64}$.

$$
\begin{array}{cccccc}
\angle & \mathsf{C} & \int & \mathsf{1} & \mathsf{3} & \mathsf{1} \\
\hline
2 & 4 & 8 & 16 & 32 & 64
\end{array}
$$

Horus Eye fractions above their modern equivalent.

These symbols were once thought of as being cursive versions of the parts of the hieroglyph of the Eye of Horus. Horus, as I've already related, was the son of Osiris, once king of the gods. Osiris was killed by his brother Set. When Horus grew up, he challenged Set. During their battle, Set shredded one of Horus's eyes, despite which Horus still won the battle. Thoth, the god of wisdom, restored the eye, and Horus gave this eye to his dead father, now lord of the underworld, so he could watch the land of the living. Horus's good eye is the sun, and the one he gave away is the moon.

The hieroglyph of the Eye of Horus.

In the above example we took $\overline{5}\,\overline{10}$ hekat, converted it into 96 ro, which in turn was converted into $\overline{4}\,\overline{32}\,\overline{64}$ hekat with 1 ro remaining. In Horus Eye fractions it would be written as ⊂ 𝟑 𝟏 hekat, 1 ro. We can use this double conversion to turn standard Egyptian fractions into Horus Eye fractions. You just need to remember that when converting back, you simply divide by 2.

EXAMPLE: Convert $\overline{\overline{3}}\,\overline{20}$ hekat into Horus Eye fractions.

SOLUTION: First convert into ro by multiplying by 320.

$\overline{3}\,\overline{20}$	?
1	320
$\overline{3}$	$213\,\overline{3}$ ✓
$\overline{10}$	32
$\overline{20}$	16 ✓
	$229\,\overline{3}$

$\overline{3}\,\overline{20}$ hekat is $229\,\overline{3}$ ro.

Then convert back to hekat by dividing by 320.

?	$229\,\overline{3}$
1	320
$\overline{2}$	160 ✓
$\overline{4}$	80
$\overline{8}$	40 ✓
$\overline{16}$	20 ✓
$\overline{32}$	10
$\overline{64}$	5 ✓
	$\overline{2}\,\overline{8}\,\overline{16}\,\overline{64}$ rem. $4\,\overline{3}$

$229\,\overline{3}$ ro is $\overline{2}\,\overline{8}\,\overline{16}\,\overline{64}$ hekat and $4\,\overline{3}$ ro.

Now write the hekat fractions as Horus Eye fractions. ∠ ʃ 11 hekat and $4\,\overline{3}$ ro.

PRACTICE: Convert $\overline{3}\,\overline{4}\,\overline{5}$ hekat into Horus Eye fractions.

ANSWER: $250\,\overline{\overline{3}}$ hekat is ∠ ⊂ 3 hekat and $\overline{\overline{3}}$ ro.

Consider the binary fractional value 0.101011. The value of each digit beyond the decimal point is half of the one before it. The place in front of the decimal point has a value of 1, so the next place has a value of $\overline{2}$. There is a 1 here, so our fraction starts off with a half. The next place is half of $\overline{2}$, so it has a value of $\overline{4}$. There is a 0 here, and we ignore it. The next place is half of $\overline{4}$ and there is a 1 there, so it contributes a value of $\overline{8}$. The final three places have values of $\overline{16}$, $\overline{32}$, and $\overline{64}$. There are 1s in the last two, so they contribute a total of $\overline{32}\,\overline{64}$. This means that the base-two decimal has a combined value of $\overline{2}\,\overline{8}\,\overline{32}\,\overline{64}$. This process

can be best understood visually as follows. Notice how easy the conversion from binary to Horus Eye fractions is.

$$0\;.\;1\quad 0\quad 1\quad 0\quad 1\quad 1$$

0.101011 = $\overline{2}\,\overline{8}\,\overline{32}\,\overline{64}$ = ∠ ʃ 3 1.

We need to realize that Horus Eye fractions and binary fractions are almost equivalent. A 1 in the appropriate spot tells us to use the corresponding fraction, and a 0 tells us to ignore it. It is now trivial to convert any binary fraction with at most six digits into Horus Eye fractions and vice versa.

EXAMPLE: Convert 0.111001_2 into standard Egyptian and Horus Eye fractions.

SOLUTION:

$$0\;.\;1\quad 1\quad 1\quad 0\quad 0\quad 1$$

So $0.111001_2 = \overline{2}\,\overline{4}\,\overline{8}\,\overline{64}$ = ∠ ⊂ ʃ 1.

PRACTICE: Convert 0.011011_2 into standard Egyptian and Horus Eye fractions.

ANSWER: $\overline{4}\,\overline{8}\,\overline{32}\,\overline{64}$ = ⊂ ʃ 3 1

EXAMPLE: Convert ∠ ⊂ 3 1 into standard Egyptian fractions and binary.

$$0\;.\;1\quad 1\quad 0\quad 0\quad 1\quad 1$$

So ∠ ⊂ 3 1 = $\overline{2}\,\overline{4}\,\overline{32}\,\overline{64}$ = 0.110011_2.

PRACTICE: Convert ꟼⲘⲖ into standard Egyptian fractions and binary.

ANSWER: $\overline{4}\ \overline{8}\ \overline{16}\ \overline{64} = 0.011101_2$

Some of you might be thinking that this isn't equivalent to binary, which can be used to express values with any number of digits. Since the Egyptian fractions only go to $\overline{64}$, in some sense you would be right. However, take your calculator and type "1 ÷ 3." You almost certainly would see something like "0.33333333." Your calculator, just like Egyptian Horus fractions, stops after a fixed number of digits. It has to. An important point I keep making is that most fractions can't be expressed in a useable way in decimal because they often go on forever. Decimal can only approximate these, and the Egyptian Horus-fraction system reflects this.

Horus Eye fractions are trivial to add. When you add two different fractions, just leave them be. So, for example, $\overline{2} + \overline{8}$ is just $\overline{2}\ \overline{8}$. When you have to add equal fractions, you simply double them.

The operations of doubling and halving are trivial in this system. For example, when we double $\overline{32}$ we get $\overline{16}$ and in general, doubling or halving one of these fractions begets another. If we double $\overline{4}\ \overline{8}\ \overline{32}$, we get $\overline{2}\ \overline{4}\ \overline{16}$. Note what happens pictorially when we think of them as calculi.

$$2\times \boxed{\overline{2}}\ \boxed{\overline{4}}\ \boxed{\overline{8}}\ \boxed{\overline{16}}\ \boxed{\overline{32}}\ \boxed{\overline{64}}$$

$$= \boxed{\overline{2}}\ \boxed{\overline{4}}\ \boxed{\overline{8}}\ \boxed{\overline{16}}\ \boxed{\overline{32}}\ \boxed{\overline{64}}$$

Multiplying by 2 "shifts" Horus fractions left.

The pattern of calculi shifts left. This is just what happens when we multiply by 10 in our base-ten system. For example, 10×12.438 is 124.38. If we think of the decimal point as being in a fixed location, each digit marched one spot left. Similarly, since Horus Eye fractions mimic binary, multiplication by two shifts these numbers left one position. Division by two has the same effect except that the numbers shift right.

Consider the following multiplication of $3\ \overline{2}\ \overline{8}$ by $3\ \overline{4}\ \overline{8}$. I'm going to use only Horus Eye fractions but I'm going to express them as traditional Egyptian fractions for ease of translation.

	$3\ \overline{4}\ \overline{8}$	?	
1)	1	$3\ \overline{2}\ \overline{8}$ ✓	
2)	2	$7\ \overline{4}$ ✓	(1)×2
3)	$\overline{2}$	$1\ \overline{2}\ \overline{4}\ \overline{16}$	(1)÷2
4)	$\overline{4}$	$\overline{2}\ \overline{4}\ \overline{8}\ \overline{32}$ ✓	(3)÷2
5)	$\overline{8}$	$\overline{4}\ \overline{8}\ \overline{16}\ \overline{64}$ ✓	(4)÷2
		$12\ \overline{8}\ \overline{16}\ \overline{32}\ \overline{64}$	

I'm going to rearrange the above multiplication and make one significant change. At the point where I start dividing by 2, instead of going down, I'm going to go up above the 1.

	$3\ \overline{4}\ \overline{8}$	?
	$\overline{8}$	$\overline{4}\ \overline{8}\ \overline{16}\ \overline{64}$ ✓
	$\overline{4}$	$\overline{2}\ \overline{4}\ \overline{8}\ \overline{32}$ ✓
	$\overline{2}$	$1\ \overline{2}\ \overline{4}\ \overline{16}$
	1	$3\ \overline{2}\ \overline{8}$ ✓
	2	$7\ \overline{4}$ ✓
		$12\ \overline{8}\ \overline{16}\ \overline{32}\ \overline{64}$

Now let's portray this pictorially as calculi. Since I'm trying to mimic binary, I'm going to express whole numbers as sums of the powers of 2. For example, 7 will be represented with one each of 1-, 2-, and 4-valued calculi.

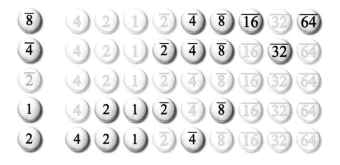

Now let's convert these to binary and multiply them in the traditional way. Note that I'm going to use a calculi motif for some of the numbers so that you can see the connection to the following multiplication and the previous example.

$$
\begin{array}{r}
1\ 1\ .\ 1\ 0\ 1 \\
\times\ 1\ 1\ .\ 0\ 1\ 1 \\
\hline
0\ 0\ 0\ .\ 0\ 1\ 1\ 1\ 0\ 1 \\
0\ 0\ 0\ .\ 1\ 1\ 1\ 0\ 1\ 0 \\
0\ 0\ 0\ .\ 0\ 0\ 0\ 0\ 0\ 0 \\
0\ 1\ 1\ .\ 1\ 0\ 1\ 0\ 0\ 0 \\
1\ 1\ 1\ .\ 0\ 1\ 0\ 0\ 0\ 0 \\
\hline
1\ 1\ 0\ 0\ .\ 0\ 0\ 1\ 1\ 1\ 1 \\
\end{array}
$$

If we portray the rows between the two lines in the above multiplication as calculi, we get the following:

Compare the pattern of the calculi of the two token diagrams above. They're exactly the same except that in one the calculi are labeled in Horus Eye fractions and in the other in binary. For all practical purposes the two systems are the same.

So if the Egyptians knew and used a system similar to the decimal system, why didn't they use it for everything? It's hard to say for certain, but we've already seen some of the flaws of the decimal system. Consider its inability to perform certain divisions precisely. If we use Horus Eye fractions and only doubling and halving, we run into the exact same problem. Let's divide 1 by 3 using Horus Eye fractions. Since the only prime factor in base 2 is 2, division by 3 should lead to infinite answers.

Knowing that $\overline{2}\ \overline{4}$ is close to 1, we complete $\overline{2}\ \overline{4}$ to 1, getting $\overline{4}$. We now continue the division, trying to make $\overline{4}$ on the right.

Now we complete $\overline{8}\ \overline{16}$ to $\overline{4}$, giving $\overline{16}$ remaining.

Completing $\overline{32}\ \overline{64}$ to $\overline{16}$, we get $\overline{64}$ remaining. We've run out of the official six Horus Eye fractions, but if there were more, this process would continue forever, giving the answer $\overline{4}\ \overline{16}\ \overline{64}\ldots$, each fraction being a quarter of the one before it. In binary this is $0.01010101\ldots$.

We can no longer assume that Egyptian mathematics was "primitive" compared to decimal. They knew of decimal-like systems and deliberately chose not to use them for most applications. Whether or not you agree with their choice, their decision was not made out of ignorance.

EPILOGUE

Who Wins?

Thousands of years ago the isolation of the Egyptian civilization led them to a decidedly unique method of doing mathematics. It carefully balanced the two primary requirements of number systems: the ability to approximate values and the need for exact values. We, on the other hand, have had to learn two different systems that each provides for one of the needs. This comes at a great expense, requiring many years to master.

Many mathematicians have cast a brief glance at the Egyptian mathematics and have been perplexed. Lacking the skill and finesse this flexible system requires, they run into difficulty. Rather than acknowledge their own lack of skill and patience, they attribute their problems to the Egyptian system itself. As a result, they automatically assume that their own system is the pinnacle of mathematical achievement and issue judgments based solely on a system's similarity to their own.

So which system is better? I hope I've left you with the knowledge that this is not an easy question. "Better" is a subjective term. What is better depends upon what you want out of a system. Do you want easy approximations, exact values, or some compromise between the two? How long are you willing to take to learn the system? Do you want it to abstract to variable-filled equations? Are you going to work with whole numbers, fractions, or irrationals? Do you like fixed procedures guaranteed to produce a solution, or do you want the possibility of a sharp wit leading to a short, inspired solution? Do you want to be able to memorize a multiplication table, or do you mind carrying around tables for use in calculation? Do

you insist on numbers being expressed in only one way? Do you need to know which god is carrying time, or do you want a consistent base for ease of calculation?

There are good reasons we use the two systems that we do today. Theoreticians need fractions to hold their variables, and business people need decimals to estimate their bottom line. I suspect that as much as I love Egyptian mathematics, it would not suffice for the needs of the modern world. However, we have to wonder if something is lost. The procedural methods of modern calculation turn our children into mathematical automatons, memorizing without understanding. The flexibility of the Egyptian system keeps its practitioners sharp, forcing them to think that in every problem there is a mystery to be unwrapped.

PRACTICE SOLUTIONS

Many Egyptian problems have more than one solution. You should not automatically assume that if your answers don't look like the following solutions that they are wrong. If you have a calculator, find a decimal approximation for your and my solutions and see if they are the same.

NUMBERS

Thoth, Scribes, and Bureaucracy

PRACTICE: Add IIII eeeeeeı + IIIIIIIIĪĪeeeeeeeeII.

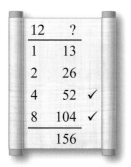

ANSWER: ᔑĪĪĪeeeIII

Bread, Beer, and Pesu

PRACTICE: Multiply 13 by 12 as an Egyptian would.

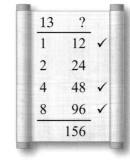

Simplicity

PRACTICE: Divide 187 by 17 and 100 by 21.

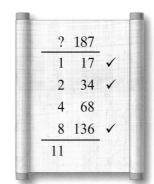

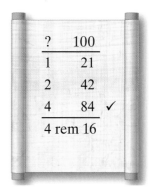

FRACTIONS

The Best Thing since Sliced Bread

PRACTICE: Divide 2 by 5 using "sliced bread." Make your first cut in thirds.

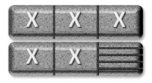

ANSWER: $\overline{3}\,\overline{15}$

PRACTICE: Divide 4 by 18 using "sliced bread." Make your first cut in fifths.

ANSWER: $\overline{5}\,\overline{45}$

PRACTICE: Divide 5 by 7 in two ways using "sliced bread."

or

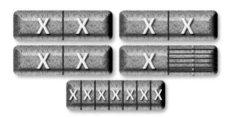

ANSWER: $\overline{2}\ 6\ 24\ \overline{168}$ or $\overline{2}\ 7\ \overline{14}$

OPERATIONS

Memorization and Triangles

PRACTICE: Find half of 58.

$$
\begin{aligned}
58 &= 40 + 18 \\
\overline{2} \times 58 &= 20 + 9 = 29
\end{aligned}
$$

ANSWER: 29

PRACTICE: Find half of 258 by breaking up the number.

$$
\begin{aligned}
258 &= 200 + 40 + 18 \\
\overline{2} \times 258 &= 100 + 20 + 9 = 129
\end{aligned}
$$

ANSWER: 129

PRACTICE: Find half of 8743 by breaking up the number.

$$
\begin{aligned}
8743 &= 8000 + 600 + 140 + 2 + 1 \\
\overline{2} \times 8743 &= 4000 + 300 + 70 + 1 + \overline{2} \\
&= 4371\,\overline{2}
\end{aligned}
$$

ANSWER: $4371\,\overline{2}$

PRACTICE: Repeat for 951.

$$
\begin{aligned}
951 &= 800 + 140 + 10 + 1 \\
\overline{2} \times 951 &= 400 + 70 + 5 + \overline{2} \\
&= 475\,\overline{2}
\end{aligned}
$$

ANSWER: $475\,\overline{2}$

PRACTICE: Find the area of a triangle that has a base of 9 and a height of 10.

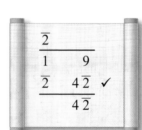

ANSWER: 45

PRACTICE: Repeat for a base of 23 and a height of 11.

ANSWER: $126\,\overline{2}$

Circles in the Sand

PRACTICE: Find the circumference of a circle of diameter 14.

$3\,\overline{8}$	
1	14 ✓
2	28 ✓
$\overline{2}$	7
$\overline{4}$	$3\,\overline{2}$
$\overline{8}$	$1\,\overline{2}\,\overline{4}$ ✓
	$43\,\overline{2}\,\overline{4}$

ANSWER: $43\,\overline{2}\,\overline{4}$

PRACTICE: Repeat for diameter 31.

$3\,\overline{8}$	
1	31 ✓
2	62 ✓
$\overline{2}$	$15\,\overline{2}$
$\overline{4}$	$7\,\overline{2}\,\overline{4}$
$\overline{8}$	$3\,\overline{2}\,\overline{4}\,\overline{8}$ ✓
	$96\,\overline{2}\,\overline{4}\,\overline{8}$

ANSWER: $96\,\overline{2}\,\overline{4}\,\overline{8}$

PRACTICE: Multiply 48 by $1\,\overline{8}\,\overline{32}$.

$1\,\overline{8}\,\overline{32}$	
1	48 ✓
$\overline{2}$	24
$\overline{4}$	12
$\overline{8}$	6 ✓
$\overline{16}$	3
$\overline{32}$	$1\,\overline{2}$ ✓
	$55\,\overline{2}$

ANSWER: $55\,\overline{2}$

Ahmose's Table

PRACTICE: Multiply $3\,\overline{20}$ by 6.

6	
1	$3\,\overline{20}$
2	$6\,\overline{10}$ ✓
4	$12\,\overline{5}$ ✓
	$18\,\overline{5}\,\overline{10}$

ANSWER: $18\,\overline{5}\,\overline{10}$

PRACTICE: Multiply $6\,\overline{14}$ by $3\,\overline{4}$.

$3\,\overline{4}$	
1	$6\,\overline{14}$ ✓
2	$12\,\overline{7}$ ✓
$\overline{2}$	$3\,\overline{28}$
$\overline{4}$	$1\,\overline{2}\,\overline{56}$ ✓
	$19\,\overline{2}\,\overline{7}\,\overline{14}\,\overline{56}$

ANSWER: $19\,\overline{2}\,\overline{7}\,\overline{14}\,\overline{56}$

PRACTICE: Multiply $3\,\overline{60}$ by 16.

16	
1	$3\,\overline{60}$
2	$6\,\overline{30}$
4	$12\,\overline{15}$
8	$24\,\overline{10}\,\overline{30}$
16	$48\,\overline{5}\,\overline{15}$ ✓
	$48\,\overline{5}\,\overline{15}$

ANSWER: $48\,\overline{5}\,\overline{15}$

Pyramids and Seked

PRACTICE: How many palms would you need to push a $3\,\overline{4}$-cubit block back if the seked were 7?

7	
1	$3\,\overline{4}$ ✓
2	$6\,\overline{2}$ ✓
8	13 ✓
	$22\,\overline{2}\,\overline{4}$

ANSWER: $22\,\overline{2}\,\overline{4}$ palms

PRACTICE: Divide 23 by 10

	23
1	10
2	20 ✓
$\overline{10}$	1 ✓
$\overline{5}$	2 ✓
	$2\,\overline{5}\,\overline{10}$

ANSWER: $2\,\overline{5}\,\overline{10}$

PRACTICE: Divide 68 by 12.

	68
1	12 ✓
2	24
4	48 ✓
$\overline{12}$	1
$\overline{6}$	2
$\overline{3}$	4
$\overline{\overline{3}}$	8 ✓
	$5\,\overline{\overline{3}}$

ANSWER: $5\,\overline{\overline{3}}$ or $5\,\overline{2}\,\overline{6}$

PRACTICE: Is a pyramid with height of 24 and a base of 52 safe?

$$\text{Run} = \text{half } 52 = 26, \quad \text{Rise} = 24$$

	26
1	24 ✓
$\overline{24}$	1
$\overline{12}$	2 ✓
	$1\,\overline{12}$

	7
1	$1\,\overline{12}$ ✓
2	$2\,\overline{6}$ ✓
4	$4\,\overline{3}$ ✓
	$7\,\overline{3}\,\overline{6}\,\overline{12}$

ANSWER: Yes, since its seked, $7\,\overline{3}\,\overline{6}\,\overline{12}$, is greater than $5\,\overline{2}$.

The Night Watchmen's Shares

PRACTICE: What is $\overline{\overline{3}}$ of 15?

SOLUTION: Since $10 + 5 = 15$, $\overline{\overline{3}}$ of 15 is 10.

ANSWER: 10

PRACTICE: Find $\overline{\overline{3}}$ of 24 and 39,156.

SOLUTION: Since $16 + 8 = 24$, $\overline{\overline{3}}$ of 24 is 16.
$\overline{\overline{3}} \times (30{,}000 + 9{,}000 + 150 + 6)$
$= 20{,}000 + 6{,}000 + 100 + 4 = 26{,}104$

ANSWERS: 16 and 26,104

PRACTICE: Find $\overline{\overline{3}}$ of 192 and 294.

SOLUTION: $\overline{\overline{3}} \times 192 = \overline{\overline{3}}(180 + 12) = 120 + 8 = 128$
$\overline{\overline{3}} \times 294 = \overline{\overline{3}}(270 + 24) = 180 + 16 = 196$.

ANSWERS: 128 and 196

PRACTICE: Find $\overline{\overline{3}}$ of 36,168 and 574,329.

SOLUTION: $\overline{\overline{3}} \times 36{,}168 = \overline{\overline{3}}(30{,}000 + 6{,}000 + 150 + 18)$
$= 20{,}000 + 4{,}000 + 100 + 12 = 24{,}112$
$\overline{\overline{3}} \times 574{,}329$
$= \overline{\overline{3}}(300{,}000 + 270{,}000 + 3{,}000 + 1{,}200 + 120 + 9)$
$= 200{,}000 + 180{,}000 + 2{,}000 + 800 + 80 + 6$
$= 382{,}886$

ANSWERS: 24,112 and 382,886.

PRACTICE: Find $\bar{\bar{3}}$ of 9,454 and 41,627.

SOLUTION: $\bar{\bar{3}} \times 9{,}454 = \bar{\bar{3}}(9{,}000+300+150+3+1) = 6{,}000+200+100+2+\bar{\bar{3}} = 6{,}302\,\bar{\bar{3}}$
$\bar{\bar{3}} \times 41{,}627 = \bar{\bar{3}}(30{,}000+9{,}000+2{,}400+210+15+2) = 20{,}000+6{,}000+1{,}600+140+10+1\,\bar{3} = 27{,}751\,\bar{3}$

ANSWERS: $6302\,\bar{\bar{3}}$ and $27{,}751\,\bar{3}$

Beer—Good to the Last Drop

PRACTICE: Find $\bar{\bar{3}}$ of $\overline{30}$.

SOLUTION: $\bar{\bar{3}} \times \overline{30} = \overline{30+15} = \overline{45}$

ANSWER: $\overline{45}$

PRACTICE: Take $\bar{\bar{3}}$ of $\overline{11}$.

SOLUTION: $\bar{\bar{3}} \times \overline{11} = \overline{2\times11}\ \overline{6\times11} = \overline{22}\ \overline{66}$

ANSWER: $\overline{22}\ \overline{66}$

PRACTICE: If a 1 share is $\overline{5}\ \overline{20}$, find the night watchman's share.

SOLUTION: $\bar{\bar{3}} \times \overline{5}\ \overline{20} = \overline{2\times5}\ \overline{6\times5} + \overline{20+10} = \overline{10}\ \overline{30}\ \overline{30} = \overline{10}\ \overline{15}$

ANSWER: $\overline{10}\ \overline{15}$

PRACTICE: If a 1 share is $\bar{\bar{3}}\ \overline{6}$, find the night watchman's share.

SOLUTION: $\bar{\bar{3}} \times \bar{\bar{3}}\ \overline{6} = \overline{3}\ \overline{9}\ \overline{9} = \overline{3}\ \overline{6}\ \overline{18}$

ANSWER: $\overline{3}\ \overline{6}\ \overline{18}$, or equivalently, $\overline{2}\ \overline{18}$

The Perfect Woman

PRACTICE: Divide the following by 3 using the Egyptian trick.

- 24
- 52
- $35\,\overline{9}$

ANSWERS: 8, $17\,\overline{3}$, and $11\,\bar{\bar{3}}\ \overline{36}\ \overline{108}$

$\bar{\bar{3}}$	?
1	24
$\overline{3}$	16
$\overline{3}$	8 ✓
	8

$\bar{\bar{3}}$	?
1	52
$\overline{3}$	$34\,\overline{3}$
$\overline{3}$	$17\,\overline{3}$ ✓
	$17\,\overline{3}$

$\bar{\bar{3}}$	?
1	$35\,\overline{9}$
$\overline{3}$	$23\,\overline{3}\ \overline{18}\ \overline{54}$
$\overline{3}$	$11\,\overline{2}\ \overline{6}\ \overline{36}\ \overline{108}$ ✓
	$11\,\bar{\bar{3}}\ \overline{36}\ \overline{108}$

PRACTICE: Divide 9 and $27\,\overline{2}$ by 18.

	18	?	
1)	1	9	
2)	$\bar{\bar{3}}$	6	$(1)\times\bar{\bar{3}}$
3)	$\overline{3}$	3	$(2)\div2$
4)	$\overline{6}\ \overline{18}$	2	$(3)\times\bar{\bar{3}}$
5)	$\overline{9}$	1	$(4)\div2$
6)	$\overline{18}$	2 ✓	$(5)\div2$
		$\overline{2}$	

	18	?	
1)	1	$27\,\overline{2}$	
2)	$\bar{\bar{3}}$	$18\,\overline{3}$	$(1)\times\bar{\bar{3}}$
3)	$\overline{3}$	$9\,\overline{6}$	$(2)\div2$
4)	$\overline{6}\ \overline{18}$	$6\,\overline{9}$	$(3)\times\bar{\bar{3}}$
5)	$\overline{9}$	$3\,\overline{18}$	$(4)\div2$
6)	$\overline{18}$	$1\,\overline{2}\ \overline{36}$ ✓	$(5)\div2$
		$1\,\overline{2}\ \overline{36}$	

ANSWERS: $\overline{2}$ and $1\,\overline{2}\,\overline{36}$

The Dock Enlistees

PRACTICE: Find the volume of a 3-by-4-by-5 block and a 7-by-3-by-1 $\overline{2}$ block.

3	?
1	4 ✓
2	8 ✓
	12

5	?
1	12 ✓
2	24
4	48 ✓
	60

7	?
1	3 ✓
2	6 ✓
4	12 ✓
	21

$1\,\overline{2}$	?
1	21 ✓
$\overline{2}$	10 $\overline{2}$ ✓
	31 $\overline{2}$

ANSWERS: 60 and $31\,\overline{2}$

PRACTICE: What is $53\,\overline{2}\,\overline{31} \div 10$?

SOLUTION: $53\,\overline{2}\,\overline{31} \div 10 = (50+3+\overline{2}+\overline{31}) \div 10$
$= 5+\overline{5}\,\overline{10}+\overline{20}+\overline{310}$
$= 5\,\overline{5}\,\overline{10}\,\overline{20}\,\overline{310}$

ANSWER: $5\,\overline{5}\,\overline{10}\,\overline{20}\,\overline{310}$

PRACTICE: Bricks are piled in a 3-by-2 $\overline{2}$-by-7 cubic cubit block. How many enlistees are needed to load the bricks?

SOLUTION: Volume $= 3 \times 2\,\overline{2} \times 7$.

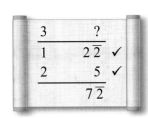

3	?
1	2 $\overline{2}$ ✓
2	5 ✓
	7 $\overline{2}$

7	?
1	7 $\overline{2}$ ✓
2	15 ✓
4	30 ✓
	52 $\overline{2}$

Number of enlistees = volume ÷ 10

$52\,\overline{2} \div 10 = (50+2+\overline{2}) \div 10 = 5\,\overline{5}\,\overline{20}$
ANSWER: $5\,\overline{5}\,\overline{20}$ or $5\,\overline{4}$

SIMPLIFICATION

Pedagogy and Pizza

PRACTICE: Use inches and feet to simplify $\overline{3}\,\overline{4}\,\overline{12}$ as feet.

$\overline{3}\,\overline{4}\,\overline{12}$ feet $= 4+3+1$ inches
$= 8$ inches
$= \overline{\overline{3}}$ feet

ANSWER: $\overline{\overline{3}}$ feet

PRACTICE: Use dollars and pennies to simplify $\overline{20}\,\overline{25}\,\overline{50}$ dollars.

$\overline{20}\,\overline{25}\,\overline{50}$ dollars $= 5+4+2$ pennies
$= 11$ pennies
$= 10+1$ pennies
$= \overline{10}\,\overline{100}$ dollars

ANSWER: $\overline{10}\,\overline{100}$

PRACTICE: Add $\overline{3}, \overline{5}, \overline{6}$, and $\overline{30}$ as parts of 30.

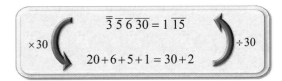

$\overline{\overline{3}}\,\overline{5}\,\overline{6}\,\overline{30} = 1\,\overline{15}$
×30   ÷30
$20+6+5+1 = 30+2$

ANSWER: $1\,\overline{15}$

PRACTICE: Add $\overline{4}, \overline{5}, \overline{10}$, and $\overline{20}$ as parts of 20.

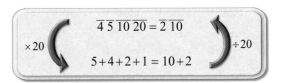

$\overline{4}\,\overline{5}\,\overline{10}\,\overline{20} = \overline{2}\,\overline{10}$
×20   ÷20
$5+4+2+1 = 10+2$

ANSWER: $\overline{2}\,\overline{10}$

The Easy Life

PRACTICE: Simplify the following:

$\overline{12}\,\overline{36}$
$\overline{20}\,\overline{180}$

SOLUTION:

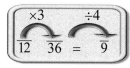 

ANSWERS: $\overline{9}$, $\overline{18}$

PRACTICE: Multiply and use the G rule to simplify your answer.

$5\times\overline{20}$
$7\times\overline{24}$

ANSWERS:

$\overline{4}$
$\overline{4}\,\overline{24}$ or $\overline{6}\,\overline{8}$

5	?	
1	$\overline{20}$	✓
2	$\overline{10}$	
4	$\overline{5}$	✓
$\overline{5}\,\overline{20}$		

$\overline{5}\,\overline{20} = (\overline{5}\,\overline{20}) = \overline{4}$

7	?	
1	$\overline{24}$	✓
2	$\overline{12}$	✓
4	$\overline{6}$	✓
$\overline{6}\,\overline{12}\,\overline{24}$		

$\overline{6}\,\overline{12}\,\overline{24} = (\overline{6}\,\overline{12})\,\overline{24}$
$= \overline{4}\,\overline{24}$

or

$\overline{6}\,\overline{12}\,\overline{24} = \overline{6}\,(\overline{12}\,\overline{24})$
$= \overline{6}\,\overline{8}$

The Suave Scribe

PRACTICE: Multiply then simplify using $\overline{7}\,\overline{14}\,\overline{28} = \overline{4}$

$5\times\overline{7}\,\overline{28}$
$5\,\overline{4}\times\overline{7}$

5	?	
1	$\overline{7}\,\overline{28}$	✓
2	$\overline{4}\,\overline{14}\,\overline{28}$	
4	$\overline{2}\,\overline{7}\,\overline{14}$	✓
$\overline{2}\,\overline{7}\,\overline{7}\,\overline{14}\,\overline{28}$		

$\overline{2}\,\overline{7}\,(\overline{7}\,\overline{14}\,\overline{28}) = \overline{2}\,\overline{4}\,\overline{7}$

$5\,\overline{4}$	?	
1	$\overline{7}$	✓
2	$\overline{4}\,\overline{28}$	
4	$\overline{2}\,\overline{14}$	✓
$\overline{2}$	$\overline{14}$	✓
$\overline{4}$	$\overline{28}$	✓
$\overline{2}\,\overline{7}\,\overline{14}\,\overline{28}$		

$\overline{2}\,(\overline{7}\,\overline{14}\,\overline{28}) = \overline{2}\,\overline{4}$

ANSWERS:

$\overline{2}\,\overline{4}\,\overline{7}$
$\overline{2}\,\overline{4}$

PRACTICE: Multiply then simplify using $\overline{18}\,\overline{27}\,\overline{54} = \overline{9}$

$\overline{36}\,\overline{54}\times 6$

6	?	
1	$\overline{36}\,\overline{54}$	
2	$\overline{18}\,\overline{27}$	✓
4	$\overline{9}\,\overline{18}\,\overline{54}$	✓
$\overline{6}\,\overline{9}$		

$\overline{9}\,\overline{18}\,(\overline{18}\,\overline{27}\,\overline{54}) = (\overline{9}\,\overline{9})\,\overline{18}$
$= \overline{6}\,(\overline{18}\,\overline{18}) = \overline{6}\,\overline{9}$

ANSWER: $\overline{6}\,\overline{9}$

PRACTICE: Simplify $\overline{12}\,\overline{18}\,\overline{36}$.

SOLUTION: Since $12 \times 1\,\overline{2} = 18$ and 18×2 is 36, $\overline{12}\,\overline{18}\,\overline{36} = \overline{12 \div 2} = \overline{6}$.

ANSWER: $\overline{6}$

Self-Sufficiency

PRACTICE: Make an identity for the simplification of $\overline{5}\,\overline{10}\,\overline{30}$.

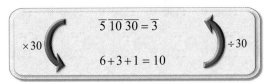

$$\overline{5}\,\overline{10}\,\overline{30} = \overline{3}$$

$\times 30$ · · · $6+3+1 = 10$ · · · $\div 30$

ANSWER: $\overline{3}$

PRACTICE: Using the whole parts of 30, find all identities that sum to $\overline{3}$.

$\overline{3}$ is 10 parts of 30. Since $6+3+1$ and $5+3+2$ both equal 10, we get $\overline{5}\,\overline{10}\,\overline{30}$ and $\overline{6}\,\overline{10}\,\overline{15}$.

ANSWERS: $\overline{5}\,\overline{10}\,\overline{30}$ and $\overline{6}\,\overline{10}\,\overline{15}$

PRACTICE: Using the whole parts of 30, find expressions for $3 \div 10$ and $8 \div 10$.

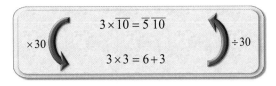

$$3 \times \overline{10} = \overline{5}\,\overline{10}$$

$\times 30$ · · · $3 \times 3 = 6+3$ · · · $\div 30$

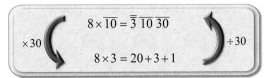

$$8 \times \overline{10} = \overline{3}\,\overline{10}\,\overline{30}$$

$\times 30$ · · · $8 \times 3 = 20+3+1$ · · · $\div 30$

ANSWERS: $\overline{5}\,\overline{10}$ and $\overline{3}\,\overline{10}\,\overline{30}$

Two Choices

PRACTICE: Find an expression for $2 \times \overline{15}$ using the factors 3 and 1.

SOLUTION:

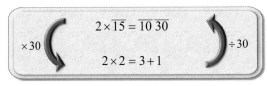

$$2 \times \overline{15} = \overline{10}\,\overline{30}$$

$\times 30$ · · · $2 \times 2 = 3+1$ · · · $\div 30$

ANSWER: $\overline{10}\,\overline{30}$

PRACTICE: Find an expression for $2 \times \overline{15}$ using the factors 15 and 5.

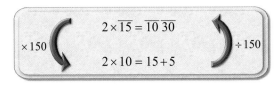

$$2 \times \overline{15} = \overline{10}\,\overline{30}$$

$\times 150$ · · · $2 \times 10 = 15+5$ · · · $\div 150$

ANSWER: $\overline{10}\,\overline{30}$

Points of View

PRACTICE: Using algebra, prove that $3 \times \overline{12} = \overline{4}$.

PROOF: $3 \times \overline{12} = 3 \times \frac{1}{12} = \frac{3}{1} \times \frac{1}{12} = \frac{3 \times 1}{1 \times 12} = \frac{3}{12} = \frac{1}{4} = \overline{4}$

PRACTICE: Prove $\overline{5x}\,\overline{4 \times 5 \times x} = \overline{4x}$

PROOF: $\overline{5x}\,\overline{4 \times 5 \times x} = \frac{1}{5x} + \frac{1}{20x} = \frac{4}{20x} + \frac{1}{20x} = \frac{5}{20x} = \frac{1}{4x} = \overline{4x}$

TECHNIQUES AND STRATEGIES

Precision, Pyramids, and Pesu

PRACTICE: Add

$8\,\overline{3}\,\overline{15}\,\overline{120} + 2\,\overline{10}\,\overline{120} + 9\,\overline{3}\,\overline{15}$.

$(8+2+9)+(\overline{\overline{3}}\ 3)\ \overline{10}\ (\overline{15}\ 15)\ (\overline{120}\ 120)$

$= 19+1\ (\overline{10}\ 10)\ (\overline{30}\ 60)$

$= 20\ (\overline{5}\ 20)$

$= 20\ \overline{4}$

ANSWER: 20 $\overline{4}$

The Solar Eye

PRACTICE: Build a $\overline{56}$ in the first column when the first row is $\overline{7}$, 12 $\overline{5}$.

1)	$\overline{7}$	12 $\overline{5}$	
2)	$\overline{14}$	6 $\overline{10}$	(1)÷2
3)	$\overline{28}$	3 $\overline{20}$	(2)÷2
4)	$\overline{56}$	1 $\overline{2}$ 40	(3)÷2

ANSWER: The fourth row should be $\overline{56}$, 1 $\overline{2}$ 40.

PRACTICE: Build a $\overline{15}$ in the first column when the first row is $\overline{5}$, 11 $\overline{2}$.

1)	$\overline{5}$	11 $\overline{2}$	
2)	$\overline{10}$	5 $\overline{2}$ $\overline{4}$	(1)÷2
3)	$\overline{15}$	3 $\overline{3}$ $\overline{3}$ 6	(2)×$\overline{\overline{3}}$

ANSWER: The third row should be $\overline{15}$, 3 $\overline{\overline{3}}$ 6.

PRACTICE: Build a $\overline{25}$ in the first column when the first row is $\overline{5}$, 7 $\overline{4}$.

1)	$\overline{5}$	7 $\overline{4}$	
2)	$\overline{50}$	$\overline{\overline{3}}$ $\overline{30}$ $\overline{40}$	(1)÷10
3)	$\overline{25}$	1 $\overline{3}$ $\overline{15}$ $\overline{20}$	(2)×2

ANSWER: The third row should be $\overline{25}$, 1 $\overline{3}$ $\overline{15}$ $\overline{20}$.

PRACTICE: Build a $\overline{300}$ in the first column given a first row of $\overline{20}$, 15 $\overline{3}$.

1)	$\overline{20}$	15 $\overline{3}$	
2)	$\overline{200}$	1 $\overline{2}$ 30	(1)÷10
3)	$\overline{300}$	$\overline{\overline{3}}$ $\overline{3}$ 45	(2)×$\overline{\overline{3}}$

ANSWER: The third row should be $\overline{300}$, 1 $\overline{45}$.

PRACTICE: Build a $\overline{33}$ in the first column given a first row of $\overline{8}$, 11.

1)	$\overline{8}$	11	
2)	$\overline{11}$	8	(1)↔
3)	$\overline{22}$	4	(2)÷2
4)	$\overline{33}$	2 $\overline{3}$	(3)×$\overline{\overline{3}}$

ANSWER: The fourth row should be $\overline{33}$, 2 $\overline{\overline{3}}$.

PRACTICE: Prove 2÷27 is $\overline{18}$ $\overline{54}$.

	$\overline{18}$ $\overline{54}$	?	
1)	1	27	
2)	$\overline{2}$	13 $\overline{2}$	(1)÷2
3)	$\overline{3}$	9	(2)×$\overline{\overline{3}}$
4)	$\overline{6}$	4 $\overline{2}$	(3)÷2
5)	$\overline{9}$	3	(4)×$\overline{\overline{3}}$
6)	$\overline{18}$	1 $\overline{2}$✓	(5)÷2
7)	$\overline{27}$	1	(1)↔
8)	$\overline{54}$	$\overline{2}$✓	(7)÷2
		2	

PRACTICE: Prove 2÷17 is $\overline{12}$ $\overline{51}$ $\overline{68}$.

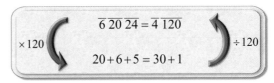

	$\overline{12\ 51\ 68}$	?	
1)	1	17	
2)	$\overline{2}$	$8\,\overline{2}$	$(1)\div2$
3)	$\overline{3}$	$5\,\overline{3}$	$(2)\times\overline{\overline{3}}$
4)	$\overline{6}$	$2\,\overline{2}\,\overline{3}$	$(3)\div2$
5)	$\overline{12}$	$1\,\overline{4}\,\overline{6}$✓	$(4)\div2$
6)	$\overline{17}$	1	$(1)\leftrightarrow$
7)	$\overline{34}$	$\overline{2}$	$(6)\div2$
8)	$\overline{51}$	$\overline{3}$✓	$(7)\times\overline{\overline{3}}$
9)	$\overline{68}$	$\overline{4}$✓	$(7)\div2$
		$1\,\overline{4}\,\overline{6}\,\overline{3}\,\overline{4}$	

$$1\,(\overline{4}\,\overline{4})(\overline{3}\,\overline{6}) = 1\,\overline{2}\,\overline{2} = 2$$

Choose Your Salvation

PRACTICE: Simplify the following or explain why there is no need.

• $\overline{5}\ \overline{21}\ \overline{40}$

21 is more than three times 5 so the first pair are fine. While 40 is roughly twice 21, it's the second pair, so it is fine.

• $\overline{3}\ \overline{4}\ \overline{36}$

4 is much less than three times 3 so we will use parts of 36, the LCM.

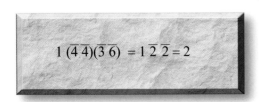

So $\overline{3}\ \overline{4}\ \overline{36}$ simplifies to $\overline{2}\ \overline{9}$.

• $\overline{6}\ \overline{20}\ \overline{24}$

While 20 is more than three times 6, the 20 and 24 are too close. Simplify with parts of 120.

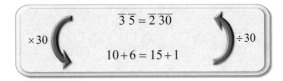

So $\overline{6}\ \overline{20}\ \overline{24}$ simplifies to $\overline{4}\ \overline{120}$.

PRACTICE: Simplify $\overline{3}\ \overline{5}$ so the solution starts with $\overline{2}$.

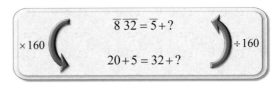

ANSWER: $\overline{2}\ \overline{30}$

PRACTICE: Try to simplify $\overline{8}\ \overline{32}$ starting with $\overline{5}$. What goes wrong and why?

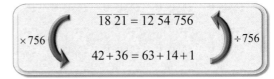

There's nothing you can add to 32 to get 25.

PRACTICE: Find a good approximation for $\overline{33}\ \overline{37}$.

SOLUTION: The even numbers between the two are 34 and 36. 36 is as close to 37 as 33, so half, $\overline{18}$, is a decent approximation.

ANSWER: $\overline{18}$

PRACTICE: Consider the simplification of $\overline{18}\ \overline{21}$. Which is the best fraction to start with, $\overline{11}$, $\overline{12}$, or $\overline{13}$? Explain and simplify with this fraction.

SOLUTION: $\overline{12}$ is the best since it has many small factors and shares a 3 with both $\overline{18}$ and $\overline{21}$ and a 2 with $\overline{18}$.

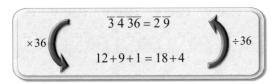

ANSWER: $\overline{12}\ \overline{54}\ \overline{756}$, $\overline{12}\ \overline{63}\ \overline{252}$, or $\overline{12}\ \overline{84}\ \overline{126}$

Easy Units

PRACTICE: Add $\overline{10}\ \overline{14}$ as parts of 10×14. Use Egyptian multiplication tricks to figure out their parts and the parts of the final solution. Include a multiplication-like document to show their derivation.

SOLUTION:

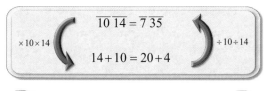

1)	$\overline{10}$	14	
2)	$\overline{14}$	10	(1)$\leftrightarrow$
3)	$\overline{7}$	20	(2)$\times2$
4)	$\overline{70}$	2	(3)$\div10$
5)	$\overline{35}$	4	(4)$\times2$

ANSWER: $\overline{7}\ \overline{35}$

PRACTICE: Construct Ahmose's doubling of $\overline{11}$ starting with the fraction $\overline{6}$. Do all parts as a table.

SOLUTION:

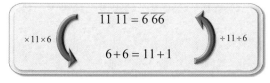

1)	$\overline{11}$	6	
2)	$\overline{6}$	11	(1)$\leftrightarrow$
3)	$\overline{22}$	3	(1)$\div2$
4)	$\overline{33}$	2	(3)$\times\overline{\overline{3}}$
5)	$\overline{66}$	1	(4)$\div2$

ANSWER: $\overline{6}\ \overline{66}$

PRACTICE: Repeat for $\overline{13}$ starting with an $\overline{8}$.

SOLUTION:

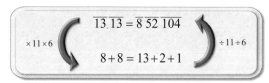

1)	$\overline{13}$	8	
2)	$\overline{8}$	13	(1)$\leftrightarrow$
3)	$\overline{26}$	4	(1)$\div2$
4)	$\overline{52}$	2	(3)$\div2$
5)	$\overline{104}$	1	(4)$\div2$

ANSWER: $\overline{8}\ \overline{52}\ \overline{104}$

Remapping a Nation

PRACTICE: Find the area of a trapezoid with a height of 9 and bases of 10 and 13. Give the answer in cubits and cubit strips.

SOLUTION: Add the bases to get 23. Take half to get $11\ \overline{2}$. Multiply $11\ \overline{2}$ by the height of 9 to get 100.

9	?	
1	$11\ \overline{2}$	✓
2	23	
4	46	
8	92	✓
	$103\ \overline{2}$	

This is the answer in square cubits. Multiply by $\overline{100}$, the number of cubit strips in a square cubit, to get the answer, $\overline{40}\ \overline{100}$, in cubit strips.

$$\begin{array}{l|l}
\overline{100} & \quad ? \\
\hline
1 & 103\ \overline{2} \\
\overline{10} & 10\ 5\ \overline{10}\ \overline{20} \\
\overline{100} & 1\ \overline{50}\ \overline{100}\ \overline{200} \quad \checkmark \\
\hline
& 1\ \overline{40}\ \overline{100}
\end{array}$$

$1\ \overline{50}\ \overline{100}\ \overline{200} = 1\ (\overline{50}\ \overline{200})\ \overline{100} = 1\ \overline{40}\ \overline{100}$

ANSWERS: $103\ \overline{2}$ cubits and $1\ \overline{40}\ \overline{100}$ cubit strips

PRACTICE: Starting with the given rows, create the last number on the right in four lines or less.

• 11, 90 to 3

1)	11	90	
2)	$1\ \overline{10}$	9	$(1)\div10$
3)	$\overline{\overline{3}}\ \overline{15}$	6	$(2)\times\overline{\overline{3}}$
4)	$\overline{3}\ \overline{30}$	3	$(3)\div2$

• 1, 45 to 31

1)	1	45	
2)	$\overline{\overline{3}}$	30	$(1)\times\overline{\overline{3}}$
3)	$\overline{45}$	1	$(1)\leftrightarrow$
4)	$\overline{\overline{3}}\ \overline{45}$	31	$(2)+(3)$

• 11, 18 to 45

1)	11	18	
2)	22	36	$(1)\times2$
3)	$5\ \overline{2}$	9	$(1)\div2$
4)	$27\ \overline{2}$	45	$(2)+(3)$

• 12, 9 to $\overline{36}$

1)	12	9	
2)	$\overline{9}$	12	$(1)\leftrightarrow$
3)	$\overline{18}$	24	$(2)\div2$
4)	$\overline{27}$	36	$(3)\times\overline{\overline{3}}$

• $\overline{4}$, 50 to 58

1)	$\overline{4}$	50	
2)	$\overline{50}$	4	$(1)\leftrightarrow$
3)	$\overline{25}$	8	$(2)\times2$
4)	$\overline{4}\ \overline{25}$	58	$(1)+(3)$

• 4, 150 to 5

1)	4	150	
2)	$\overline{3}\ \overline{15}$	15	$(1)\div10$
3)	$\overline{5}\ \overline{15}$	10	$(2)\times\overline{\overline{3}}$
4)	$\overline{10}\ \overline{30}$	5	$(3)\div2$

• $\overline{6}$, 80 to 89

1)	$\overline{6}$	80	
2)	$\overline{80}$	6	$(1)\leftrightarrow$
3)	$\overline{160}$	3	$(2)\div2$
4)	$\overline{6}\ \overline{80}\ \overline{160}$	89	$(1)+(2)+(3)$

• 10, 36 to 48

1)	10	36	
2)	$6\ \overline{\overline{3}}$	24	$(1)\times\overline{\overline{3}}$
3)	$13\ \overline{3}$	48	$(2)\times2$

• $\overline{6}$, 40 to 100

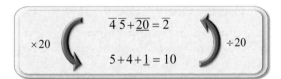

1)	$\overline{6}$	40	
2)	$\overline{3}$	80	$(1)\times 2$
3)	$\overline{12}$	20	$(1)\div 2$
4)	$\overline{3}\ \overline{12}$	100	$(2)+(3)$

• 20, 12 to $\overline{45}$

1)	20	12	
2)	$\overline{12}$	$\overline{20}$	$(1)\leftrightarrow$
3)	$\overline{18}$	$\overline{30}$	$(2)\times\overline{\overline{3}}$
4)	$\overline{27}$	$\overline{45}$	$(3)\times\overline{\overline{3}}$

What Is an Nb.t?

PRACTICE: Complete $\overline{4}\ \overline{5}$ to $\overline{2}$.

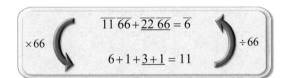

$$\overline{4}\,\overline{5}+\underline{\overline{20}}=\overline{2}$$
$$5+4+\underline{1}=10$$
$\times 20$ $\div 20$

ANSWER: $\overline{20}$

PRACTICE: Complete $\overline{11}\ \overline{66}$ to $\overline{6}$.

$$\overline{11}\,\overline{66}+\underline{\overline{22}\,\overline{66}}=\overline{6}$$
$$6+1+\underline{3+1}=11$$
$\times 66$ $\div 66$

ANSWER: $\overline{22}\,\overline{66}$

PRACTICE: Complete $\overline{3}\ \overline{5}$ to 1.

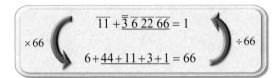

$$\overline{3}\,\overline{5}+\underline{\overline{3}\,\overline{10}\,\overline{30}}=1$$
$$10+6+\underline{10+3+1}=30$$
$\times 30$ $\div 30$

ANSWER: $\overline{3}\ \overline{10}\ \overline{30}$

PRACTICE: Complete $\overline{11}$ to 1.

$$\overline{11}+\underline{\overline{3}\,\overline{6}\,\overline{22}\,\overline{66}}=1$$
$$6+\underline{44+11+3+1}=66$$
$\times 66$ $\div 66$

ANSWER: $\overline{3}\ \overline{6}\ \overline{11}\ \overline{22}$

PRACTICE: Complete $8\,\overline{2}\ \overline{10}$ to $12\,\overline{2}\,\overline{5}$.

This is the same as completing $\overline{10}$ to $4\,\overline{5}$, which can be computed by completing $\overline{10}$ to $\overline{5}$ and adding 4.

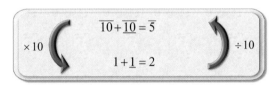

$$\overline{10}+\underline{\overline{10}}=\overline{5}$$
$$1+\underline{1}=2$$
$\times 10$ $\div 10$

ANSWER: $4\ \overline{10}$

Don't Disturb My Circles

PRACTICE: Estimate the area of a circle of diameter 27.

SOLUTION: First calculate $\overline{9}$ of 27.

$\overline{9}$	?
1	27
$\overline{2}$	$13\,\overline{2}$
$\overline{3}$	9
$\overline{6}$	$4\,\overline{2}$
$\overline{9}$	3 ✓
	3

Complete 3 to get 27.
The answer is 24.
Square 24.

24	?
1	24
2	48
4	96
8	192 ✓
16	384 ✓
	576

576 is the area of the circle.
Redo with the square-first method.

27	?
1	27 ✓
2	54 ✓
4	108
8	216 ✓
16	432 ✓
	729

$\overline{2}\ \overline{4}\ \overline{32}$	?
1	729
$\overline{2}$	$364\,\overline{2}$ ✓
$\overline{4}$	$182\,\overline{4}$ ✓
$\overline{8}$	$91\,\overline{8}$
$\overline{16}$	$45\,\overline{2}\ \overline{16}$
$\overline{32}$	$22\,\overline{2}\ \overline{4}\ \overline{32}$ ✓
	$569\,\overline{2}\ \overline{32}$

ANSWERS: $576,\ 569\,\overline{2}\ \overline{32}$

PRACTICE: Find the area of a circle of radius 12. Then redo using the square-first method.

SOLUTION: First calculate $\overline{9}$ of 12.

$\overline{9}$	?
1	12
$\overline{2}$	6
$\overline{3}$	4
$\overline{6}$	2
$\overline{9}$	$1\,\overline{3}$ ✓
	$1\,\overline{3}$

Complete $1\,\overline{3}$ to get 12.
The answer is $10\,\overline{\overline{3}}$.
Square $10\,\overline{\overline{3}}$.

$10\,\overline{\overline{3}}$	?
1	$10\,\overline{\overline{3}}$
2	$21\,\overline{3}$ ✓
4	$42\,\overline{\overline{3}}$
8	$85\,\overline{3}$ ✓
$\overline{\overline{3}}$	$6\,\overline{\overline{3}}\ \overline{3}\ \overline{9}$ ✓
	$113\,\overline{\overline{3}}\ \overline{9}$

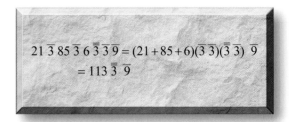

$$21\,\overline{3}\ 85\,\overline{3}\ 6\,\overline{\overline{3}}\ \overline{3}\ \overline{9} = (21 + 85 + 6)(\overline{3}\,\overline{3})(\overline{3}\,\overline{3})\ \overline{9}$$
$$= 113\,\overline{\overline{3}}\ \overline{9}$$

$113\,\overline{\overline{3}}\ \overline{9}$ is the area of the circle.
Redo with the square-first method.

12	?
1	12
2	24
4	48 ✓
8	96 ✓
	144

$\overline{2}\ \overline{4}\ \overline{32}$	?
1	144
$\overline{2}$	72 ✓
$\overline{4}$	36 ✓
$\overline{8}$	18
$\overline{16}$	9
$\overline{32}$	$4\,\overline{2}$ ✓
	$112\,\overline{2}$

ANSWERS: $113\,\overline{\overline{3}}\ \overline{9},\ 112\,\overline{2}$

PRACTICE: Find the area of a circle of radius 7. Then redo using the square-first method.

SOLUTION: First calculate $\overline{9}$ of 7.

$\overline{9}$	?
1	7
$\overline{2}$	3 $\overline{2}$
$\overline{3}$	2 $\overline{3}$
$\overline{6}$	1 $\overline{6}$
$\overline{9}$	$\overline{\overline{3}}$ $\overline{9}$ ✓
	$\overline{\overline{3}}$ $\overline{9}$

Complete $\overline{\overline{3}}\,\overline{9}$ to get 7.

This can be computed by completing $\overline{\overline{3}}\,\overline{9}$ to 1 and adding 6.

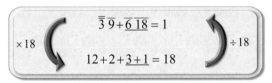

$$\overline{\overline{3}}\,\overline{9} + \overline{6}\,\overline{18} = 1$$
$$\times 18 \qquad\qquad \div 18$$
$$12 + 2 + \underline{3+1} = 18$$

The answer is $6\,\overline{6}\,\overline{18}$.

Square $6\,\overline{6}\,\overline{18}$.

$6\,\overline{6}\,\overline{18}$	?
1	$6\,\overline{6}\,\overline{18}$
2	$12\,\overline{\overline{3}}\,\overline{9}$ ✓
4	$24\,\overline{\overline{3}}\,\overline{6}\,\overline{18}$ ✓
$\overline{2}$	$3\,\overline{9}$
$\overline{3}$	$2\,\overline{18}\,\overline{54}$
$\overline{6}$	$1\,\overline{36}\,\overline{108}$ ✓
$\overline{12}$	$\overline{2}\,\overline{72}\,\overline{216}$
$\overline{18}$	$\overline{3}\,\overline{108}\,\overline{324}$ ✓

$12\,\overline{3}\,\overline{9}\ 24\,\overline{\overline{3}}\,\overline{6}\,\overline{18}\ 1\,\overline{36}\,\overline{108}\ \overline{3}\,\overline{108}\,\overline{324}$

$= (12+24+1)(\overline{3}\,\overline{\overline{3}})(\overline{3}\,\overline{6})(\overline{9}\,\overline{18})\,\overline{36}\,(\overline{108}\,\overline{108})\,\overline{324}$

$= 37 + 1 + \overline{2}\,\overline{6}\,\overline{36}\,\overline{54}\,\overline{324}$

$= (37+1)\,(\overline{2}\,\overline{6})\,\overline{36}\,\overline{54}\,\overline{324}$

$= 38\,\overline{\overline{3}}\,\overline{36}\,\overline{54}\,\overline{324}$

Simplify $\overline{36}\,\overline{54}\,\overline{324}$ as parts of 324.

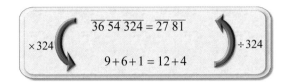

$$\overline{36}\,\overline{54}\,\overline{324} = \overline{27}\,\overline{81}$$
$$\times 324 \qquad\qquad \div 324$$
$$9 + 6 + 1 = 12 + 4$$

Redo with the square-first method.

7	?
1	7 ✓
2	14 ✓
4	28 ✓
	49

$2\,\overline{4}\,\overline{32}$	?
1	49
$\overline{2}$	$24\,\overline{2}$ ✓
$\overline{4}$	$12\,\overline{4}$ ✓
$\overline{8}$	$6\,\overline{8}$
$\overline{16}$	$3\,\overline{16}$
$\overline{32}$	$1\,\overline{2}\,\overline{32}$ ✓
	$38\,\overline{4}\,\overline{32}$

ANSWERS: $38\,\overline{\overline{3}}\,\overline{27}\,\overline{81}$, $38\,\overline{4}\,\overline{32}$

I've Got Nothing

PRACTICE: Divide 15 by 42 in four lines.

?	15
1	42
$\overline{2}$	21
$\overline{3}$	14 ✓
$\overline{42}$	1 ✓
$\overline{3}\,\overline{42}$	

ANSWER: $\overline{3}\,\overline{42}$

PRACTICE: Divide 16 by 22.

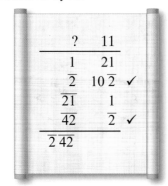

?	16	
1	22	
$\overline{\overline{3}}$	14 $\overline{\overline{3}}$	✓
$\overline{22}$	1	✓
$\overline{33}$	$\overline{3}$	
$\overline{66}$	$\overline{3}$	✓
$\overline{\overline{3}}\ \overline{22}\ \overline{66}$		

ANSWER: $\overline{\overline{3}}\ \overline{22}\ \overline{66}$

PRACTICE: Divide 11 by 21.

?	11	
1	21	
$\overline{2}$	10 $\overline{2}$	✓
$\overline{21}$	1	
$\overline{42}$	$\overline{2}$	✓
$\overline{2}\ \overline{42}$		

ANSWER: $\overline{2}\ \overline{42}$

PRACTICE: Divide 31 by 2000.

?	31	
1	2000	
$\overline{10}$	200	
$\overline{100}$	20	✓
$\overline{200}$	10	✓
$\overline{2000}$	1	✓
$\overline{100}\ \overline{200}\ \overline{2000}$		

ANSWER: $\overline{100}\ \overline{200}\ \overline{2000}$

PRACTICE: Divide 66 by 840.

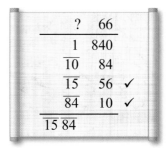

?	66	
1	840	
$\overline{10}$	84	
$\overline{15}$	56	✓
$\overline{84}$	10	✓
$\overline{15}\ 84$		

ANSWER: $\overline{15}\ 84$

What Lies Ahead

PRACTICE: Divide 14 by 11 using completions.

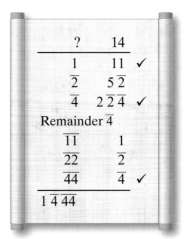

?	14	
1	11	✓
$\overline{2}$	5 $\overline{2}$	
$\overline{4}$	2 $\overline{2}$ $\overline{4}$	✓
Remainder $\overline{4}$		
$\overline{11}$	1	
$\overline{22}$	$\overline{2}$	
$\overline{44}$	$\overline{4}$	✓
1 $\overline{4}$ $\overline{44}$		

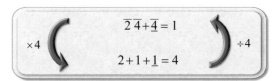

$\times 4$ $\overline{2}\ \overline{4} + \overline{4} = 1$ $\div 4$

$2 + 1 + \underline{1} = 4$

ANSWER: 1 $\overline{4}$ $\overline{44}$

PRACTICE: Divide 8 $\overline{6}$ by 23 using completions.

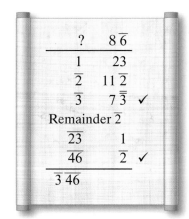

$$\overline{3}+\underline{2}=1\,\overline{6}$$
×6 ↺ ↻ ÷6
$$4+\underline{3}=6+1$$

ANSWER: $\overline{3}\,\overline{46}$

Payday

PRACTICE: Finish the following multiplication by forming a completion and finishing in parts. Divide $8\,\overline{5}$ by $2\,\overline{3}$.

?	$8\,\overline{5}$	
1	$2\,\overline{3}$	✓
2	$4\,\overline{\overline{3}}$	✓
4	$1\,\overline{6}$	✓
Rem.		

SOLUTION:

?	$8\,\overline{5}$	
1	$2\,\overline{3}$	✓
2	$4\,\overline{\overline{3}}$	✓
$\overline{2}$	$1\,\overline{6}$	✓
Rem. 1 of 30 parts		
1	70	
$\overline{70}$	1	✓
$3\,\overline{2}\,\overline{70}$		

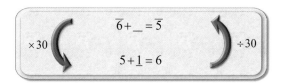

ANSWER: $7\,\overline{70}$

MISCELLANY

That's a Lot of Stone

PRACTICE: Consider a truncated pyramid with a 9-by-9 lower base, a 4-by-4 upper base, and a height of 8. What's the volume?

Solution: Take the 4 and square it to get 16. Take the 9 and square it to get 81. Multiply the 4 and 9 to get 36.

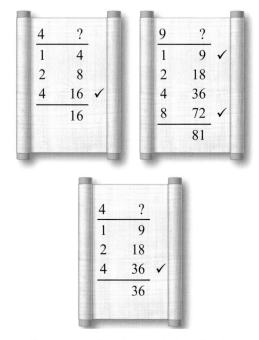

The squares and products of the base lengths.

Add the 16, 81, and 36 to get 133. A third of the height is $2\,\overline{\overline{3}}$. Multiply $2\,\overline{\overline{3}}$ by 133 to get $354\,\overline{\overline{3}}$, the volume.

$\overline{3}$	?
1	8
$\overline{3}$	$5\,\overline{3}$
$\overline{3}$	$2\,\overline{3}$ ✓
	$2\,\overline{\overline{3}}$

$2\,\overline{3}$	?
1	133
2	266 ✓
$\overline{3}$	$88\,\overline{3}$ ✓
	$354\,\overline{3}$

A third the height and the answer times the sum.

ANSWER: $354\,\overline{\overline{3}}$

Going to St. Ives

PRACTICE: Given that $7 + 49 + 343 + 2{,}401 + 16{,}807 = 19{,}607$, use the Egyptian trick to find $7 + \ldots + 16{,}807 + 117{,}649$.

SOLUTION: $7 \times (19{,}607 + 1) = 7 \times 19{,}608 = 137{,}256$

ANSWER: 137,256

Ahah!

PRACTICE: A number and its $\overline{2}$ is added and it is 16. What is the number?

SOLUTION: Assume the number is 2.

Calculate 2 and its $\overline{2}$ to get 3.

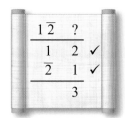

2 and its $\overline{2}$ is 3.

Scale 3 to 16 by calculating $16 \div 3$, which is $5\ \overline{3}$.

?	16
1	3 ✓
2	6
4	12 ✓
$\overline{3}$	1 ✓
	$5\,\overline{3}$

$16 \div 3$ gives the scaling factor.

Scale up the assumed answer of 2 by a factor of $5\,\overline{3}$ to get the answer $10\,\overline{\overline{3}}$.

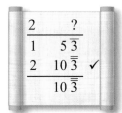

2	?
1	$5\,\overline{3}$
2	$10\,\overline{\overline{3}}$ ✓
	$10\,\overline{3}$

Scale up the false answer by $5\,\overline{3}$.

ANSWER: $10\,\overline{\overline{3}}$

PRACTICE: A ninth is removed from a number and 20 remains. What is the number?

SOLUTION: Assume the number is 9, its ninth is 1 and when removed leaves 8.

Scale 8 to 20 by calculating $20 \div 8$, which is $5\ \overline{3}$.

?	20
1	8
2	16 ✓
$\overline{2}$	4 ✓
$2\,\overline{2}$	

$20 \div 8$ gives the scaling factor.

Scale up the assumed answer of 9 by a factor of $2\,\overline{2}$ to get the answer $22\,\overline{2}$.

9	?	
1	$2\overline{2}$	✓
2	5	
4	10	
8	20	✓
	$22\overline{2}$	

Scale up the false answer by $2\overline{2}$.

ANSWER: $22\overline{2}$

Useless or Theoretical

PRACTICE: Five people get 35 loaves of bread, and each person gets 3 more than the person before. How much does each person get?

SOLUTION: The middle man gets the average of $35 \div 5$.

?	35	
1	5	✓
2	10	✓
4	20	✓
7		

The middle man gets 7. The two above each get 3 more, 10 and 13. The two below each get 3 less, 4 and 1.

ANSWERS: 1, 4, 7, 10, and 13

PRACTICE: There are six people who get 33 loaves of bread. If each one gets $1\overline{2}$ more than the one before, how much does each person get?

SOLUTION: The average is $33 \div 6$, which is $5\overline{2}$.

?	33	
1	6	
$\overline{2}$	3	✓
5	30	✓
$5\overline{2}$		

The middle is $2\ \overline{2}$ jumps of $1\ \overline{2}$ from the start.

$2\overline{2}$	?	
1	$1\overline{2}$	
2	3	✓
$\overline{2}$	$2\overline{4}$	✓
	$3\overline{2}\overline{4}$	

Since $2\overline{2} \times 1\overline{2}$ is $3\overline{2}\overline{4}$, the first person gets $5\overline{2} - 3\overline{2}\overline{4}$. Complete $\overline{4}$ to 1 and add 1, which is $1\overline{2}\overline{4}$.

$$\overline{4} + \overline{2}\,\overline{4} = 1$$
$$\times 4 \qquad\qquad \div 4$$
$$1 + \underline{2} + \underline{1} = 4$$

Adding $1\overline{2}$ to get the rest, the wages are $1\overline{2}\ \overline{4}, 3\overline{4}, 4\overline{2}$ $\overline{4}, 6\overline{4}, 7\ \overline{2}\ \overline{4}$, and $9\overline{4}$.

ANSWERS: $1\overline{2}\ \overline{4}, 3\overline{4}, 4\overline{2}\ \overline{4}, 6\overline{4}, 7\ \overline{2}\ \overline{4}$, and $9\overline{4}$.

Too Easy Is Just Right

REDO PRACTICE: Ninety loaves of bread get distributed between five people in arithmetic progression. If the upper three get 4 times as much as the lower two, how much does each person get?

SOLUTION: Assume they get a total of 60 loaves. The middle person gets $60 \div 5$, or 12 loaves.

?	60	
1	5	
10	50	✓
2	10	✓
12		

The lower two get one part while the upper three get four, so the lower two get a fifth of 60. So the lower two get a total of $60 \div 5$, which is we already know is 12. The "middle person" of the two gets $12 \div 2$, which is 6. The "middle person" of the first two makes 6 less than the middle man of all five and is $1\,\overline{2}$ jumps away, and hence each jump is $6 \div 1\,\overline{2}$, or 4.

?	6
1	$1\,\overline{2}$
2	3
4	6 ✓
4	

Adding and subtracting 4 from the middle person, we get 4, 8, 12, 16, and 20 loaves. We need to scale up our answer to 90 loaves of bread by multiplying by $90 \div 60$, which is $1\,\overline{2}$.

?	90
1	60 ✓
$\overline{2}$	30 ✓
$1\,\overline{2}$	

Now we multiply $1\,\overline{2}$ by each answer to get the adjusted shares.

$1\,\overline{2}$	
1	4 ✓
$\overline{2}$	2 ✓
6	

$1\,\overline{2}$	
1	8 ✓
$\overline{2}$	4 ✓
12	

$1\,\overline{2}$	
1	12 ✓
$\overline{2}$	6 ✓
18	

$1\,\overline{2}$	
1	16 ✓
$\overline{2}$	8 ✓
24	

$1\,\overline{2}$	
1	20 ✓
$\overline{2}$	10 ✓
30	

Hence they get 6, 12, 18, 24, and 30.

ANSWERS: 6, 12, 18, 24, and 30

PRACTICE: Sixty-six loaves of bread get distributed between five people in arithmetic progression. If the upper two get two times as much as the lower three, how much does each person get?

SOLUTION: Assume they get a total of 60 loaves. The middle person gets $60 \div 5$, or 12 loaves.

?	60
1	5
10	50 ✓
2	10 ✓
12	

The lower three get one part while the upper two get two, so the lower two get a third of 60. So the lower two get a total of $60 \div 3$, which is 20.

$\overline{3}$	?
1	60
$\overline{3}$	40
$\overline{\overline{3}}$	20 ✓
20	

The middle person of the three gets $20 \div 3$, which is $6\,\overline{\overline{3}}$. The middle person of the first three makes $5\,\overline{\overline{3}}$ less than the middle person of all five and is 1 jump away, and hence each jump is $5\,\overline{\overline{3}} \div 1$, or $5\,\overline{\overline{3}}$.

Adding and subtracting $5\,\overline{\overline{3}}$ from the middle person, we get $1\,\overline{3}$, $6\,\overline{\overline{3}}$, 12, $17\,\overline{3}$, and $22\,\overline{\overline{3}}$ loaves. We need to scale up our answer to 66 loaves of bread by multiplying by $66 \div 60$, which is $1\,\overline{10}$.

?	66
1	60 ✓
$\overline{10}$	6 ✓
$1\,\overline{10}$	

Now we multiply $1\,\overline{10}$ by each answer to get the adjusted shares. (Add sums as parts of 30.)

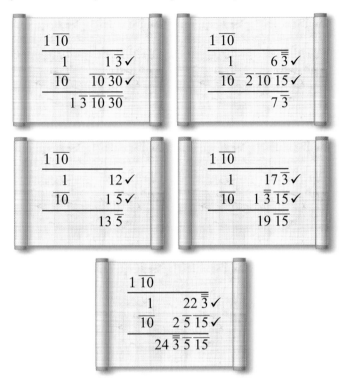

$$1\,\overline{10}$$
$$1 \qquad 1\,\overline{3}\checkmark$$
$$\overline{10} \qquad \overline{10}\,\overline{30}\checkmark$$
$$1\,\overline{3}\,\overline{10}\,\overline{30}$$

$$1\,\overline{10}$$
$$1 \qquad 6\,\overline{\overline{3}}\checkmark$$
$$\overline{10} \qquad 2\,\overline{10}\,\overline{15}\checkmark$$
$$7\,\overline{3}$$

$$1\,\overline{10}$$
$$1 \qquad 12\checkmark$$
$$\overline{10} \qquad 1\,\overline{5}\checkmark$$
$$13\,\overline{5}$$

$$1\,\overline{10}$$
$$1 \qquad 17\,\overline{3}\checkmark$$
$$\overline{10} \qquad 1\,\overline{\overline{3}}\,\overline{15}\checkmark$$
$$19\,\overline{15}$$

$$1\,\overline{10}$$
$$1 \qquad 22\,\overline{\overline{3}}\checkmark$$
$$\overline{10} \qquad 2\,\overline{5}\,\overline{15}\checkmark$$
$$24\,\overline{\overline{3}}\,\overline{5}\,\overline{15}$$

Hence they get $1\,\overline{3}\,\overline{10}\,\overline{30}, 7\,\overline{3}, 13\,\overline{5}, 19\,\overline{15},$ and $24\,\overline{\overline{3}}\,\overline{5}\,\overline{15}$.

ANSWERS: $1\,\overline{3}\,\overline{10}\,\overline{30}, 7\,\overline{3}, 13\,\overline{5}, 19\,\overline{15},$ and $24\,\overline{\overline{3}}\,\overline{5}\,\overline{15}$

The Pylons of the Night

PRACTICE: Represent the $\overline{49}$ node of an odd-fraction family tree.

SOLUTION: $2\times\overline{49} = \overline{28}\,\overline{196}, 2\times\overline{28} = \overline{14}, 2\times\overline{14} = \overline{7}, 2\times\overline{196} = \overline{98}, 2\times\overline{98} = \overline{49}$

ANSWER:

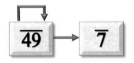

PRACTICE: Find the complete odd-fraction family tree for $\overline{21}$.

SOLUTION: $2\times\overline{21} = \overline{14}\,\overline{42}, 2\times\overline{14} = \overline{7}, 2\times\overline{42} = \overline{21}, 2\times\overline{7} = \overline{4}\,\overline{28}, 2\times\overline{4} = \overline{2} = *, 2\times\overline{28} = \overline{14}$

ANSWER:

PRACTICE: Use the previous answer to find the complete odd-fraction family tree for $\overline{63}$.

SOLUTION: $2\times\overline{63} = \overline{42}\,\overline{126}$, above we know 42 goes to the $\overline{21}$ tree. $2\times\overline{126} = \overline{63}$.

ANSWER:

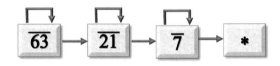

PRACTICE: Draw the augmented complete family tree of $\overline{75}$ and show that it contains the fractions on the left-hand side of $\overline{25}\,\overline{50}\,\overline{150} = \overline{15}$.

SOLUTION: $2\times\overline{75} = \overline{50}\,\overline{150}, 2\times\overline{50} = \overline{25}, 2\times\overline{150} = \overline{75}, 2\times\overline{25} = \overline{15}\,\overline{75}, 2\times\overline{15} = \overline{10}\,\overline{30}, 2\times\overline{10} = \overline{5}, 2\times\overline{30} = \overline{15}, 2\times\overline{5} = \overline{3}\,\overline{15}$

ANSWER:

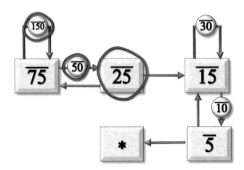

BASE-BASED MATHEMATICS

Coins and Large Quantities

PRACTICE: Express 107_{10} in base five and base-five calculi.

SOLUTIONS:

4   1 2_5

ANSWER: 412_5

PRACTICE: Convert 321_5 into base ten.

SOLUTION: Three quarters, 2 nickels, and 1 penny is worth $(3 \times 25) + (2 \times 5) + 1 = 86_{10}$

ANSWER: 86_{10}

PRACTICE: Using penny, nickel, and quarter calculi, add $214_5 + 123_5$.

SOLUTION:

ANSWER: 342_5

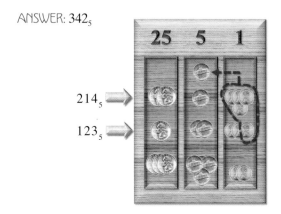

Mayans, Gods, and Numbers

PRACTICE: Convert the following tokens into base four and base ten.

SOLUTION: In base four: 2131_4.

In base ten:

$$(2 \times 64) + (1 \times 16) + (3 \times 4) + 1 = 128 + 16 + 12 + 1 = 157_{10}$$

ANSWERS: 2131_4 and 157_{10}

PRACTICE: Using base-seven calculi, add $351_7 + 132_7$.

SOLUTION:

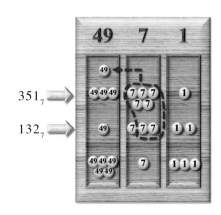

ANSWER: 513_7

PRACTICE: Using base-three calculi, add $120_3 + 211_3$.

SOLUTION:

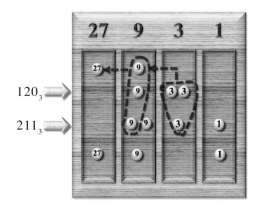

ANSWER: 1101_3

Romans ≠ Mathematicians

PRACTICE: Add CXXVIII to LXXXXVIIII.

SOLUTION:

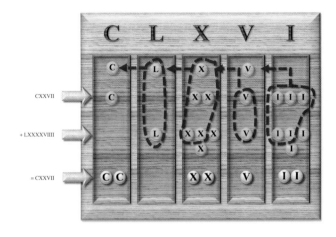

ANSWER: CCXXVII

The Legacy of Sumer

PRACTICE: Add ⊚◔◑◑●●●◦◑ + ◑◑◑◔●●●◖◦◦◦

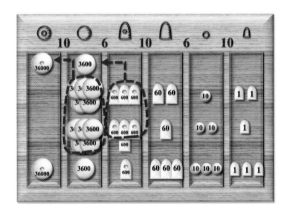

ANSWER: ⊚◔◖●●●◦◑◑

The Stylus Is Mightier Than the Spear

PRACTICE: Convert the Sumerian calculi ⊚◑◑◑◔●●●◑ into Babylonian cuneiform.

SOLUTION: ⊚◑◑◑◔ = 15 = ⟨𝗬
🛆◖◖◖ = 23 = ⟪𝖨𝖨𝖨 and ◖◖ = 2 = 𝖨𝖨,
so the answer is ⟨𝗬 ⟪𝖨𝖨𝖨 𝖨𝖨.

ANSWER: ⟨𝗬 ⟪𝖨𝖨𝖨 𝖨𝖨

PRACTICE: Convert ⟨𝖨𝖨 ⟪𝖨𝖨𝖨 ⟪𝗛 into base ten.

SOLUTION: The digits are 12, 43, and 29 so it's equal to
$(12 \times 60^2) + (43 \times 60) + 29 = 43,\!200 + 2,\!580 + 29 = 45,\!809_{10}$.

ANSWER: $45,\!809_{10}$

The Bull of Heaven

PRACTICE: Translate $31\frac{2}{3}$ into Babylonian.

SOLUTION: $31\frac{2}{3} = 31\frac{40}{60} = $ ⟪𝖨.⟪𝗬

ANSWER: ⟪𝖨.⟪𝗬

PRACTICE: How many degrees is 𝖨.⟨𝖨𝖨 watches?

SOLUTION: 𝖨.⟨𝖨𝖨 $\times 60 = 1\frac{12}{60} \times 60$
$= (1 \times 60) + (\frac{12}{60} \times 60)$
$= 60 + 12$
$= 72$

ANSWER: 72 degrees

The Tablets of Nippur

PRACTICE: Multiply ⟪𝖨𝗬 by ⟪𝗛.

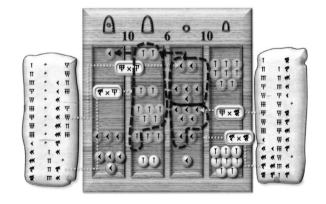

ANSWER: ⟪𝖨𝖨 ⟨𝗬

The Diagonal and the Soul

PRACTICE: Find the length of the diagonal of a square whose side is ⟨𝖧. Use common sense to place the decimal point.

ANSWER: ⟨𝖧.ππ ⟨𝗛 𝗦

PRACTICE: Multiply ⟨πππ by ⟨𝖧 𝗪 𝖧.

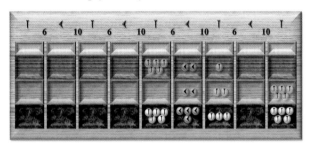

ANSWER: 𝖧 𝗦ππ 𝖧

The Ignorant, Liars, and the Insane

PRACTICE: Multiply 𝗦ππ ⟨𝗛 by ππ ⟨𝗦ᵀ 𝗦𝖧.

ANSWER: ππ ⟨ππ ⟨𝗛 πππ ⟨𝖧

JUDGMENT DAY

The Root's All Evil

PRACTICE: Estimate the square root of 102.

SOLUTION: $102 = 100 + 2 = 10^2 + 2$
  So $\sqrt{102} \approx 10^{2/2 \times 10} = 10\frac{1}{10}$.

ANSWER: 10

PRACTICE: Divide 𝗦πππ by ⟨ππ knowing that the reciprocal of ⟨ππ is 𝖧.

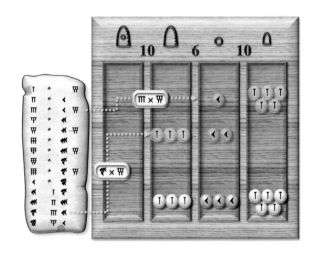

ANSWER: πππ.⟨⟨𝖧

PRACTICE: Use the above table to divide ⟨ by ⟨𝗛.

The ⟨ row is ⟨⟨𝖧 ⟨⟨⟨. Since $10 \div 16$ is just under 1, the decimal point goes in front.

ANSWER: .⟨⟨𝗛 ⟨⟨⟨

I Ching and the Digital Age

PRACTICE: What is 1000110_2 in base ten?

SOLUTION: $(1 \times 64) + (0 \times 32) + (0 \times 16) + (0 \times 8) + (1 \times 4) + (1 \times 2) + (0 \times 1) = 70_{10}$

PRACTICE: Multiply 1110_2 by 101_2 using binary calculi.

ANSWER: 1000110_2

PRACTICE: Use a one-column Egyptian style table to convert 55 into binary.

55		
1	✓	1
2	✓	1
4	✓	1
8		0
16	✓	1
32	✓	1

ANSWER: 110111_2

PRACTICE: Convert $\overline{3}\,\overline{4}\,\overline{5}$ hekat into Horus Eye fractions.

SOLUTION: First convert into ro by multiplying by 320.

$\overline{3}\,\overline{4}\,\overline{5}$	?	
1	320	
$\overline{2}$	160	
$\overline{3}$	$106\,\overline{\overline{3}}$	✓
$\overline{4}$	80	✓
$\overline{10}$	32	
$\overline{5}$	64	✓
	$250\,\overline{\overline{3}}$	

$\overline{3}\,\overline{4}\,\overline{5}$ hekat is $250\,\overline{\overline{3}}$ ro.

Then convert back to hekat by dividing by 320.

?	$250\,\overline{\overline{3}}$	
1	320	
$\overline{2}$	160	✓
$\overline{4}$	80	✓
$\overline{8}$	40	
$\overline{16}$	20	
$\overline{32}$	10	✓
$\overline{64}$	5	
$\overline{2}\,\overline{4}\,\overline{32}$ rem. $\overline{\overline{3}}$		

$250\,\overline{\overline{3}}$ ro is $\overline{2}\,\overline{4}\,\overline{32}$ hekat and $\overline{\overline{3}}$ ro.

Now write the hekat fractions as Horus Eye fractions.

• ∠⊂3 hekat and $\overline{\overline{3}}$ ro.

ANSWER: $250\,\overline{\overline{3}}$ hekat is ∠⊂3 hekat and $\overline{\overline{3}}$ ro.

INDEX